Year 4B

A Guide to Teaching for Mastery

Series Editor: Tony Staneff

Contents

Introduction

Foreword by the series editor and author, Tony Staneff

For far too long in the UK, maths has been feared by learners – and by many teachers, too. As a result, most learners consistently underachieve. More crucially, negative beliefs about ability, aptitude and the nature of maths are entrenched in children's thinking from an early age.

Yet, as someone who has loved maths all my life, I've always believed that every child has the capacity to succeed in maths. I've also had the great pleasure of leading teams and departments who share that belief and passion. Teaching for mastery, as practised in China and other South-East Asian jurisdictions since the 1980s, has confirmed my conviction that maths really is for everyone and not just those who have a special talent. In recent years, my team and I at Trinity Academy, Halifax, have had the privilege of researching with and working alongside some of the finest mastery practitioners from the UK and beyond, whose impact on learners' confidence, achievement and attitude is an inspiration.

The mastery approach recognises the value of developing the power to think rather than just do. It also recognises the value of making a coherent journey in which whole-class groups tackle concepts in very small steps, one by one. You cannot build securely on loose foundations – and it is just the same with maths: by creating a solid foundation of deep understanding, our children's skills and confidence will be strong and secure. What's more, the mindset of learner and teacher alike is fundamental: everyone can do maths … EVERYONE CAN!

I am proud to have been part of the extensive team responsible for turning the best of the world's practice, research, insights, and shared experiences into *Power Maths*, a unique teaching and learning resource developed especially for UK classrooms. *Power Maths* embodies our vision to help and support primary maths teachers to transform every child's mathematical and personal development. 'Everyone can!' has become our mantra and our passion, and we hope it will be yours, too.

Now, explore and enjoy all the resources you need to teach for mastery, and please get back to us with your *Power Maths* experiences and stories!

What is *Power Maths*?

Created especially for UK primary schools, and aligned with the new National Curriculum, *Power Maths* is a whole-class, textbook-based mastery resource that empowers every child to understand and succeed. *Power Maths* rejects the notion that some people simply 'can't do' maths. Instead, it develops growth mindsets and encourages hard work, practice and a willingness to see mistakes as learning tools.

Best practice consistently shows that mastery of small, cumulative steps builds a solid foundation of deep mathematical understanding. *Power Maths* combines interactive teaching tools, high-quality textbooks and continuing professional development (CPD) to help you equip children with a deep and long lasting understanding. Based on extensive evidence, and developed in partnership with practising teachers, *Power Maths* ensures that it meets the needs of children in the UK.

Power Maths and Mastery

Power Maths makes mastery practical and achievable by providing the structures, pathways, content, tools and support you need to make it happen in your classroom.

To develop mastery in maths children need to be enabled to acquire a deep understanding of maths concepts, structures and procedures, step by step. Complex mathematical concepts are built on simpler conceptual components and when children understand every step in the learning sequence, maths becomes transparent and makes logical sense. Interactive lessons establish deep understanding in small steps, as well as effortless fluency in key facts such as tables and number bonds. The whole class works on the same content and no child is left behind.

Power Maths

- Builds every concept in small, progressive steps.
- Is built with interactive, whole-class teaching in mind.
- Provides the tools you need to develop growth mindsets.
- Helps you check understanding and ensure that every child is keeping up.
- Establishes core elements such as intelligent practice and reflection.

The *Power Maths* approach

Everyone can!

Founded on the conviction that every child can achieve, *Power Maths* enables children to build number fluency, confidence and understanding, step by step.

Child-centred learning

Children master concepts one step at a time in lessons that embrace a Concrete-Pictorial-Abstract (C-P-A) approach, avoid overload, build on prior learning and help them see patterns and connections. Same-day intervention ensures sustained progress.

Continuing professional development

Embedded teacher support and development offer every teacher the opportunity to continually improve their subject knowledge and manage whole-class teaching for mastery.

Whole-class teaching

An interactive, whole-class teaching model encourages thinking and precise mathematical language and allows children to deepen their understanding as far as they can.

Introduction to the author team

Power Maths arises from the work of maths mastery experts who are committed to proving that, given the right mastery mindset and approach, **everyone can do maths**. Based on robust research and best practice from around the world, *Power Maths* was developed in partnership with a group of UK teachers to make sure that it not only meets our children's wide-ranging needs but also aligns with the National Curriculum in England.

Tony Staneff, Series Editor and author

Vice Principal at Trinity Academy, Halifax, Tony also leads a team of mastery experts who help schools across the UK to develop teaching for mastery via nationally recognised CPD courses, problem-solving and reasoning resources, schemes of work, assessment materials and other tools.

A team of experienced authors, including:

 **Josh Lury** – a specialist maths teacher, author and maths consultant with a passion for innovative and effective maths education

 **Trinity Academy, Halifax** (Michael Gosling CEO, Tony Staneff, Emily Fox, Kate Henshall, Rebecca Holland, Stephanie Kirk, Stephen Monaghan and Rachel Webster)

David Board, **Belle Cottingham**, **Jonathan East**, **Tim Handley**, **Derek Huby**, **Neil Jarrett**, **Stephen Monaghan**, **Beth Smith**, **Tim Weal**, **Paul Wrangles** – skilled maths teachers and mastery experts

Cherri Moseley – a maths author, former teacher and professional development provider

Professors Liu Jian and Zhang Dan, Series Consultants and authors, and their team of mastery expert authors:

Wei Huinv, Huang Lihua, Zhu Dejiang, Zhu Yuhong, Hou Huiying, Yin Lili, Zhang Jing, Zhou Da and Liu Qimeng

Used by over 20 million children, Professor Liu Jian's textbook programme is one of the most popular in China. He and his author team are highly experienced in intelligent practice and in embedding key maths concepts using a C-P-A approach.

A group of 15 teachers and maths co-ordinators

We have consulted our teacher group throughout the development of *Power Maths* to ensure we are meeting their real needs in the classroom.

Your *Power Maths* resources

To help you teach for mastery, *Power Maths* comprises a variety of high-quality resources.

Pupil Textbooks

Discover, Share, and Think together sections promote discussion and introduce mathematical ideas logically, so that children understand more easily.

Using a Concrete-Pictorial-Abstract approach, clear mathematical models help children to make connections and grasp concepts.

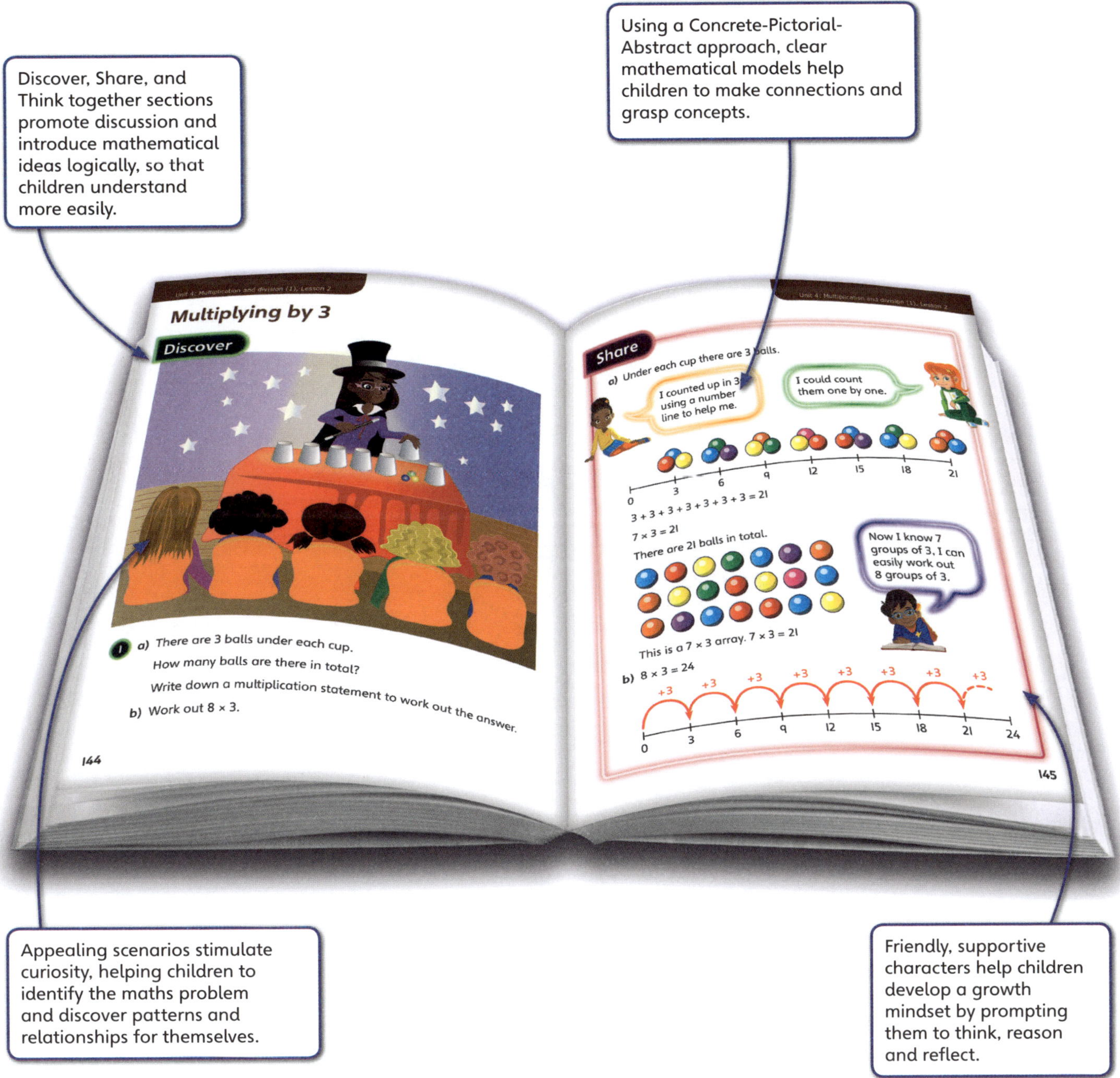

Appealing scenarios stimulate curiosity, helping children to identify the maths problem and discover patterns and relationships for themselves.

Friendly, supportive characters help children develop a growth mindset by prompting them to think, reason and reflect.

The coherent *Power Maths* lesson structure carries through into the vibrant, high-quality textbooks. Setting out the core learning objectives for each class, the lesson structure follows a carefully mapped journey through the curriculum and supports children on their journey to deeper understanding.

Pupil Practice Books

The Practice Books offer just the right amount of intelligent practice for children to complete independently in the final section of each lesson.

The practice questions are for everyone – each question varies one small element to move children on in their thinking. Look at the different parts in question ❶!

Calculations are connected so that children think about the underlying concept. In question ❸, children have to write out the calculation to find the answer. Concepts are presented differently again in question ❹ to challenge children.

Practice questions are finely tuned to move children forward in their thinking and to reveal misconceptions.

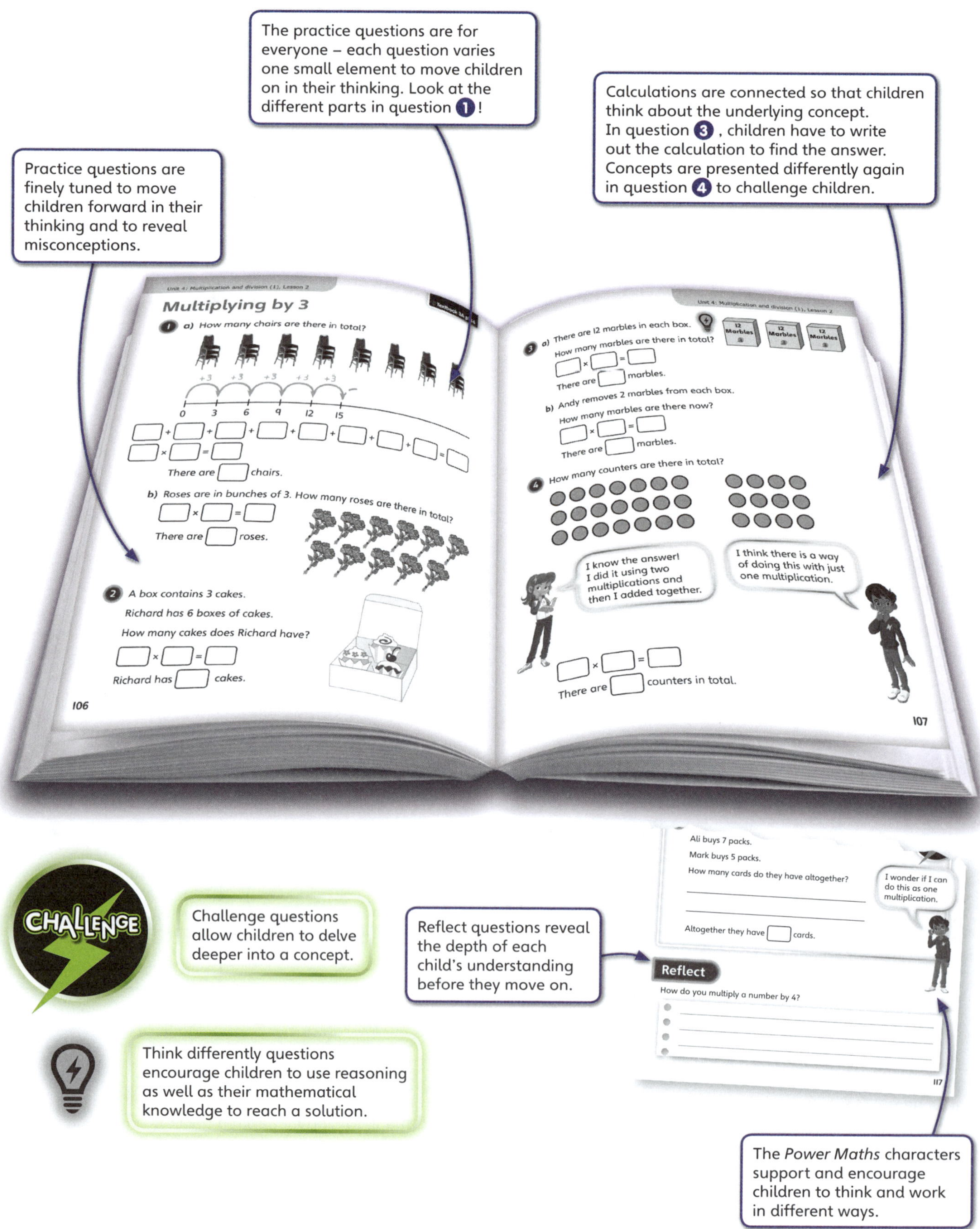

Challenge questions allow children to delve deeper into a concept.

Reflect questions reveal the depth of each child's understanding before they move on.

Think differently questions encourage children to use reasoning as well as their mathematical knowledge to reach a solution.

The *Power Maths* characters support and encourage children to think and work in different ways.

Online subscriptions

The online subscription will give you access to additional resources.

eTextbooks

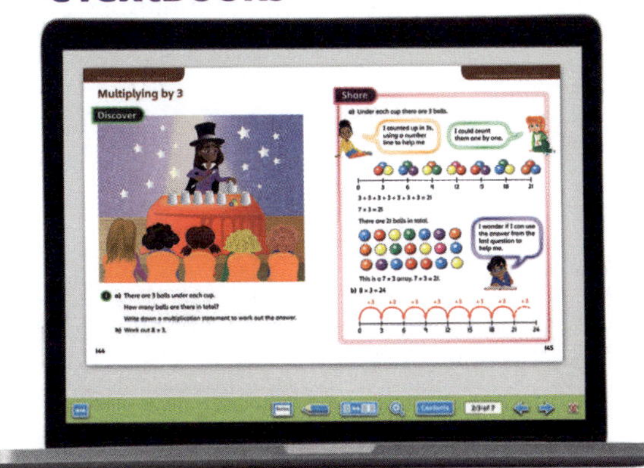

Digital versions of *Power Maths* Textbooks allow class groups to share and discuss questions, solutions and strategies. They allow you to project key structures and representations at the front of the class, to ensure all children are focusing on the same concept.

Teaching tools

Here you will find interactive versions of key *Power Maths* structures and representations.

Power Ups

Use this series of daily activities to promote and check number fluency.

Online versions of Teacher Guide pages

PDF pages give support at both unit and lesson levels. You will also find help with key strategies and templates for tracking progress.

Unit videos

Watch the professional development videos at the start of each unit to help you teach with confidence. The videos explore common misconceptions in the unit, and include intervention suggestions as well as suggestions on what to look out for when assessing mastery in your children.

End of unit Strengthen and Deepen materials

Each Strengthen activity at the end of every unit addresses a key misconception and can be used to support children who need it. The Deepen activities are designed to be 'Low Threshold High Ceiling' and will challenge those children who can understand more deeply. These resources will help you ensure that every child understands and will help you keep the class moving forward together. These printable activities provide an optional resource bank for use after the assessment stage.

Underpinning all of these resources, *Power Maths* is infused throughout with continual professional development, supporting you at every step.

The *Power Maths* teaching model

At the heart of *Power Maths* is a clearly structured teaching and learning process that helps you make certain that every child masters each maths concept securely and deeply. For each year group, the curriculum is broken down into core concepts, taught in units. A unit divides into smaller learning steps – lessons. Step by step, strong foundations of cumulative knowledge and understanding are built.

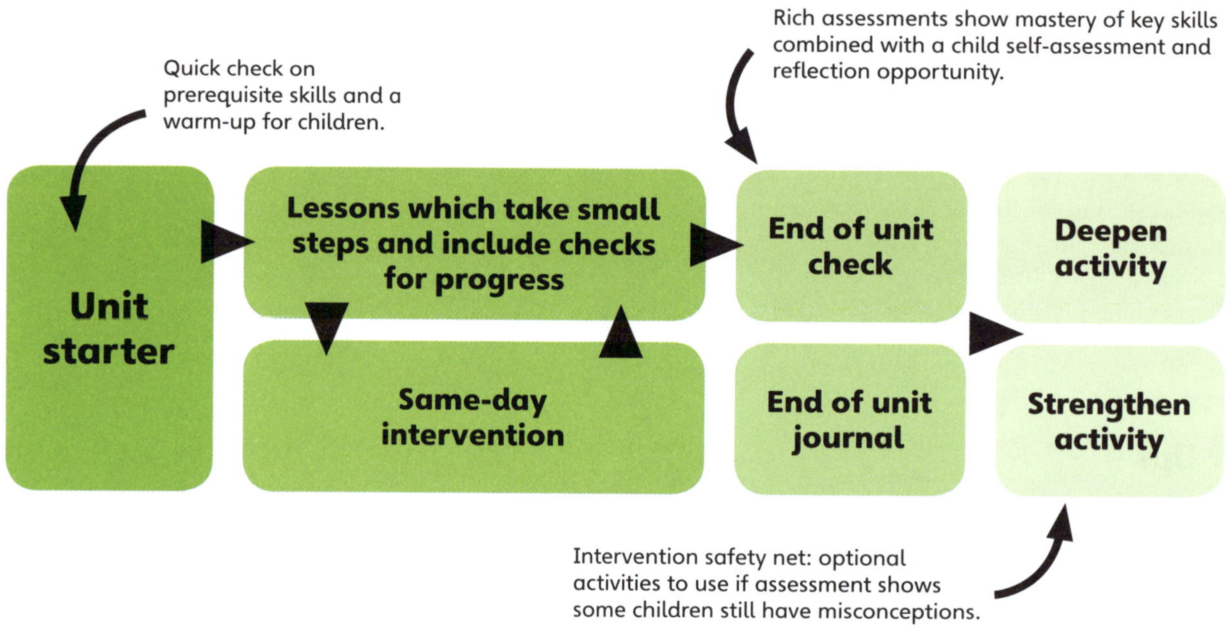

Unit starter

Each unit begins with a unit starter, which introduces the learning context along with key mathematical vocabulary, structures and representations.

- The Textbooks include a check on readiness and a warm-up task for children to complete.

- Your Teacher Guide gives support right from the start on important structures and representations, mathematical language, common misconceptions and intervention strategies.

- Unit-specific videos develop your subject knowledge and insights so you feel confident and fully equipped to teach each new unit. These are available via the online subscription.

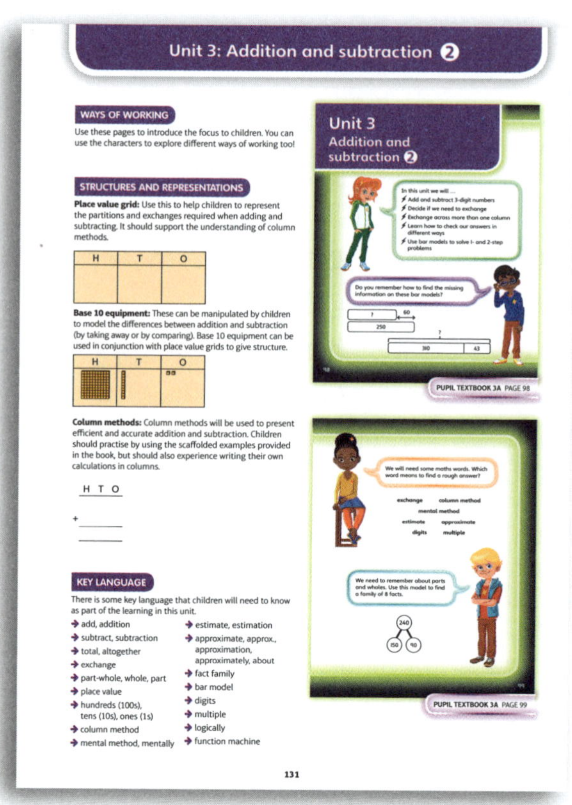

Lesson

Once a unit has been introduced, it is time to start teaching the series of lessons.

- Each lesson is scaffolded with Textbook and Practice Book activities and always begins with a Power Up activity (available via online subscription).

- *Power Maths* identifies lesson by lesson what concepts are to be taught.

- Your Teacher Guide offers lots of support for you to get the most from every child in every lesson. As well as highlighting key points, tricky areas and how to handle them, you will also find question prompts to check on understanding and clarification on why particular activities and questions are used.

Same-day intervention

Same-day interventions are vital in order to keep the class progressing together. Therefore, *Power Maths* provides plenty of support throughout the journey.

- Intervention is focused on keeping up now, not catching up later, so interventions should happen as soon as they are needed.

- Practice questions are designed to bring misconceptions to the surface, allowing you to identify these easily as you circulate during independent practice time.

- Child-friendly assessment questions in the Teacher Guide help you identify easily which children need to strengthen their understanding.

End of unit check and journal

At the end of a unit, summative assessment tasks reveal essential information on each child's understanding. An End of unit check in the Pupil Textbook lets you see which children have mastered the key concepts, which children have not and where their misconceptions lie. The Practice Book includes an End of unit journal in which children can reflect on what they have learnt.
Each unit also offers Strengthen and Deepen activities, available via the online subscription.

The Teacher Guide offers support with handling misconceptions.

The End of unit check presents six to nine multiple-choice questions. These questions are designed to reveal misconceptions and help you target areas that need strengthening.

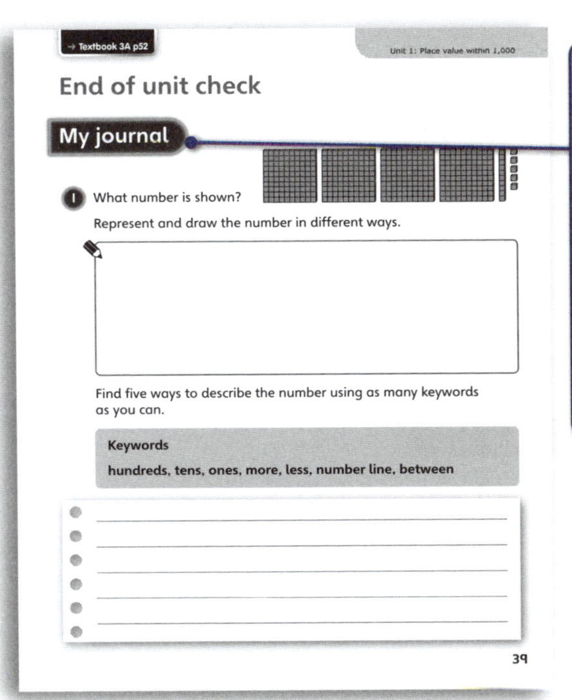

The End of unit journal is an opportunity for children to test out their learning and reflect on how they feel about it. Tackling the 'journal' problem reveals whether a child understands the concept deeply enough to move on to the next unit.

In KS2, the End of unit assessment will also include one SATs-style question.

The *Power Maths* lesson sequence

At the heart of *Power Maths* is a unique lesson sequence designed to empower children to understand core concepts and grow in confidence. Embracing the National Centre for Excellence in the Teaching of Mathematics' (NCETM's) definition of mastery, the sequence guides and shapes every *Power Maths* lesson you teach.

Flexibility is built into the *Power Maths* programme so there is no one-to-one mapping of lessons and concepts meaning you can pace your teaching according to your class. While some children will need to spend longer on a particular concept (through interventions or additional lessons), others will reach deeper levels of understanding. However, it is important that the class moves forward together through the termly schedules.

Power Up 5 minutes

Each lesson begins with a Power Up activity (available via the online subscription) which supports fluency in key number facts.

The whole-class approach depends on fluency, so the Power Up is a powerful and essential activity.

TOP TIP

If the class is struggling with the task, revisit it later and check understanding.

Power Ups reinforce key skills such as times-tables, number bonds and working with place value.

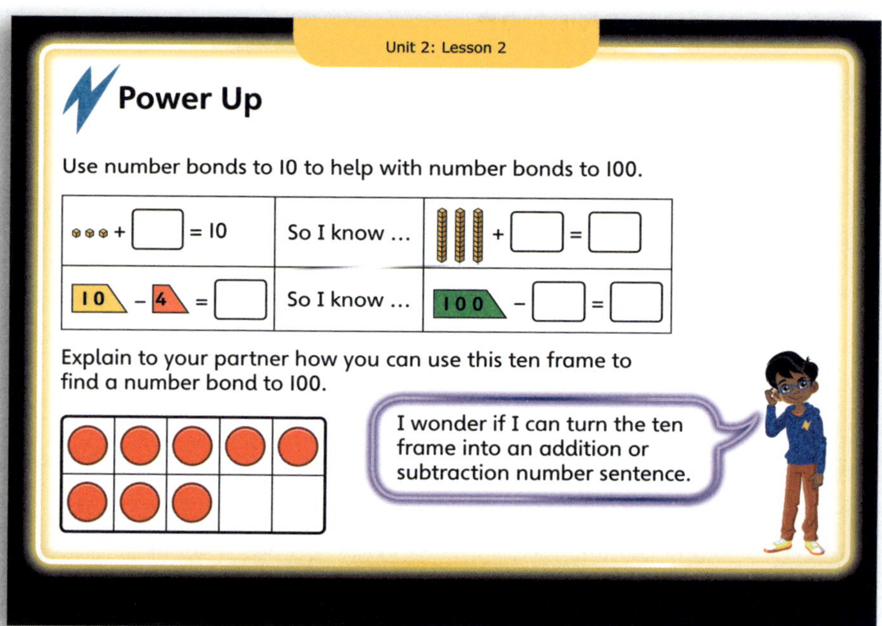

Discover 10 minutes

A practical, real-life problem arouses curiosity. Children find the maths through story-telling.

A real-life scenario is provided for the Discover section but feel free to build upon these with your own examples that are more relevant to your class.

TOP TIP

Discover works best when run at tables, in pairs with concrete objects.

Question **1** a) tackles the key concept and question **1** b) digs a little deeper. Children have time to explore, play and discuss possible strategies.

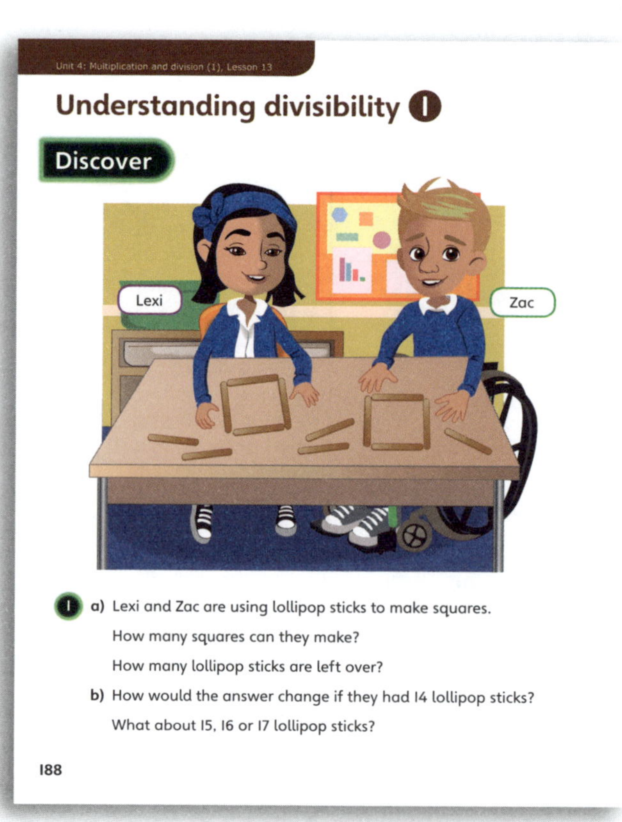

1 a) Lexi and Zac are using lollipop sticks to make squares.

How many squares can they make?

How many lollipop sticks are left over?

b) How would the answer change if they had 14 lollipop sticks?

What about 15, 16 or 17 lollipop sticks?

188

Share

Teacher-led, this interactive section follows the Discover activity and highlights the variety of methods that can be used to solve a single problem.

TOP TIP

Bring children to the front (or onto the carpet if you have this area) to discuss their methods. Pairs sharing a textbook is a great format for this!

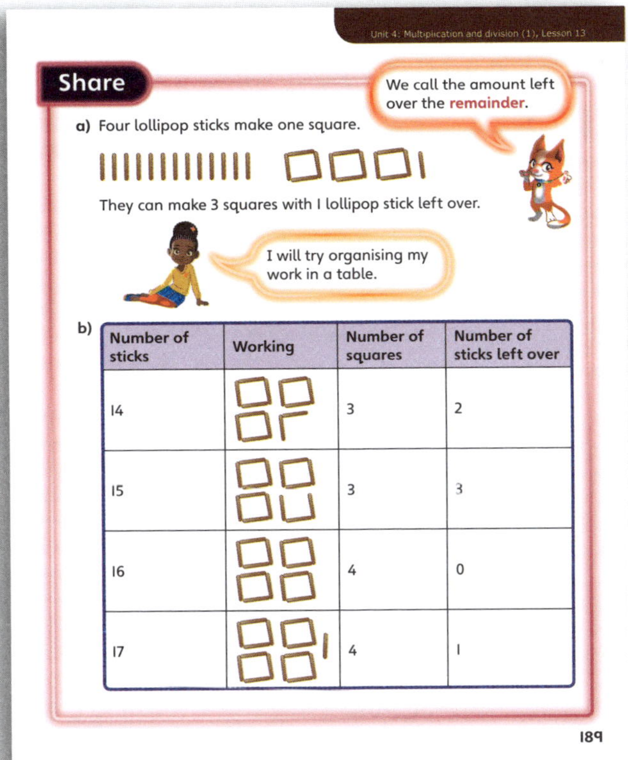

b)

Number of sticks	Working	Number of squares	Number of sticks left over
14		3	2
15		3	3
16		4	0
17		4	1

Your Teacher Guide gives target questions for children. The online toolkit provides interactive structures and representations to link concrete and pictorial to abstract concepts.

TOP TIP

Bring children to the front to share and celebrate their solutions and strategies.

Think together

10 minutes

Children work in groups on the carpet or at tables, using their textbooks or eBooks.

TOP TIP

Make sure children have mini whiteboards or pads to write on if they are not at their tables.

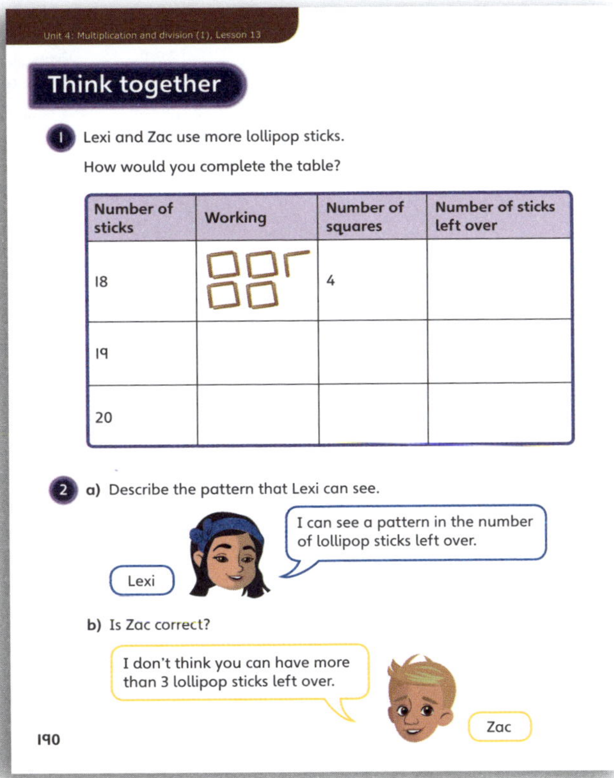

Number of sticks	Working	Number of squares	Number of sticks left over
18		4	
19			
20			

Using the Teacher Guide, model question 1 for your class.

Question 2 is less structured. Children will need to think together in their groups, then discuss their methods and solutions as a class.

In questions 3 and 4 children try working out the answer independently. The openness of the challenge question helps to check depth of understanding.

Practice — 15 minutes

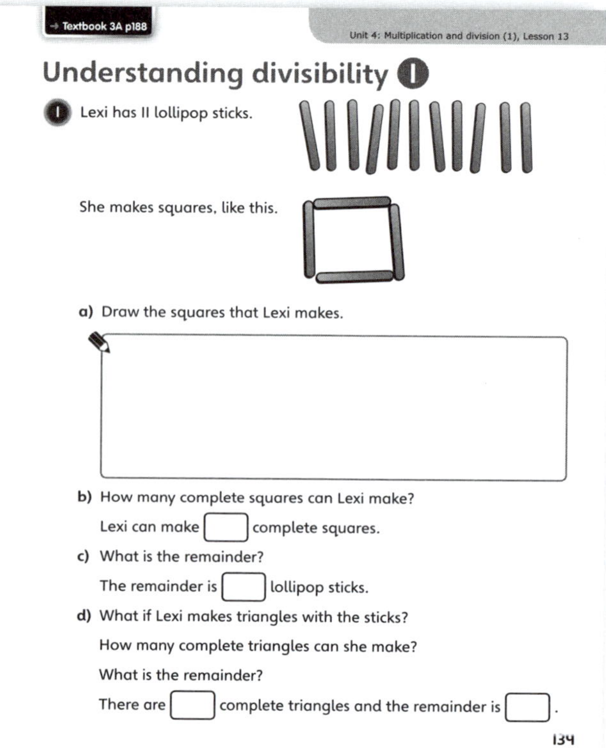

Reflect — 5 minutes

14

Using the *Power Maths* Teacher Guide

Think of your Teacher Guides as *Power Maths* handbooks that will guide, support and inspire your day-to-day teaching. Clear and concise, and illustrated with helpful examples, your Teacher Guides will help you make the best possible use of every individual lesson. They also provide wrap-around professional development, enhancing your own subject knowledge and helping you to grow in confidence about moving your children forward together.

There is a Teacher Guide per year group for every term with unit and lesson level guidance and support.

Tips and advice on key elements such as C-P-A approaches, misconceptions, language, modelling growth mindsets and same-day intervention.

Annotations for every Pupil Textbook and Practice Book page, providing prompts for key questions to ask to expose understanding and explanations as to why key questions have been chosen.

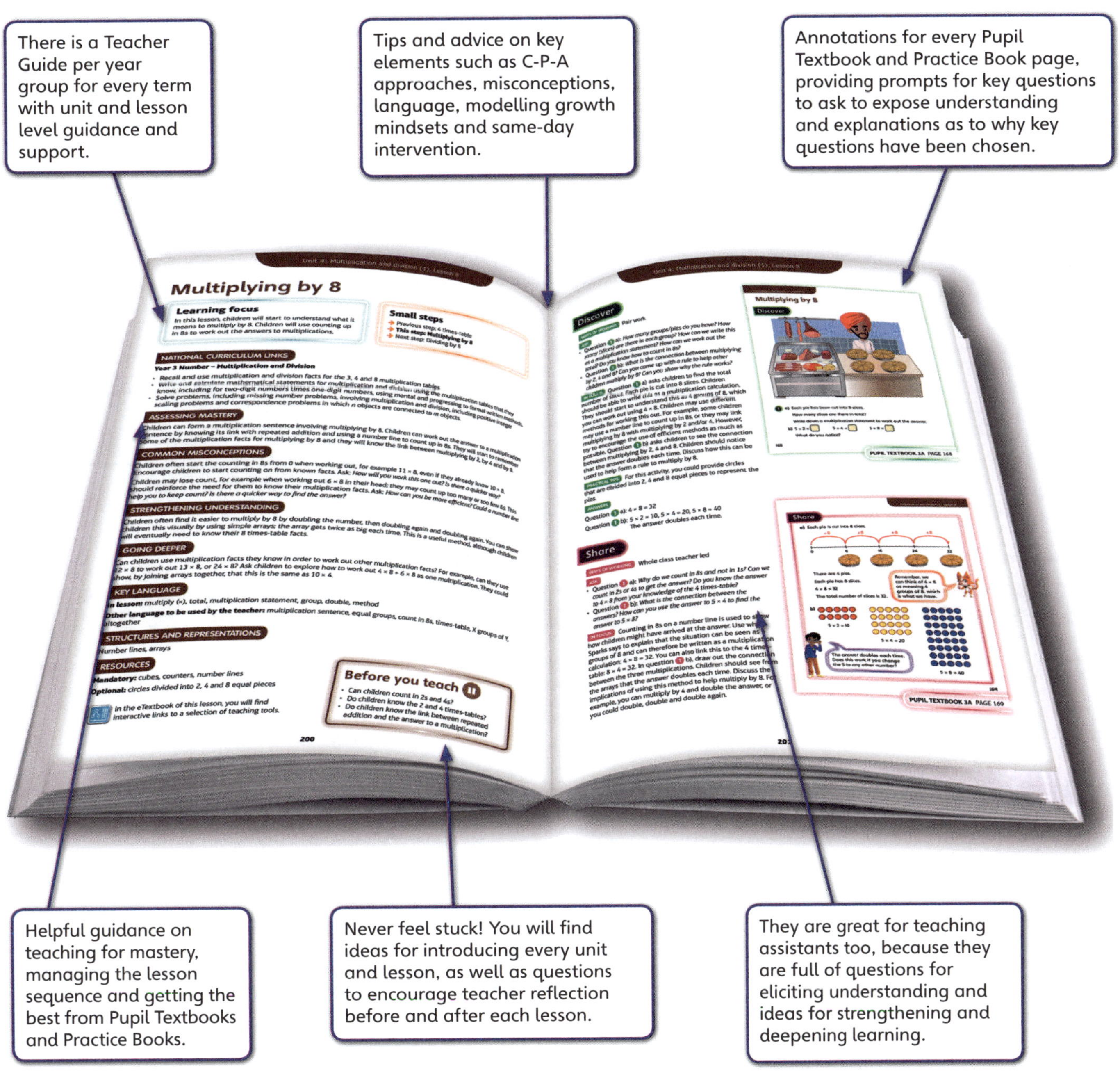

Helpful guidance on teaching for mastery, managing the lesson sequence and getting the best from Pupil Textbooks and Practice Books.

Never feel stuck! You will find ideas for introducing every unit and lesson, as well as questions to encourage teacher reflection before and after each lesson.

They are great for teaching assistants too, because they are full of questions for eliciting understanding and ideas for strengthening and deepening learning.

At the end of each unit, your Teacher Guide helps you identify who has fully grasped the concept, who has not and how to move every child forward. This is covered later in the Assessment strategies section.

Power Maths Year 4, yearly overview

Textbook	Strand	Unit		Number of Lessons
Textbook A / Practice Book A (Term 1)	Number – number and place value	1	Place value – 4-digit numbers (1)	9
	Number – number and place value	2	Place value – 4-digit numbers (2)	9
	Number – addition and subtraction	3	Addition and subtraction	15
	Measurement	4	Measure – perimeter	5
	Number – multiplication and division	5	Multiplication and division (1)	11
Textbook B / Practice Book B (Term 2)	Number – multiplication and division	6	Multiplication and division (2)	15
	Measurement	7	Measure – area	5
	Number – fractions (including decimals)	8	Fractions (1)	7
	Number – fractions (including decimals)	9	Fractions (2)	8
	Number – fractions (including decimals)	10	Decimals (1)	10
Textbook C / Practice Book C (Term 3)	Number – fractions (including decimals)	11	Decimals (2)	7
	Measurement	12	Money	9
	Measurement	13	Time	5
	Statistics	14	Statistics	5
	Geometry – properties of shapes	15	Geometry – angles and 2D shapes	10
	Geometry – position and direction	16	Geometry – position and direction	6

Power Maths Year 4, Textbook 4B (Term 2) overview

Strand 1	Strand 2	Unit		Lesson number	Lesson title	NC Objective 1	NC Objective 2
Number – multiplication and division	Year 5 – number – multiplication and division	Unit 6	Multiplication and division (2)	1	Problem solving – addition and multiplication	Solve problems involving multiplying and adding, including using the distributive law to multiply two digit numbers by one digit, integer scaling problems and harder correspondence problems such as n objects are connected to m objects	Solve problems involving addition, subtraction, multiplication and division and a combination of these, including understanding the meaning of the equals sign
Number – multiplication and division	Year 5 – number – multiplication and division	Unit 6	Multiplication and division (2)	2	Problem solving – mixed problems	Solve problems involving multiplying and adding, including using the distributive law to multiply two digit numbers by one digit, integer scaling problems and harder correspondence problems such as n objects are connected to m objects	Solve problems involving addition, subtraction, multiplication and division and a combination of these, including understanding the meaning of the equals sign
Number – multiplication and division		Unit 6	Multiplication and division (2)	3	Using written methods to multiply	Multiply two-digit and three-digit numbers by a one-digit number using formal written layout	
Number – multiplication and division		Unit 6	Multiplication and division (2)	4	Multiplying a 2-digit number by a 1-digit number	Multiply two-digit and three-digit numbers by a one-digit number using formal written layout	
Number – multiplication and division		Unit 6	Multiplication and division (2)	5	Multiplying a 3-digit number by a 1-digit number	Multiply two-digit and three-digit numbers by a one-digit number using formal written layout	
Number – multiplication and division		Unit 6	Multiplication and division (2)	6	Problem solving – multiplication	Solve problems involving multiplying and adding, including using the distributive law to multiply two digit numbers by one digit, integer scaling problems and harder correspondence problems such as n objects are connected to m objects	Multiply two-digit and three-digit numbers by a one-digit number using formal written layout

Strand 1	Strand 2	Unit		Lesson number	Lesson title	NC Objective 1	NC Objective 2
Number – multiplication and division		Unit 6	Multiplication and division (2)	7	Multiplying more than two numbers (1)	Solve problems involving multiplying and adding, including using the distributive law to multiply two digit numbers by one digit, integer scaling problems and harder correspondence problems such as *n* objects are connected to *m* objects	
Number – multiplication and division		Unit 6	Multiplication and division (2)	8	Multiplying more than two numbers (2)	Recognise and use factor pairs and commutativity in mental calculations	
Number – multiplication and division		Unit 6	Multiplication and division (2)	9	Problem solving – mixed correspondence problems	Recognise and use factor pairs and commutativity in mental calculations	
Number – multiplication and division		Unit 6	Multiplication and division (2)	10	Dividing a 2-digit number by a 1-digit number (1)	Recognise and use factor pairs and commutativity in mental calculations	Solve problems involving multiplying and adding, including using the distributive law to multiply two digit numbers by one digit, integer scaling problems and harder correspondence problems such as *n* objects are connected to *m* objects
Number – multiplication and division		Unit 6	Multiplication and division (2)	11	Division with remainders	Multiply two-digit and three-digit numbers by a one-digit number using formal written layout	Use place value, known and derived facts to multiply and divide mentally, including: multiplying by 0 and 1; dividing by 1; multiplying together three numbers
Number – multiplication and division		Unit 6	Multiplication and division (2)	12	Dividing a 2-digit number by a 1-digit number (2)	Use place value, known and derived facts to multiply and divide mentally, including: multiplying by 0 and 1; dividing by 1; multiplying together three numbers	
Number – multiplication and division		Unit 6	Multiplication and division (2)	13	Dividing a 2-digit number by a 1-digit number (3)	Use place value, known and derived facts to multiply and divide mentally, including: multiplying by 0 and 1; dividing by 1; multiplying together three numbers	Multiply two-digit and three-digit numbers by a one-digit number using formal written layout
Number – multiplication and division		Unit 6	Multiplication and division (2)	14	Dividing a 3-digit number by a 1-digit number	Use place value, known and derived facts to multiply and divide mentally, including: multiplying by 0 and 1; dividing by 1; multiplying together three numbers	
Number – multiplication and division		Unit 6	Multiplication and division (2)	15	Problem solving – division	Solve problems involving multiplying and adding, including using the distributive law to multiply two digit numbers by one digit, integer scaling problems and harder correspondence problems such as *n* objects are connected to *m* objects	
Measurement		Unit 7	Measure – area	1	What is area?	Find the area of rectilinear shapes by counting squares	Estimate, compare and calculate different measures, including money in pounds and pence
Measurement		Unit 7	Measure – area	2	Counting squares (1)	Find the area of rectilinear shapes by counting squares	
Measurement		Unit 7	Measure – area	3	Counting squares (2)	Find the area of rectilinear shapes by counting squares	
Measurement		Unit 7	Measure – area	4	Making shapes	Find the area of rectilinear shapes by counting squares	
Measurement		Unit 7	Measure – area	5	Comparing area	Estimate, compare and calculate different measures, including money in pounds and pence	
Number – fractions (including decimals)		Unit 8	Fractions (1)	1	Tenths and hundredths (1)	Count up and down in hundredths; recognise that hundredths arise when dividing an object by one hundred and dividing tenths by ten	
Number – fractions (including decimals)		Unit 8	Fractions (1)	2	Tenths and hundredths (2)	Count up and down in hundredths; recognise that hundredths arise when dividing an object by one hundred and dividing tenths by ten	

Strand 1	Strand 2	Unit		Lesson number	Lesson title	NC Objective 1	NC Objective 2
Number – fractions (including decimals)		Unit 8	Fractions (1)	3	Equivalent fractions (1)	Recognise and show, using diagrams, families of common equivalent fractions	
Number – fractions (including decimals)		Unit 8	Fractions (1)	4	Equivalent fractions (2)	Recognise and show, using diagrams, families of common equivalent fractions	
Number – fractions (including decimals)		Unit 8	Fractions (1)	5	Simplifying fractions	Recognise and show, using diagrams, families of common equivalent fractions	
Number – fractions (including decimals)		Unit 8	Fractions (1)	6	Fractions greater than 1 (1)	Solve problems involving increasingly harder fractions to calculate quantities, and fractions to divide quantities, including non-unit fractions where the answer is a whole number	
Number – fractions (including decimals)		Unit 8	Fractions (1)	7	Fractions greater than 1 (2)	Solve problems involving increasingly harder fractions to calculate quantities, and fractions to divide quantities, including non-unit fractions where the answer is a whole number	
Number – fractions (including decimals)		Unit 9	Fractions (2)	1	Adding fractions	Add and subtract fractions with the same denominator	
Number fractions (including decimals)		Unit 9	Fractions (2)	2	Subtracting fractions (1)	Add and subtract fractions with the same denominator	Solve problems involving increasingly harder fractions to calculate quantities, and fractions to divide quantities, including non-unit fractions where the answer is a whole number
Number – fractions (including decimals)		Unit 9	Fractions (2)	3	Subtracting fractions (2)	Add and subtract fractions with the same denominator	Solve problems involving increasingly harder fractions to calculate quantities, and fractions to divide quantities, including non-unit fractions where the answer is a whole number
Number – fractions (including decimals)		Unit 9	Fractions (2)	4	Problem solving – adding and subtracting fractions (1)	Solve problems involving increasingly harder fractions to calculate quantities, and fractions to divide quantities, including non-unit fractions where the answer is a whole number	
Number – fractions (including decimals)		Unit 9	Fractions (2)	5	Problem solving – adding and subtracting fractions (2)	Solve problems involving increasingly harder fractions to calculate quantities, and fractions to divide quantities, including non-unit fractions where the answer is a whole number	
Number – fractions (including decimals)		Unit 9	Fractions (2)	6	Calculating fractions of a quantity	Solve problems involving increasingly harder fractions to calculate quantities, and fractions to divide quantities, including non-unit fractions where the answer is a whole number	
Number – fractions (including decimals)		Unit 9	Fractions (2)	7	Problem solving – fraction of a quantity (1)	Solve problems involving increasingly harder fractions to calculate quantities, and fractions to divide quantities, including non-unit fractions where the answer is a whole number	
Number – fractions (including decimals)		Unit 9	Fractions (2)	8	Problem solving – fraction of a quantity (2)	Solve problems involving increasingly harder fractions to calculate quantities, and fractions to divide quantities, including non-unit fractions where the answer is a whole number	
Number – fractions (including decimals)		Unit 10	Decimals (1)	1	Tenths (1)	Recognise and write decimal equivalents of any number of tenths or hundredths	

Strand 1	Strand 2	Unit		Lesson number	Lesson title	NC Objective 1	NC Objective 2
Number – fractions (including decimals)		Unit 10	Decimals (1)	2	Tenths (2)	Recognise and write decimal equivalents of any number of tenths or hundredths	
Number – fractions (including decimals)		Unit 10	Decimals (1)	3	Tenths (3)	Recognise and write decimal equivalents of any number of tenths or hundredths	Solve simple measure and money problems involving fractions and decimals to two decimal places
Number – fractions (including decimals)		Unit 10	Decimals (1)	4	Dividing by 10 (1)	Find the effect of dividing a one- or two-digit number by 10 and 100, identifying the value of the digits in the answer as ones, tenths and hundredths	
Number – fractions (including decimals)		Unit 10	Decimals (1)	5	Dividing by 10 (2)	Find the effect of dividing a one- or two-digit number by 10 and 100, identifying the value of the digits in the answer as ones, tenths and hundredths	
Number – fractions (including decimals)		Unit 10	Decimals (1)	6	Hundredths (1)	Recognise and write decimal equivalents of any number of tenths or hundredths	Count up and down in hundredths; recognise that hundredths arise when dividing an object by one hundred and dividing tenths by ten
Number – fractions (including decimals)		Unit 10	Decimals (1)	7	Hundredths (2)	Recognise and write decimal equivalents of any number of tenths or hundredths	Count up and down in hundredths; recognise that hundredths arise when dividing an object by one hundred and dividing tenths by ten
Number – fractions (including decimals)		Unit 10	Decimals (1)	8	Hundredths (3)	Find the effect of dividing a one- or two-digit number by 10 and 100, identifying the value of the digits in the answer as ones, tenths and hundredths	Count up and down in hundredths; recognise that hundredths arise when dividing an object by one hundred and dividing tenths by ten
Number – fractions (including decimals)		Unit 10	Decimals (1)	9	Dividing by 100	Find the effect of dividing a one- or two-digit number by 10 and 100, identifying the value of the digits in the answer as ones, tenths and hundredths	
Number – fractions (including decimals)		Unit 10	Decimals (1)	10	Dividing by 10 and 100	Find the effect of dividing a one- or two-digit number by 10 and 100, identifying the value of the digits in the answer as ones, tenths and hundredths	

Mindset: an introduction

Global research and best practice deliver the same message: learning is greatly affected by what learners perceive they can or cannot do. What is more, it is also shaped by what their parents, carers and teachers perceive they can do. Mindset – the thinking that determines our beliefs and behaviours – therefore has a fundamental impact on teaching and learning.

Everyone can!

Power Maths and mastery methods focus on the distinction between 'fixed' and 'growth' mindsets (Dweck, 2007).[1] Those with a fixed mindset believe that their basic qualities (for example, intelligence, talent and ability to learn) are pre-wired or fixed: 'If you have a talent for maths, you will succeed at it. If not, too bad!' By contrast, those with a growth mindset believe that hard work, effort and commitment drive success and that 'smart' is not something you are or are not, but something you become. In short, everyone can do maths!

Key mindset strategies

A growth mindset needs to be actively nurtured and developed. *Power Maths* offers some key strategies for fostering healthy growth mindsets in your classroom.

It is okay to get it wrong

Mistakes are valuable opportunities to re-think and understand more deeply. Learning is richer when children and teachers alike focus on spotting and sharing mistakes as well as solutions.

Praise hard work

Praise is a great motivator, and by focusing on praising effort and learning rather than success, children will be more willing to try harder, take risks and persist for longer.

Mind your language!

The language we use around learners has a profound effect on their mindsets. Make a habit of using growth phrases, such as, 'Everyone can!', 'Mistakes can help you learn' and 'Just try for a little longer'. The king of them all is one little word, 'yet ... I cannot solve this ... yet!' Encourage parents and carers to use the right language too.

Build in opportunities for success

The step-by-small-step approach enables children to enjoy the experience of success. In addition, avoid ability grouping and encourage every child to answer questions and explain or demonstrate their methods to others.

[1]Dweck, C (2007) *The New Psychology of Success*, Ballantine Books: New York

The *Power Maths* characters

The *Power Maths* characters model the traits of growth mindset learners and encourage resilience by prompting and questioning children as they work. Appearing frequently in the Textbooks and Practice Books, they are your allies in teaching and discussion, helping to model methods, alternatives and misconceptions, and to pose questions. They encourage and support your children, too: they are all hardworking, enthusiastic and unafraid of making and talking about mistakes.

Meet the team!

Flexible Flo is open-minded and sometimes indecisive. She likes to think differently and come up with a variety of methods or ideas.

Determined Dexter is resolute, resilient and systematic. He concentrates hard, always tries his best and he'll never give up – even though he doesn't always choose the most efficient methods!

Curious Ash is eager, interested and inquisitive, and he loves solving puzzles and problems. Ash asks lots of questions but sometimes gets distracted.

Sparks the Cat

Brave Astrid is confident, willing to take risks and unafraid of failure. She is never scared to jump straight into a problem or question, and although she often makes simple mistakes she is happy to talk them through with others.

Mathematical language

Traditionally, we in the UK have tended to try simplifying mathematical language to make it easier for young children to understand. By contrast, evidence and experience show that by diluting the correct language, we actually mask concepts and meanings for children. We then wonder why they are confused by new and different terminology later down the line! *Power Maths* is not afraid of 'hard' words and avoids placing any barriers between children and their understanding of mathematical concepts. As a result, we need to be planned, precise and thorough in building every child's understanding of the language of maths. Throughout the Teacher Guides you will find support and guidance on how to deliver this, as well as individual explanations throughout the Pupil Textbooks.

Use the following key strategies to build children's mathematical vocabulary, understanding and confidence.

Precise and consistent

Everyone in the classroom should use the correct mathematical terms in full, every time. For example, refer to 'equal parts', not 'parts'. Used consistently, precise maths language will be a familiar and non-threatening part of children's everyday experience.

Full sentences

Teachers and children alike need to use full sentences to explain or respond. When children use complete sentences, it both reveals their understanding and embeds their knowledge.

Stem sentences

These important sentences help children express mathematical concepts accurately, and are used throughout the *Power Maths* books. Encourage children to repeat them frequently, whether working independently or with others. Examples of stem sentences are:

'4 is a part, 5 is a part, 9 is the whole.'

'There are … groups. There are … in each group.'

Key vocabulary

The unit starters highlight essential vocabulary for every lesson. In the Pupil Textbooks, characters flag new terminology and the Teacher Guide lists important mathematical language for every unit and lesson. New terms are never introduced without a clear explanation.

Mathematical signs

Mathematical signs are used early on so that children quickly become familiar with them and their meaning. Often, the *Power Maths* characters will highlight the connection between language and particular signs.

The role of talk and discussion

When children learn to talk purposefully together about maths, barriers of fear and anxiety are broken down and they grow in confidence, skills and understanding. Building a healthy culture of 'maths talk' empowers their learning from day one.

Explanation and discussion are integral to the *Power Maths* structure, so by simply following the books your lessons will stimulate structured talk. The following key 'maths talk' strategies will help you strengthen that culture and ensure that every child is included.

Sentences, not words

Encourage children to use full sentences when reasoning, explaining or discussing maths. This helps both speaker and listeners to clarify their own understanding. It also reveals whether or not the speaker truly understands, enabling you to address misconceptions as they arise.

Working together

Working with others in pairs, groups or as a whole class is a great way to support maths talk and discussion. Use different group structures to add variety and challenge. For example, children could take timed turns for talking, work independently alongside a 'discussion buddy', or perhaps play different *Power Maths* character roles within their group.

Think first – then talk

Provide clear opportunities within each lesson for children to think and reflect, so that their talk is purposeful, relevant and focused.

Give every child a voice

Where the 'hands up' model allows only the more confident child to shine, *Power Maths* involves everyone. Make sure that no child dominates and that even the shyest child is encouraged to contribute – and is praised when they do.

Assessment strategies

Teaching for mastery demands that you are confident about what each child knows and where their misconceptions lie: therefore, practical and effective assessment is vitally important.

Formative assessment within lessons

The Think together section will often reveal any confusions or insecurities: try ironing these out by doing the first Think together question as a class. For children who continue to struggle, you or your teaching assistant should provide support and enable them to move on.

Performance in Practice can be very revealing: check Practice Books and listen out both during and after practice to identify misconceptions.

The Reflect section is designed to check on the all-important depth of understanding. Be sure to review how children performed in this final stage before you teach the next lesson.

End of unit check – Textbook

Each unit concludes with a summative check to help you assess quickly and clearly each child's understanding, fluency, reasoning and problem-solving skills. In KS2 this check also contains a SATs-style question to help children become familiar with answering this type of question.

In KS2 we would suggest the End of unit check is completed independently in children's exercise books, but you can adapt this to suit the needs of your class.

End of unit check – Practice Book

The Practice Book contains further opportunities for assessment, and can be completed by children independently while you are carrying out diagnostic assessment with small groups. Your Teacher Guide will advise you on what to do if children struggle to articulate an explanation – or perhaps encourage you to write down something they have explained well. It will also offer insights into children's answers and their implications for the next learning steps. It is split into three main sections, outlined below.

My journal

My journal is designed to allow children to show their depth of understanding of the unit. It can also serve as a way of checking that children have grasped key mathematical vocabulary. Children should have some time to think about how they want to answer the question, and you could ask them to talk to a partner about their ideas. Then children should write their answer in their Practice Book.

Power check

The Power check allows children to self-assess their level of confidence on the topic by colouring in different smiley faces. You may want to introduce the faces as follows:

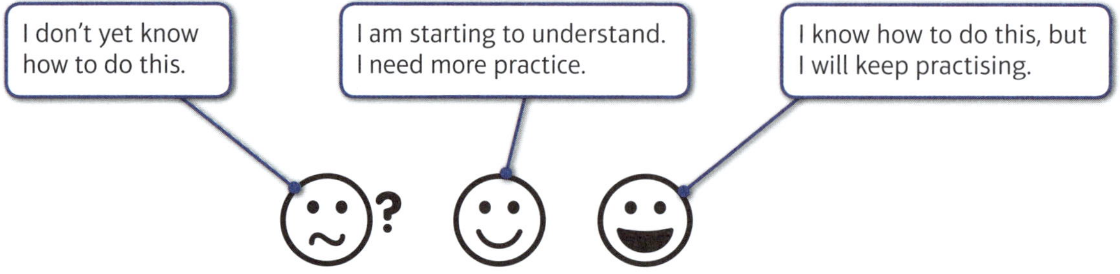

Power play or Power puzzle

Each unit ends with either a Power play or a Power puzzle. This is an activity, puzzle or game that allows children to use their new knowledge in a fun, informal way. In Key Stage 2 we have also included a deeper level to each game to help challenge those children who have grasped a concept quickly.

How to use diagnostic questions

The diagnostic questions provided in *Power Maths* Textbooks are carefully structured to identify both understanding and misconceptions (if children answer in a particular way, you will know why). The simple procedure below may be helpful:

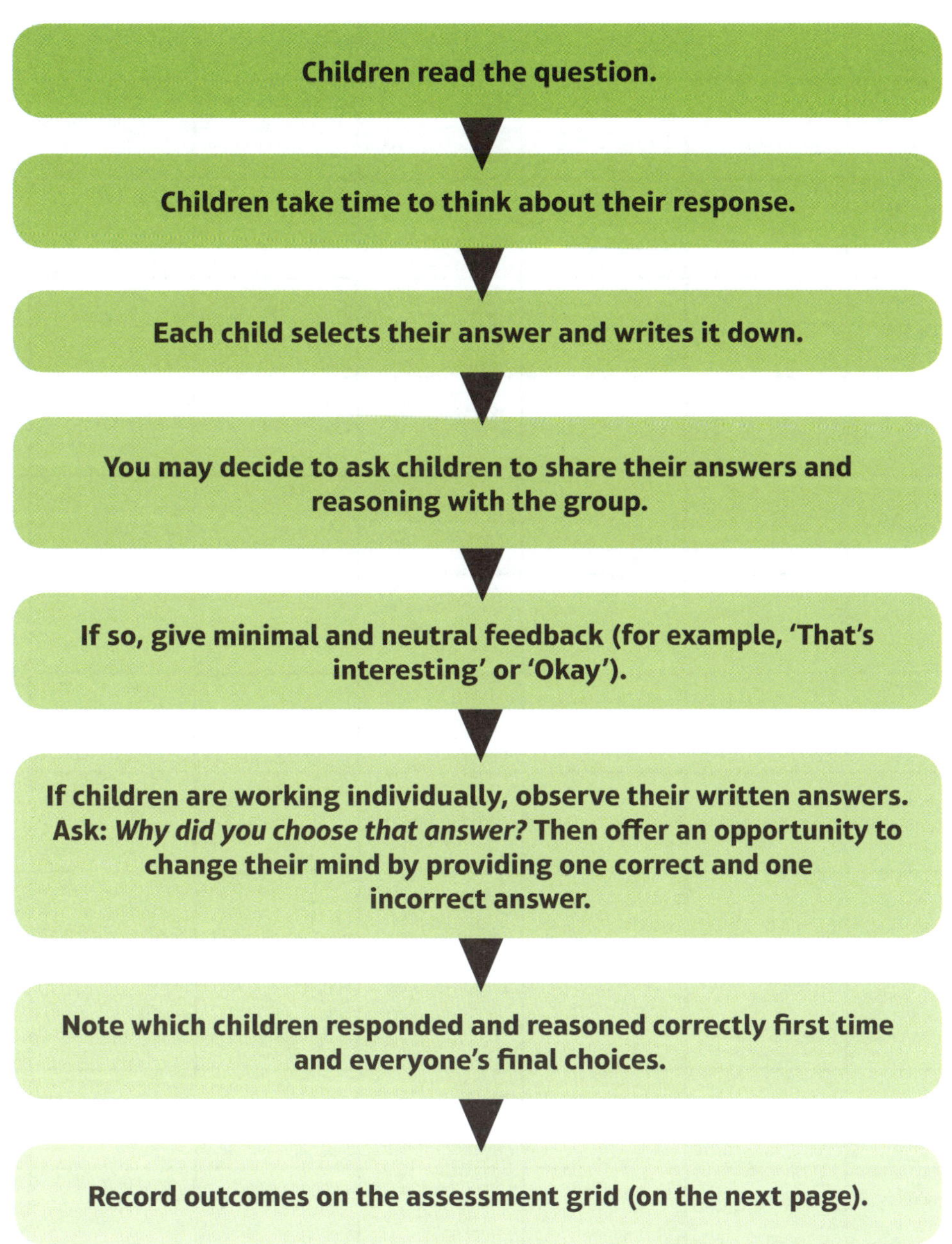

Power Maths unit assessment grid

Year ____   Unit ____ ________________________________

Record only as much information as you judge appropriate for your assessment of each child's mastery of the unit and any steps needed for intervention.

Name	Diagnostic questions	SATs-style question	My journal	Power check	Power play/puzzle	Mastery	Intervention/ Strengthen

Keeping the class together

Traditionally, children who learn quickly have been accelerated through the curriculum. As a consequence, their learning may be superficial and will lack the many benefits of enabling children to learn with and from each other.

By contrast, *Power Maths'* mastery approach values real understanding and richer, deeper learning above speed. It sees all children learning the same concept in small, cumulative steps, each finding and mastering challenge at their own level. Remember that when you teach for mastery, EVERYONE can do maths! Those who grasp a concept easily have time to explore and understand that concept at a deeper level. The whole class therefore moves through the curriculum at broadly the same pace via individual learning journeys.

For some teachers, the idea that a whole class can move forward together is revolutionary and challenging. However, the evidence of global good practice clearly shows that this approach drives engagement, confidence, motivation and success for all learners, and not just the high flyers. The strategies below will help you keep your class together on their maths journey.

Mix it up

Do not stick to set groups at each table. Every child should be working on the same concept, and mixing up the groupings widens children's opportunities for exploring, discussing and sharing their understanding with others.

Recycling questions

Reuse the Pupil Textbook and Practice Book questions with concrete materials to allow children to explore concepts and relationships and deepen their understanding. This strategy is especially useful for reinforcing learning in same-day interventions.

Strengthen at every opportunity

The next lesson in a *Power Maths* sequence always revises and builds on the previous step to help embed learning. These activities provide golden opportunities for individual children to strengthen their learning with the support of teaching assistants.

Prepare to be surprised!

Children may grasp a concept quickly or more slowly. The 'fast graspers' won't always be the same individuals, nor does the speed at which a child understands a concept predict their success in maths. Are they struggling or just working more slowly?

Depth and breadth

Just as prescribed in the National Curriculum, the goal of *Power Maths* is never to accelerate through a topic but rather to gain a clear, deep and broad understanding.

"Pupils who grasp concepts rapidly should be challenged through being offered rich and sophisticated problems before any acceleration through new content. Those who are not sufficiently fluent with earlier material should consolidate their understanding, including through additional practice, before moving on."

National Curriculum: Mathematics programmes of study: KS1 & 2, 2013

The lesson sequence offers many opportunities for you to deepen and broaden children's learning, some of which are suggested below.

Discover

As well as using the questions in the Teacher Guide, check that children are really delving into why something is true. It is not enough to simply recite facts, such as '6 + 3 = 9'. They need to be able to see why, explain it, and to demonstrate the solution in several ways.

Share

Make sure that every child is given chances to offer answers and expand their knowledge and not just those with the greatest confidence.

Think together

Encourage children to think about how they found the solution and explain it to their partner. Be sure to make concrete materials available on group tables throughout the lesson to support and reinforce learning.

Practice

Avoid any temptation to select questions according to your assessment of ability: practice questions are presented in a logical sequence and it is important that each child works through every question.

Reflect

Open-ended questions allow children to deepen their understanding as far as they can by discovering new ways of finding answers. For example, *Give me another way of working out how high the wall is … And another way?*

Online materials

For each unit you will find additional strengthening activities to support those children who need it and to deepen the understanding of those who need the additional challenge.

Same-day intervention

Since maths competence depends on mastering concepts one-by-one in a logical progression, it is important that no gaps in understanding are ever left unfilled. Same-day interventions – either within or after a lesson – are a crucial safety net for any child who has not fully made the small step covered that day. In other words, intervention is always about keeping up, not catching up, so that every child has the skills and understanding they need to tackle the next lesson. That means presenting the same problems used in the lesson, with a variety of concrete materials to help children model their solutions.

We offer two intervention strategies below, but you should feel free to choose others if they work better for your class.

Within-lesson intervention

The Think together activity will reveal those who are struggling, so when it is time for Practice, bring these children together to work with you on the first Practice questions. Observe these children carefully, ask questions, encourage them to use concrete models and check that they reach and can demonstrate their understanding.

After-lesson intervention

You might like to use Think together before an assembly, giving you or teaching assistants time to recap and expand with slow graspers during assembly time. Teaching assistants could also work with strugglers at other convenient points in the school day.

The role of practice

Practice plays a pivotal role in the *Power Maths* approach. It takes place in class groups, smaller groups, pairs and independently, so that children always have the opportunities for thinking as well as the models and support they need to practise meaningfully and with understanding.

Intelligent practice

In *Power Maths*, practice never equates to the simple repetition of a process. Instead we embrace the concept of intelligent practice, in which all children become fluent in maths through varied, frequent and thoughtful practice that deepens and embeds conceptual understanding in a logical, planned sequence. To see the difference, take a look at the following examples.

Traditional practice

- Repetition can be rote – no need for a child to think hard about what they are doing.

- Praise may be misplaced.

- Does this prove understanding?

Intelligent practice

- Varied methods – concrete, pictorial and abstract.

- Calculations expressed in different ways, requiring thought and understanding.

- Constructive feedback.

All practice questions are designed to move children on and reveal misconceptions.

Simple, logical steps build onto earlier learning.

C-P-A runs throughout – different ways of modelling and understanding the same concept.

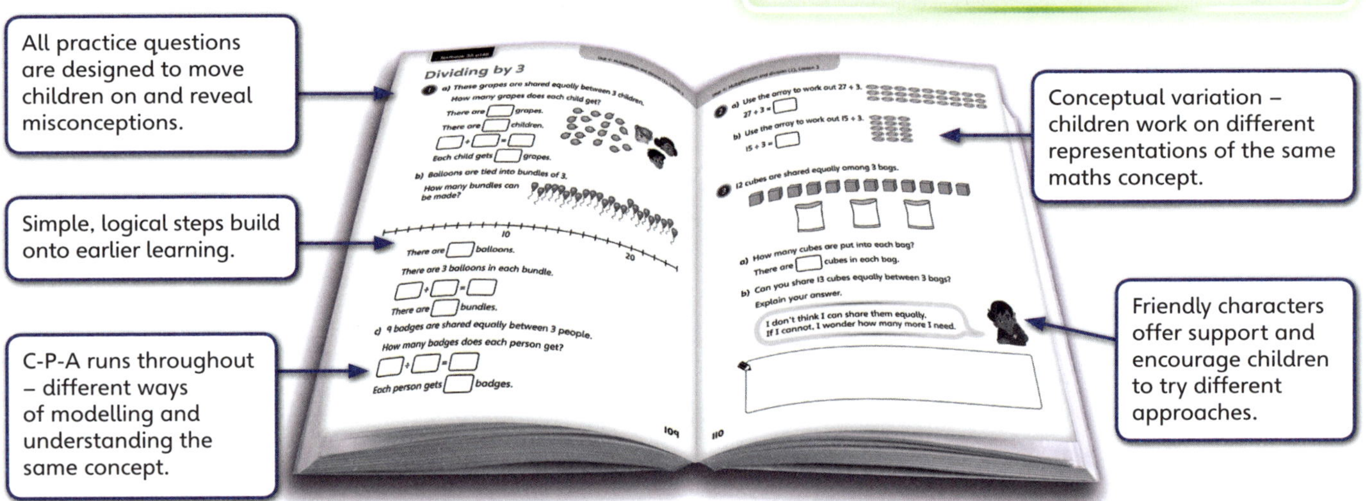

Conceptual variation – children work on different representations of the same maths concept.

Friendly characters offer support and encourage children to try different approaches.

A carefully designed progression

The Practice Books provide just the right amount of intelligent practice for children to complete independently in the final section of each lesson. It is really important that all children are exposed to the Practice questions, and that children are not directed to complete different sections. That is because each question is different and has been designed to challenge children to think about the maths they are doing. The questions become more challenging so children grasping concepts more quickly will start to slow down as they progress. Meanwhile, you have the chance to circulate and spot any misconceptions before they become barriers to further learning.

Homework and the role of carers

While *Power Maths* does not prescribe any particular homework structure, we acknowledge the potential value of practice at home. For example, practising fluency in key facts, such as number bonds and times-tables, is an ideal homework task, and carers could work through uncompleted Practice Book questions with children at either primary stage.

However, it is important to recognise that many parents and carers may themselves lack confidence in maths, and few, if any, will be familiar with mastery methods. A Parents' and Carers' Evening that helps them understand the basics of mindsets, mastery and mathematical language is a great way to ensure that children benefit from their homework. It could be a fun opportunity for children to teach their families that everyone can do maths!

Structures and representations

Unlike most other subjects, maths comprises a wide array of abstract concepts – and that is why children and adults so often find it difficult. By taking a Concrete-Pictorial-Abstract (C-P-A) approach, *Power Maths* allows children to tackle concepts in a tangible and more comfortable way.

Non-linear stages

Concrete

Replacing the traditional approach of a teacher working through a problem in front of the class, the concrete stage introduces real objects that children can use to 'do' the maths – any familiar object that a child can manipulate and move to help bring the maths to life. It is important to appreciate, however, that children must always understand the link between models and the objects they represent. For example, children need to first understand that three cakes could be represented by three pretend cakes, and then by three counters or bricks. Frequent practice helps consolidate this essential insight. Although they can be used at any time, good concrete models are an essential first step in understanding.

Pictorial

This stage uses pictorial representations of objects to let children 'see' what particular maths problems look like. It helps them make connections between the concrete and pictorial representations and the abstract maths concept. Children can also create or view a pictorial representation together, enabling discussion and comparisons. The *Power Maths* teaching tools are fantastic for this learning stage, and bar modelling is invaluable for problem solving throughout the primary curriculum.

Abstract

Our ultimate goal is for children to understand abstract mathematical concepts, signs and notation and, of course, some children will reach this stage far more quickly than others. To work with abstract concepts, a child needs to be comfortable with the meaning of, and relationships between, concrete, pictorial and abstract models and representations. The C-P-A approach is not linear, and children may need different types of models at different times. However, when a child demonstrates with concrete models and pictorial representations that they have grasped a concept, we can be confident that they are ready to explore or model it with abstract signs such as numbers and notation.

Use at any time and with any age to support understanding.

Practical aspects of *Power Maths*

One of the key underlying elements of *Power Maths* is its practical approach, allowing you to make maths real and relevant to your children, no matter their age.

Manipulatives are essential resources for both key stages and *Power Maths* encourages teachers to use these at every opportunity, and to continue the Concrete-Pictorial-Abstract approach right through to Year 6.

The Textbooks and Teacher Guides include lots of opportunities for teaching in a practical way to show children what maths means in real life.

Discover and Share

The Discover and Share sections of the Textbook give you scope to turn a real-life scenario into a practical and hands-on section of the lesson. Use these sections as inspiration to get active in the classroom. Where appropriate, use the Discover contexts as a springboard for your own examples that have particular resonance for your children – and allow them to get their hands dirty trying out the mathematics for themselves.

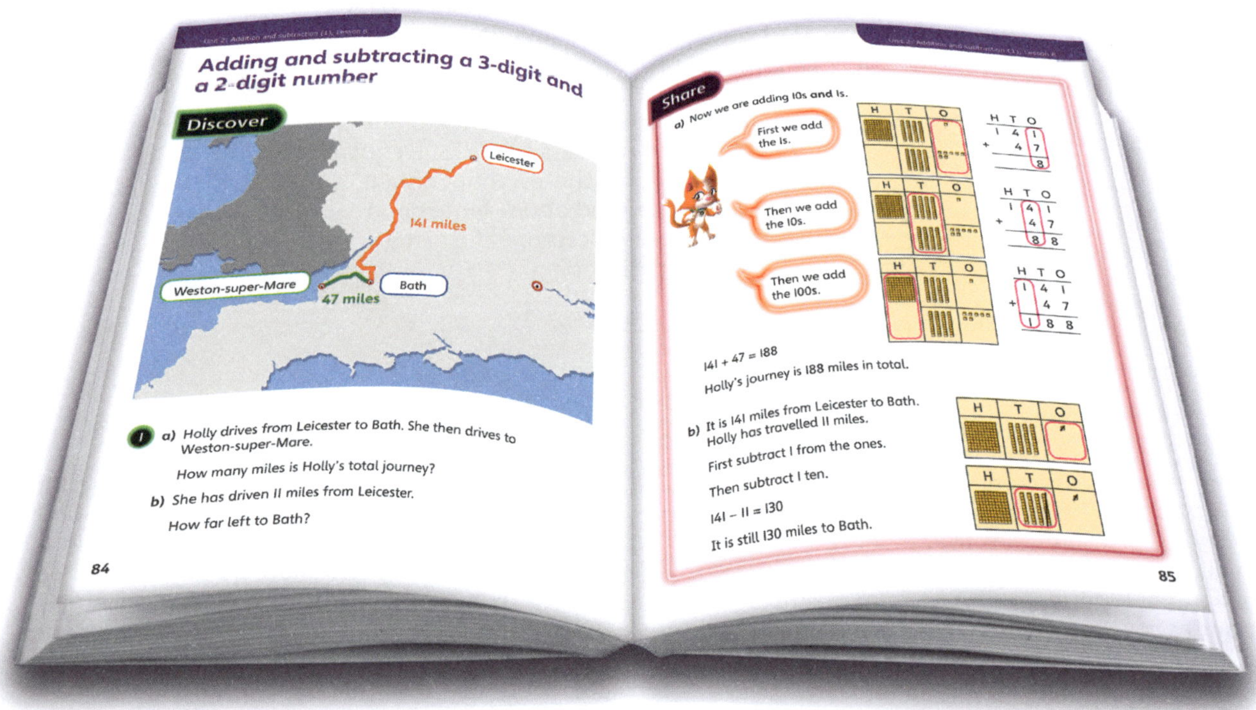

Unit videos

Every unit has a video which incorporates real-life classroom sequences.

These videos show you how the reasoning behind mathematics can be carried out in a practical manner by showing real children using various concrete and pictorial methods to come to the solution. You can see how using these practical models, such as part-whole and bar models, helps them to find and articulate their answer.

Mastery tips

Mastery Experts give anecdotal advice on where they have used hands-on and real-life elements to inspire their children.

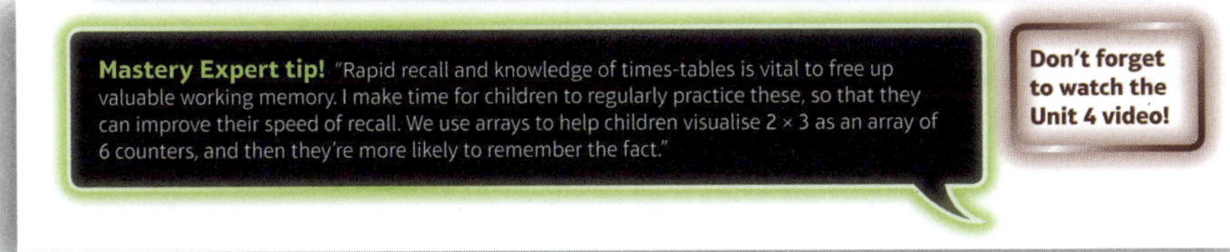

Concrete-Pictorial-Abstract (C-P-A) approach

Each Share section uses various methods to explain an answer, helping children to access abstract concepts by using concrete tools, such as counters. Remember this isn't a linear process, so even children who appear confident using the more abstract method can deepen their knowledge by exploring the concrete representations. Encourage children to use all three methods to really solidify their understanding of a concept.

Pictorial representation – drawing the problem in a logical way that helps children visualise the maths.

Concrete representation – using manipulatives to represent the problem. Encourage children to physically use resources to explore the maths.

Abstract representation – using words and calculations to represent the problem.

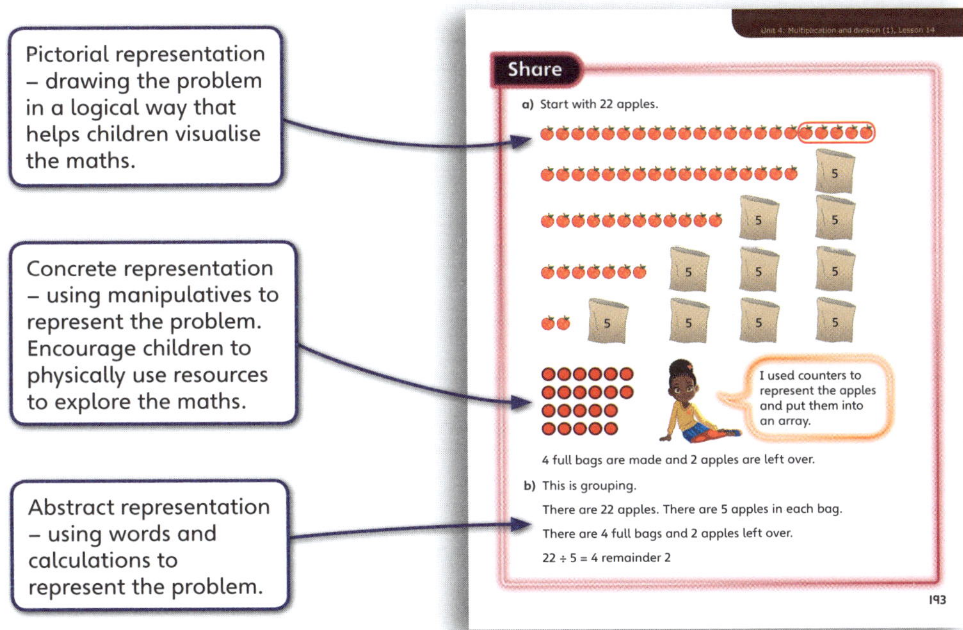

Practical tips

Every lesson suggests how to draw out the practical side of the Discover context.

You'll find these tips in the Discover section of the Teacher Guide for each lesson.

PRACTICAL TIPS You could use balls, counters or cubes under plastic cups to re-enact the artwork and help children get a feel for this activity.

Resources

Every lesson lists the practical resources you will need or might want to use. There is also a summary of all the resources used throughout the term on page 34 to help you be prepared.

RESOURCES

Mandatory: cubes, counters, number lines

Optional: balls, plastic cups

List of practical resources

Year 4B Mandatory resources

Resource	Lesson
2D shapes such as hexagons and triangles	**Unit 6** lesson 13
A variety of flat, non-standard items to measure with (plastic counters, flat coloured squares or triangles, playing cards, coins)	**Unit 7** lesson 1
Arrays	**Unit 6** lesson 12
Base 10 equipment	**Unit 6** lesson 11
Calendar	**Unit 6** lesson 13
Counters	**Unit 6** lessons 1, 3, 7, 8, 12, 15 **Unit 7** lessons 1, 4 **Unit 8** lesson 1
Cubes	**Unit 6** lessons 1, 3, 7, 8, 14, 15
Fraction cards	**Unit 9** lesson 1
Hundredth grids	**Unit 8** lesson 1 **Unit 10** lessons 6, 9
Number lines	**Unit 6** lesson 7
Paper squares	**Unit 7** lesson 4
Place value counters	**Unit 6** lessons 4, 11, 14 **Unit 10** lessons 1, 2, 3, 4, 5, 6, 7, 8, 10
Place value grid	**Unit 10** lesson 9
Print-outs of an 11 × 12 square grid	**Unit 6** lessons 14
Squared paper	**Unit 7** lessons 2, 3, 4
Square dotted paper	**Unit 7** lesson 4

Year 4B Optional resources

Resource	Lesson
4 times-table flashcards	**Unit 6** lesson 7
Base 10 equipment	**Unit 6** lesson 13 **Unit 8** lesson 1 **Unit 9** lessons 1, 2, 3, 4, 5, 6, 7, 8
Bean bags	**Unit 6** lesson 12
Bricks	**Unit 6** lesson 10
Chalk	**Unit 7** lesson 4
Counters	**Unit 6** lessons 2, 4, 5, 6, 10, 13 **Unit 7** lesson 5 **Unit 9** lesson 6, 8
Cubes	**Unit 6** lessons 2, 4, 5 ,6, 11, 12 **Unit 9** lesson 8
Dice	**Unit 10** lesson 8
Digit cards	**Unit 10** lesson 2
Fraction strips	**Unit 8** lessons 3, 4 **Unit 9** lesson 3
Fraction wall	**Unit 8** lessons 3, 4, 5
Fraction wheels	**Unit 9** lesson 4
Geoboards	**Unit 7** lesson 2
Grids	**Unit 7** lesson 5
Hundredths grid	**Unit 8** lesson 2 **Unit 10** lesson 10
Number lines	**Unit 6** lessons 2, 6 **Unit 8** lessons 2, 7 **Unit 9** lesson 7 **Unit 10** lesson 8
Paper butterflies	**Unit 6** lesson 2
Paper squares	**Unit 7** lessons 1, 5
Paper shapes, some cut into fractions	**Unit 8** lesson 6
Part-whole model	**Unit 10** lesson 6
Picture cards of buckets and spades, pictures of shorts, shirts and different-coloured socks	**Unit 6** lesson 9
Place value counters	**Unit 9** lesson 8 **Unit 10** lessons 4, 5, 9
Place value grid	**Unit 10** lesson 8
Real objects to represent fractions (of a whole)	**Unit 8** lesson 6 **Unit 9** lessons 3, 4, 6
Rulers (with mm)	**Unit 10** lesson 3
Scissors	**Unit 10** lesson 10
Strips of paper	**Unit 9** lesson 5
Ten frames	**Unit 10** lesson 4
Variety of squares and rectangles to measure (book covers, newspaper pages, paper or card)	**Unit 7** lesson 1
Wooden blocks	**Unit 6** lesson 2

Variation helps visualisation

Children find it much easier to visualise and grasp concepts if they see them presented in a number of ways, so be prepared to offer and encourage many different representations.

For example, the number six could be represented in various ways:

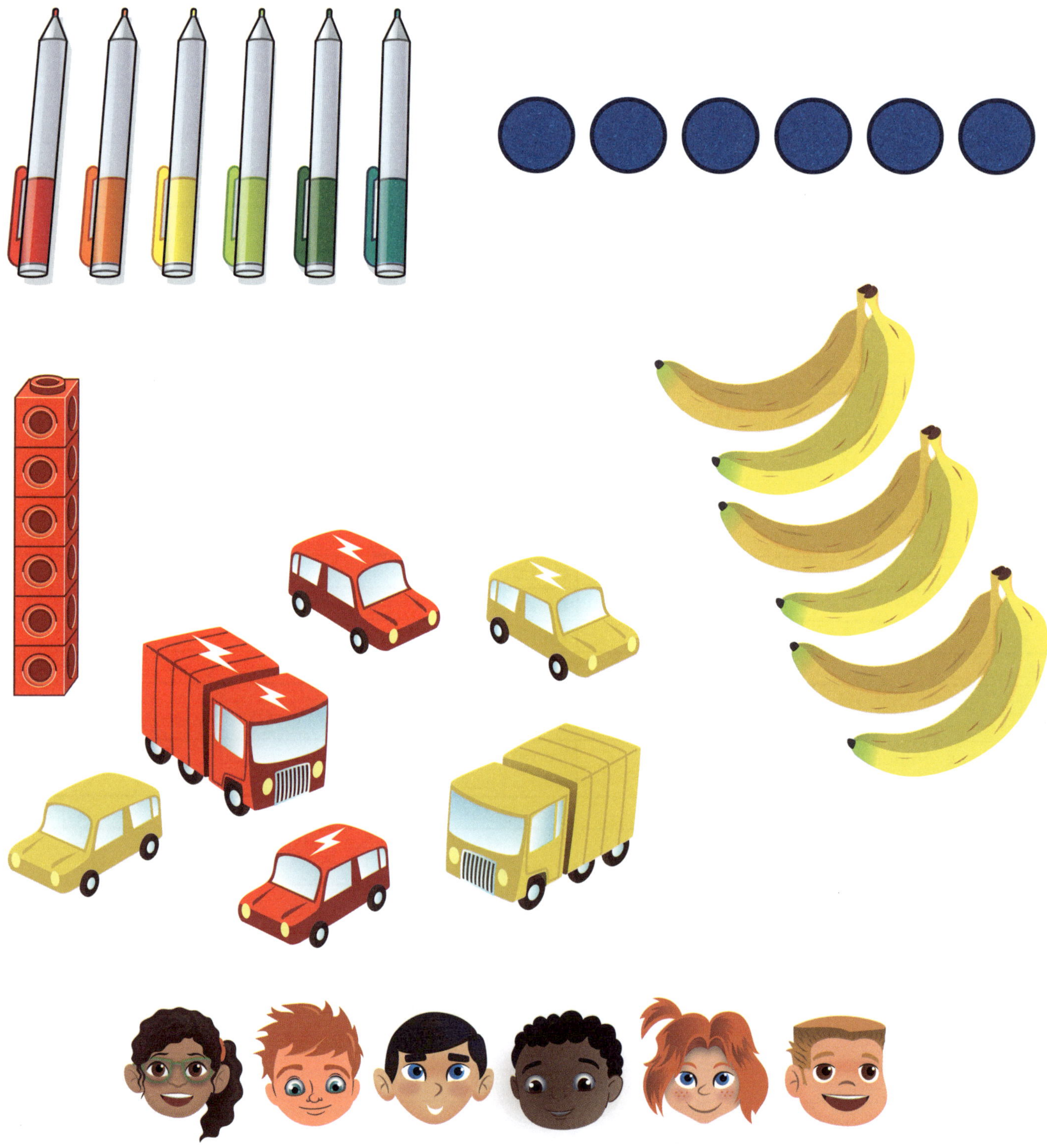

Getting started with *Power Maths*

As you prepare to put *Power Maths* into action, you might find the tips and advice below helpful.

STEP 1: Train up!

A practical, up-front, full-day professional development course will give you and your team a brilliant head-start as you begin your *Power Maths* journey. You will learn more about the ethos, how it works and why.

STEP 2: Check out the progression

Take a look at the yearly and termly overviews. Next take a look at the unit overview for the unit you are about to teach in your Teacher Guide, remembering that you can match your lessons and pacing to your class.

STEP 3: Explore the context

Take a little time to look at the context for this unit: what are the implications for the unit ahead? (Think about key language, common misunderstandings and intervention strategies, for example.) If you have the online subscription, don't forget to watch the corresponding unit video.

STEP 4: Prepare for your first lesson

Familiarise yourself with the objectives, essential questions to ask and the resources you will need. The Teacher Guide offers tips, ideas and guidance on individual lessons to help you anticipate children's misconceptions and challenge those who are ready to think more deeply.

STEP 5: Teach and reflect

Deliver your lesson – and enjoy!

Afterwards, reflect on how it went … Did you cover all five stages? Does the lesson need more time? How could you improve it? What percentage of your class do you think mastered the concept? How can you help those who didn't?

Unit 6
Multiplication and division ❷

Mastery Expert tip! "I ask children to make information posters which explain how to multiply and divide, including mental and written calculations. Each lesson, children add more information to the poster. This is very helpful for children, not only because they can refer to it throughout the lessons, but because they can also explain misconceptions and give examples of 'easy' and 'difficult' questions. Making the poster themselves helps children to take ownership of their own learning."

WHY THIS UNIT IS IMPORTANT

This unit builds on exploring written and mental calculation strategies for multiplying and dividing. Children explore in depth the distributive and associative properties of multiplication. The learning progresses from Year 3, where children used expanded methods for 2-digit × 1-digit numbers, to Year 4 where they are using the compressed single line (standard) formal multiplication. Children learn to solve more complex problems building on n objects related to m objects, find all solutions and notice how to use multiplication to solve questions. Children use partitioning to divide 2- and 3-digit numbers by a 1-digit number. They recap on the concept of a remainder after division, and move on to predicting whether a number will have a division and what the number could be if the remainder is given. Children then move on to solve simple 2-step problems that involve all of the four operations.

WHERE THIS UNIT FITS

→ Unit 5: Multiplication and division (1)
→ **Unit 6: Multiplication and division (2)**
→ Unit 7: Measure – area

This unit expands learning from Year 3, where children developed confidence in knowing when to multiply and an understanding of the difference between equal grouping and sharing. This unit also builds on what children learnt in Year 3 about remainders and on work in Year 4 Unit 5, where children learnt their multiplication facts up to 12 × 12.

Before they start this unit, it is expected that children:

- know that numbers can be partitioned in different ways
- know that multiplication can be seen as repeated addition, and division as repeated subtraction
- know that an array shows two multiplications, for example 5 × 4 = 4 × 5.

ASSESSING MASTERY

Children who have mastered this unit can show that they understand what multiplication facts mean and then apply these in order to solve problems. Children will be able to solve a range of multiplication and division multi-step problems. They are able to use key information from a question to draw a bar model and to work out whether they need to multiply or divide first. Children can use a formal written method to multiply a 2-digit and 3-digit number by a single-digit number.

COMMON MISCONCEPTIONS	STRENGTHENING UNDERSTANDING	GOING DEEPER
Children add the number of combinations rather than multiply them. For example, if they have 3 scarves and 2 hats, they think they have 5 combinations they can wear.	Children need to have a lot of practice of combining different items, such as socks, shoes, hats and scarves, before they are able to make the link between the correspondence problems and multiplying to find all possible combinations.	Start by exploring the number of possibilities arising from two choices, before adding a third. Encourage children to think of real life examples of correspondence problems: for example, there might be 1 table for every 2 children, and each child has 3 different colour pens.
When solving problems, children do not always recognise if they need to do a multiplication or division.	Use counters to represent the objects in groups and then use a bar model to represent the problem. You can use counters initially to represent the parts.	Ask children to write simple 1-step and multi-step multiplication and division problems and to explain when a problem requires more than 1 step.

WAYS OF WORKING

Use these pages to remind children of the 3, 4 and 8 times-tables. This is also a good opportunity to remind children of how to use arrays and how to apply the commutative property of multiplication.

STRUCTURES AND REPRESENTATIONS

Arrays: This model shows the total of a multiplication and reinforces commutativity. It can also be used to demonstrate sharing and grouping.

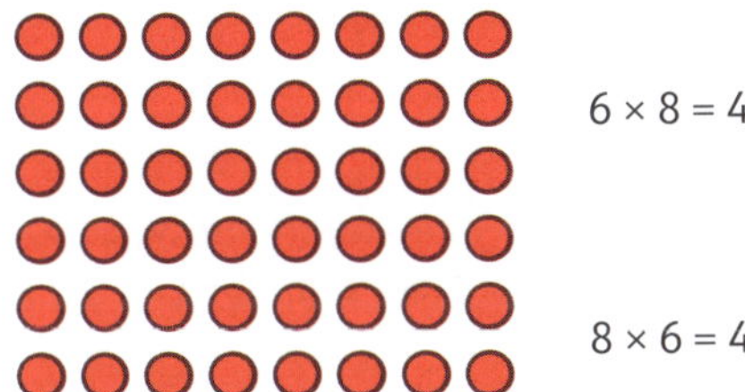

$6 \times 8 = 48$

$8 \times 6 = 48$

Bar model: This model represents the situation in multiplication and division word problems and shows the link between multiplication and repeated addition.

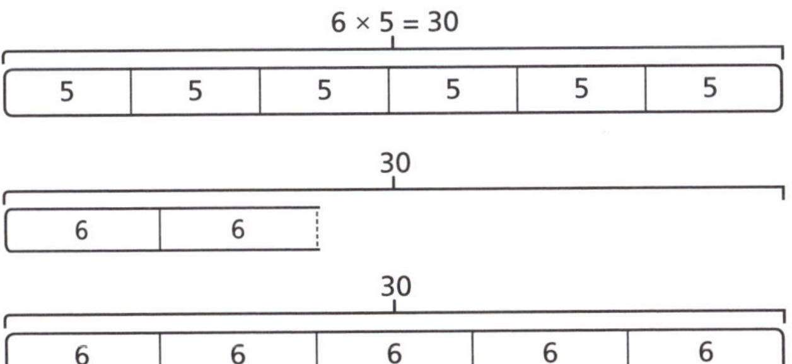

Part-whole model: This model shows how a number can be partitioned in multiplication and division problems.

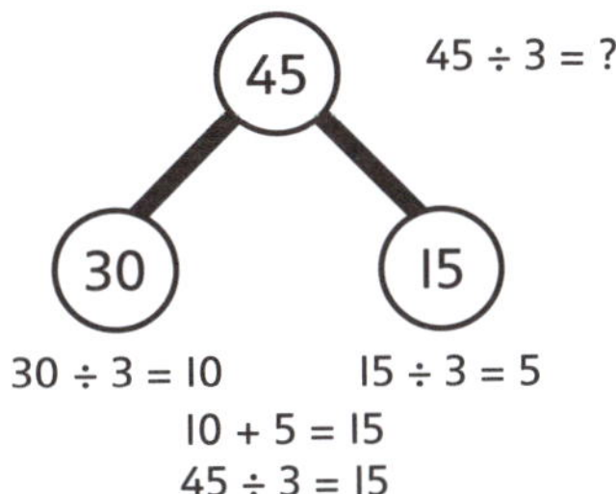

$45 \div 3 = ?$

$30 \div 3 = 10 \qquad 15 \div 3 = 5$

$10 + 5 = 15$

$45 \div 3 = 15$

KEY LANGUAGE

There is some key language that children will need to know as part of the learning in this unit.

➜ multiplication (×), multiplication statement

➜ grouping, groups, equal, total, repeated addition

➜ correspondence, multiply, divide, combinations

➜ divide (÷), division statement

➜ times-tables

➜ whole, left over, remainder

➜ one-step, two-step, multi-step

➜ array, bar model, part-whole model

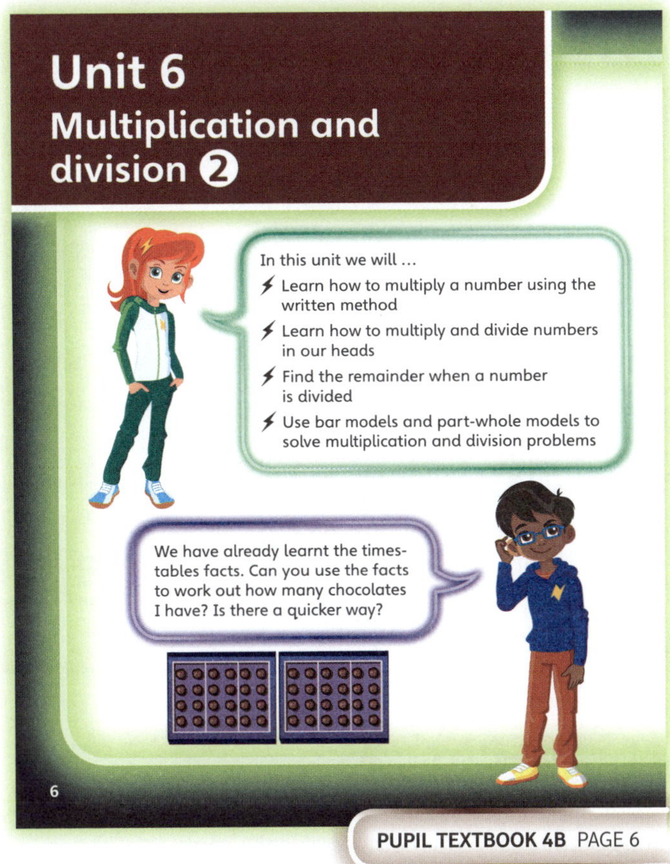

PUPIL TEXTBOOK 4B PAGE 6

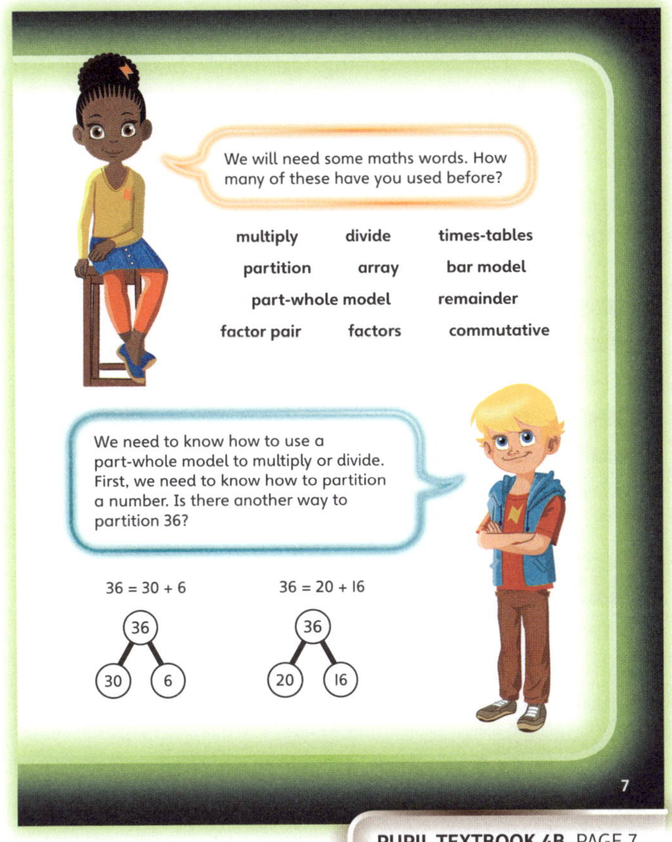

PUPIL TEXTBOOK 4B PAGE 7

Problem solving – addition and multiplication

Learning focus

In this lesson, children will learn to solve addition and multiplication problems. They will discover that multiplying a number by two numbers added together is the same as doing separate multiplications and then adding the answers (known as the distributive law).

Small steps

→ Previous step: 11 and 12 times-tables
→ **This step: Problem solving – addition and multiplication**
→ Next step: Problem solving – mixed problems

NATIONAL CURRICULUM LINKS

Year 4 Number – Multiplication and Division
Solve problems involving multiplication and addition, including using the distributive law to multiply two digit numbers by one digit, integer scaling problems and harder correspondence problems such as *n* objects are connected to *m* objects.

Year 5 Number – Multiplication and Division
Solve problems involving addition, subtraction, multiplication and division and a combination of these, including understanding the meaning of the equals sign.

ASSESSING MASTERY

Children are able to solve multi-step problems involving both addition and multiplication using two different methods. They understand that both methods give the same answer and can explain why this is the case.

COMMON MISCONCEPTIONS

Children may complete multiplication fact assessments satisfactorily, but are unable to apply their knowledge to other arithmetic and problem-solving situations. They know how to multiply 5 × 3, but fail to recognise multiplication in a problem, or to manipulate a question, so that they can solve it in different ways. Ask:

- *How many are in each row? Does each row have the same number of items? How many rows are there? Is there a quicker way to find the total?*

Children may not be able to explain why they should multiply, so when the question does not state what to do, they guess the operation they should use. Ask:

- *Can you draw me a picture? Can you use cubes/counters to explain the question?*

STRENGTHENING UNDERSTANDING

Ask children quick-fire questions to practise partitioning numbers: *Which numbers add to make 8? Which numbers add to make 9?*

GOING DEEPER

Ask children to make up problems that involve addition and multiplication. Ask: *Will you add or multiply first? Does the order matter? Why? Can they draw a bar model to represent the question?*

KEY LANGUAGE

In lesson: total, method, multiplication

Other language to be used by the teacher: times-table, multiply, multiplication sentence, grouping, groups, equal, repeated addition

STRUCTURES AND REPRESENTATIONS

arrays, part-whole model

RESOURCES

Mandatory: cubes, counters

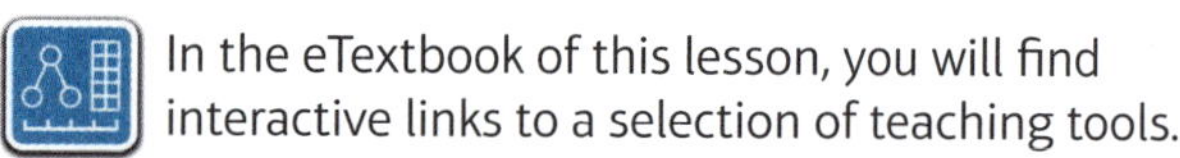 In the eTextbook of this lesson, you will find interactive links to a selection of teaching tools.

Before you teach 

- Are children able to recognise equal groups and unequal groups?
- Can children write a multiplication statement for any equal groups?

Discover

ASK

- Question **1** a): *How many rows of chairs with bows are there? How do you know? How can you use counters to show the groups in the question?*
- Question **1** b): *What does 'total' mean? How can you find it? What multiplication statements can you see? How did you work out the answers? Did you just know some of the answers?*

IN FOCUS Question **1** a) is used as a recap from Year 2 and from Year 3. Children should discuss in pairs how to find the number of chairs in each group. They may use cubes and counters or may be able to see the answer straight away. Question **1** b) demonstrates two methods for working out the total number of chairs and determines whether children can write down the corresponding multiplication sentences. It is important for children to be able to reason why the groups are added and not count the chairs or estimate.

PRACTICAL TIPS Different colour counters and cubes should be available for children to represent the objects in the different groups.

ANSWERS

Question **1** a): 12 chairs have bows. There are 20 plain chairs.

Question **1** b): 4 × 3 = 12, 4 × 5 = 20, 12 + 20 = 32
Or 3 + 5 = 8, 8 × 4 = 32
There are 32 chairs in total.

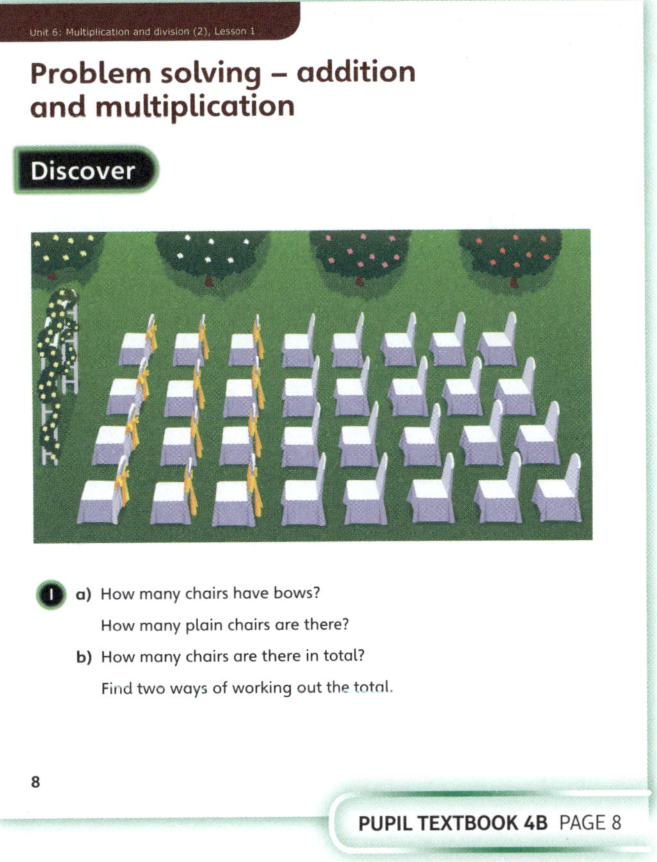

PUPIL TEXTBOOK 4B PAGE 8

Share

WAYS OF WORKING Whole class teacher led

ASK

- Question **1** a): *How many groups of chairs are there? What do the groups have in common? How do they differ? How many chairs are there in each group? What calculation can you use to find 4 rows of 3 chairs? 4 rows of 5 chairs?*
- Question **1** b): *How did you work out the total amount? Why did you add the groups? Why did you add 5 and 3?*

IN FOCUS Question **1** b) is important as it introduces children to the distributive law (that multiplying a number by two numbers added together is the same as doing separate multiplications and then adding). Children learn how to manipulate calculations so they can work them out mentally. Using the distributive law is an effective way of helping children to understand how they can manipulate difficult multiplications to make them more manageable. Discuss that 8 × 4 is the same as the total of 5 × 4 and 3 × 4. Remind children of the need to know their multiplication facts off by heart. Give children a method (for example, repeated addition, linked with counting up by the same amount each time) to help them, in case they cannot remember.

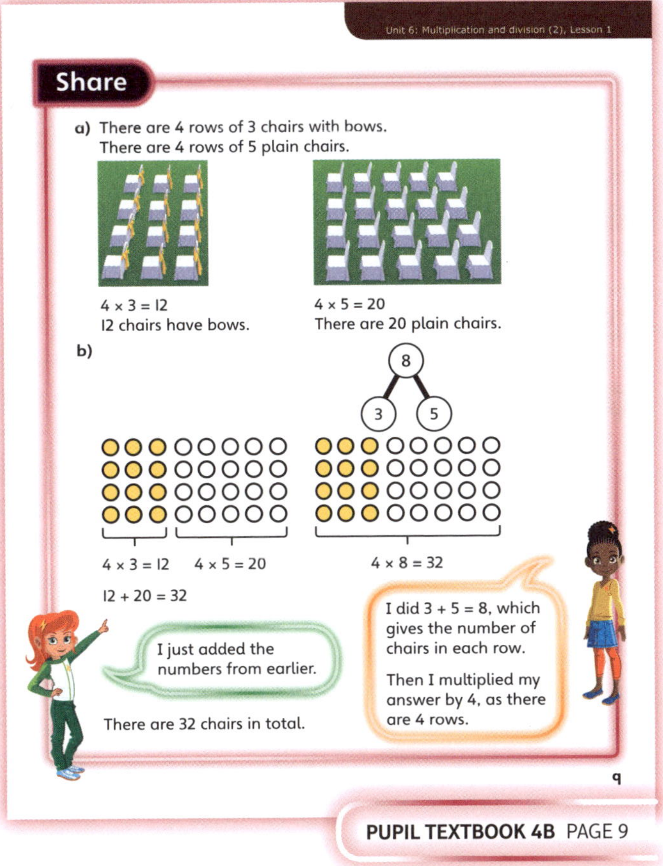

PUPIL TEXTBOOK 4B PAGE 9

Think together

 Whole class teacher led (I do, We do, You do)

ASK

- Question **1**: *How many chairs are there in each row? How many rows are there in each group? How can you find the total amount using Method 1? How can you find the total amount using Method 2? How is Method 1 different from Method 2?*
- Question **2**: *Can you use two methods to solve the question? Which method is more efficient? Which one do you prefer? Why? What multiplication statement can you write?*
- Question **3**: *What do the numbers in the multiplication statement represent? Can you use counters to explain the question? Which two numbers are the same? Which are different? Do you prefer Ash's method or Dexter's? Why?*

IN FOCUS Question **1** checks whether children understand the distributive law and can use both methods of solving the question. Question **2** reinforces the distributive law of multiplication. Ask children to discuss both methods they can use before choosing their preferred one. Question **3** starts with the abstract concept; children can use concrete and pictorial material to explain the multiplication statements. Ask children to look at the question and think of what each number represents. Which number represents the number of rows? Which number represents the number of objects in each row? Discuss that $2 \times 5 + 2 \times 7$ is the same as 2×12.

STRENGTHEN For question **1**, ask children to put counters out on their desk as the chairs are shown and ask them to show each group. Keep the two groups separate to start with, so that children can see clearly how many rows each group has and how many counter of different colours there are in each row. Then join the groups together so that each row is made of two colour counters. This may help children to see why in Method 2 they should add the number of counters in each row and then multiply by the number of rows.

DEEPEN For question **3** b), ask: *Can you see 7 rows of 5, or 7×5? Ask children to make 7×5 in different ways by partitioning 5 in different ways. Is $4 \times 7 + 1 \times 7$ the same as 5×7?*

ASSESSMENT CHECKPOINT Can children use the distributive law of multiplication to find different ways to problem solve?

ANSWERS

Question **1**: Method 1: Chairs with bows: $5 \times 3 = 15$
Plain chairs: $7 \times 3 = 21$
$15 + 21 = 36$
Method 2: $5 + 7 = 12$
$3 \times 12 = 36$
There are 36 chairs in total.

Question **2**: There are 36 cupcakes in total.

Question **3** a): $2 \times 5 + 2 \times 7 = 2 \times 12 = 24$

Question **3** b): 7 groups of 3 counters and 7 groups of 2 counters make 7 groups of 5 counters.
$7 \times 3 + 7 \times 2 = 7 \times 5$
$21 + 14 = 35$ or $7 \times 5 = 35$

Question **3** c): 2 groups of 4 and 3 groups of 2 = 7 groups of 2
$4 \times 2 + 3 \times 2 = 7 \times 2$
$8 + 6 = 14$ or $7 \times 2 = 14$

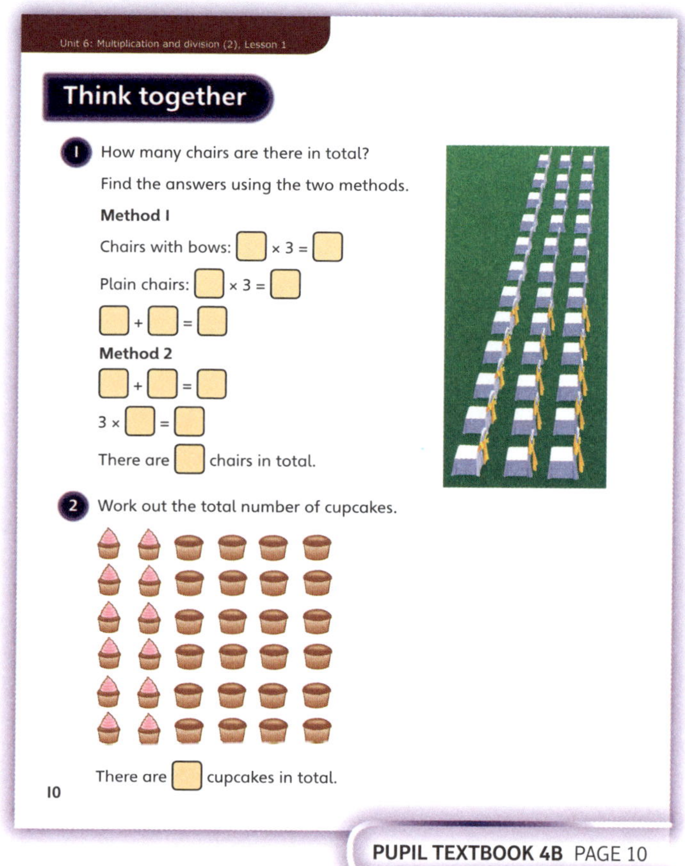

PUPIL TEXTBOOK 4B PAGE 10

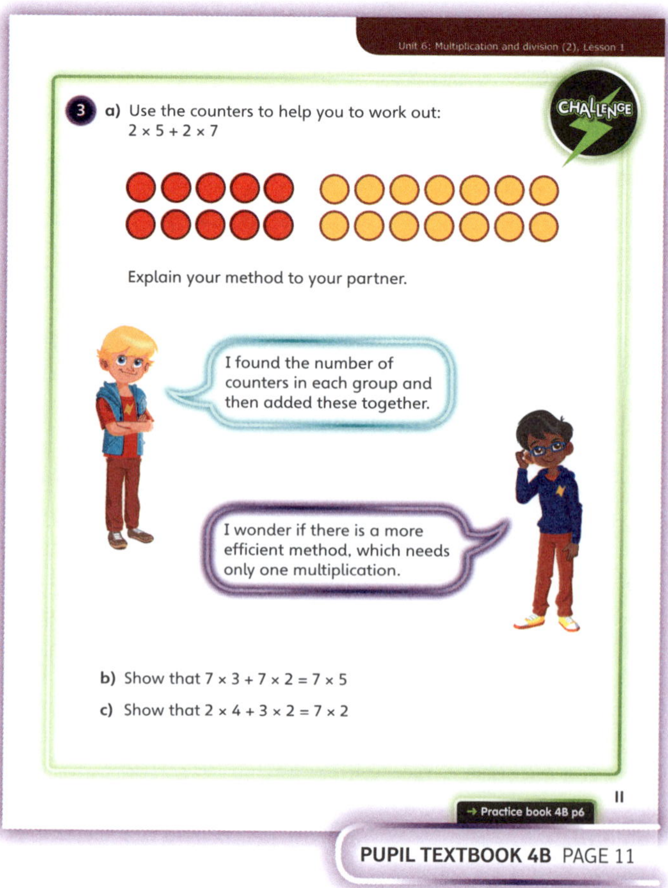

PUPIL TEXTBOOK 4B PAGE 11

Practice

WAYS OF WORKING Independent thinking

IN FOCUS Question **1** requires children to work out the number of counters, using both methods. Questions **2** and **3** require children to work out how many objects there are in total, using the distributive law of multiplication. Throughout these questions, the clear link between calculating the groups separately and joining them together is given. Children may use a number line to count on in order to work out the total. Question **6** asks children to explain the method they would use to find the total number of counters.

STRENGTHEN To support understanding of finding the total amount of objects using the distributive law of multiplication, ask 8 children to line up in 2 rows in front of the class. Ask children to write down the multiplication statement they see: 2 rows of 8, or 2 × 8. Then ask children to stay in the same row, but separate the rows into two groups. The first group has 5 children in each row and the second group has 3 children in each row. Ask children to write down the multiplication they see. Has the number of children changed? Encourage children to use multiplication facts instead of counting.

DEEPEN For question **6**, ask: *Would the answer be different if the light-grey counters were first, then the dark grey and then the white? What would the multiplication statement be?*

ASSESSMENT CHECKPOINT Can children explain how they can find the total number of counters by using two different methods? Can they explain which method is more efficient and why? Can children use repeated addition to help them to work out multiplication sentences?

ANSWERS Answers for the **Practice** part of the lesson appear in the separate **Practice and Reflect answer guide**.

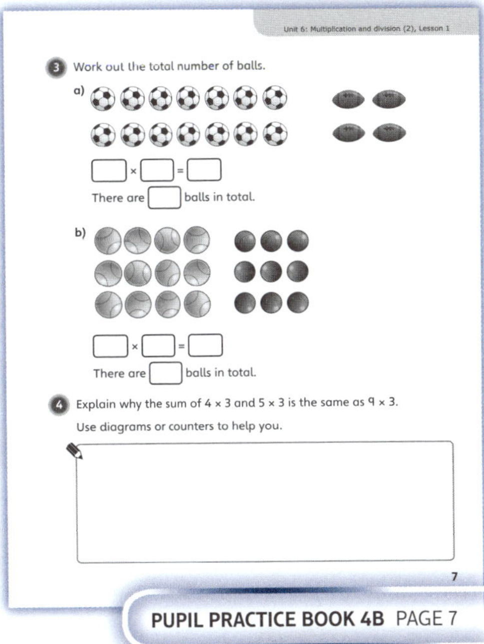

PUPIL PRACTICE BOOK 4B PAGE 6

PUPIL PRACTICE BOOK 4B PAGE 7

Reflect

WAYS OF WORKING Independent thinking

IN FOCUS This question looks to see whether children can use multiplication and addition to problem solve. Children should be able to come up with several different methods. Ask children to see how many different methods they can find. Do they all give the same answers? Why?

ASSESSMENT CHECKPOINT Check if children understand the distributive law in multiplication and check if they write the correct multiplication sentence for each group, before adding them. Can they use resources (counters, cubes) to explain why the statement is correct?

ANSWERS Answers for the **Reflect** part of the lesson appear in the separate **Practice and Reflect answer guide**.

After the lesson 

- Can children explain why groups are added together to find the total? Can they solve the question in a different way?
- Can children write a correct multiplication sentence for each group, with the correct answer?

PUPIL PRACTICE BOOK 4B PAGE 8

Problem solving – mixed problems

Learning focus

In this lesson, children will solve multi-step multiplication and division problems using bar models to expose the underlying structure.

Small steps

→ Previous step: Problem solving – addition and multiplication
→ **This step: Problem solving – mixed problems**
→ Next step: Using written methods to multiply

NATIONAL CURRICULUM LINKS

Year 4 Number – Multiplication and Division
Solve problems involving multiplication and addition, including using the distributive law to multiply two digit numbers by one digit, integer scaling problems and harder correspondence problems such as *n* objects are connected to *m* objects.

Year 5 Number – Multiplication and Division
Solve problems involving addition, subtraction, multiplication and division and a combination of these, including understanding the meaning of the equals sign.

ASSESSING MASTERY

Children will be able to solve a range of multiplication and division multi-step problems. They will be able to use key information from a question to draw a bar model and work out whether they need to multiply or divide first.

COMMON MISCONCEPTIONS

Children may guess the operation they need to use rather than understanding whether they need to multiply or divide. Drawing a bar model to represent the situation may help with this. Ask:
• *Is this multiplication or division? How do you know? What clues are in the question?*

Children may be unable to make the link between the two parts of a question. Ask:
• *Look at the two bar models. What is the same? What is different?*

STRENGTHENING UNDERSTANDING

To support children in understanding what they need to do, read the questions line by line, asking children to identify the key information. Ask: *Have you used all the information given? What does it mean? Why does the question say that? What can you find?* If necessary, ask children to represent the objects with counters or cubes. This may help them to better understand the more abstract questions. Encourage children to draw a bar model to help them see whether they need to multiply or divide.

GOING DEEPER

Encourage children to make up their own multi-step problems. Working out the answers is not as important as designing the question. If children can recognise key words used in multi-step questions, this will help them to gain a deeper understanding of problem-solving questions.

KEY LANGUAGE

In lesson: total, bar model

Other language to be used by the teacher: times-table, multiply, divide, grouping, sharing, multiplication statement, division statement, recall, array, missing value, multiplication, division, inverse operation

STRUCTURES AND REPRESENTATIONS

bar model

RESOURCES

Optional: cubes, counters, number lines, wooden blocks, paper butterflies

 In the eTextbook of this lesson, you will find interactive links to a selection of teaching tools.

Before you teach 

• Do children know the 2, 3, 4, 5, 8 and 10 times-table facts?
• Can children draw a bar model to represent a simple multiplication?
• Can children work out the answer to simple 1-step multiplication and division word problems?

Discover

WAYS OF WORKING Pair work

ASK

• Question ❶ a): *What does the picture show? How are the butterflies arranged? How many butterflies are in the tray? Is it a multiplication or division problem?*
• Question ❶ b): *How are the butterflies arranged now? Has the total number of butterflies changed? What calculation can you do to work out the answer?*

IN FOCUS Ensure that children fully understand the context. Encourage drawing: they could use crosses or circles instead of the butterflies. Some children may want to add six 8s together; help them to realise this is a multiplication. Question ❶ b) is the opposite of question ❶ a) and asks children to find the number of trays needed when the butterflies are grouped in 12s, which should help children to realise that they need to divide. Some children may count up in 12s until they get to 48. Encourage them to see the link with the division.

PRACTICAL TIPS Colourful cut-outs of butterflies could be used so that children can re-create the picture. Alternatively, they could use wooden building blocks, cubes or counters.

ANSWERS

Question ❶ a): $6 \times 8 = 48$
There are 48 butterflies in the tray.

Question ❶ b): $48 \div 12 = 4$
4 new trays are needed.

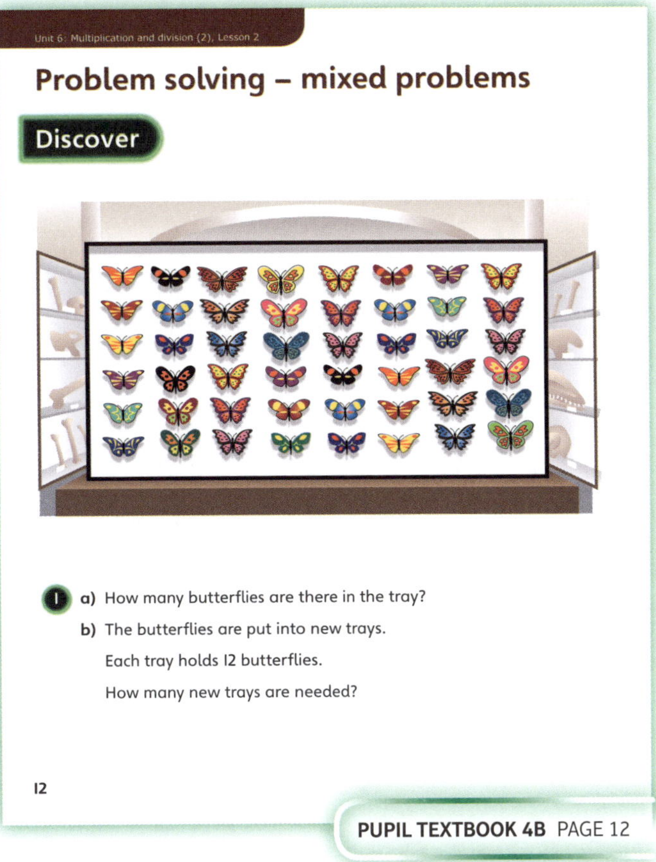

PUPIL TEXTBOOK 4B PAGE 12

Share

WAYS OF WORKING Whole class teacher led

ASK

• Question ❶ a): *How has Astrid worked out the answer to the question? How has her knowledge of arrays helped her? Why has she used a bar model? What does each part of the bar model represent? How can you use the bar model to work out if you need to do a division or multiplication?*
• Question ❶ b): *How has Dexter worked out the answer to the question? Why did he choose to divide? What is the difference between the two bar models used in the whole question?*

IN FOCUS Discuss as a class the different ways that bar models can be used to find the answers. Astrid shows how you can use a bar model to represent the situation when you need to multiply. Dexter uses a bar model when he needs to divide. Ensure children know what each part of the bar model represents and how it leads to a multiplication or division. Discuss with children how they can check that the answer is correct using the inverse operation. Highlight the fact that children need to be secure in their multiplication table facts, as well as confident in adding and subtracting in order to solve the problem.

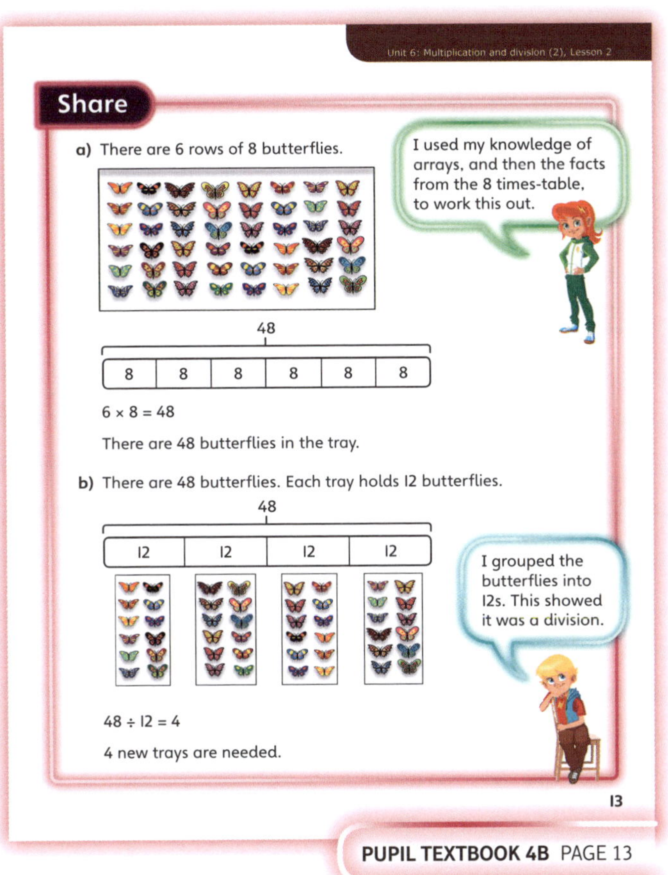

PUPIL TEXTBOOK 4B PAGE 13

Think together

WAYS OF WORKING Whole class teacher led (I do, We do, You do)

ASK

- Question ❶: *How many boxes are there? How many pencils are in each box? Why is it a multiplication? How could a bar model help you to understand that it is a multiplication? How are the pencils shared? How many classes get pencils? How can you work out how many pencils each class gets?*
- Question ❷: *How many parts are in the top bar? What does each part in the top bar represent? How many are there in total? How many parts are there in the bottom bar? What does each represent? Will you multiply or divide? Why?*
- Question ❸: *Why does Flo decide to draw bar models? Do the bar models need to be the same length? Why? What do they represent?*

IN FOCUS Question ❶ builds on the **Discover** task by asking children to find the total number of pencils in all the boxes. You may want to use blocks to help children to recognise the multiplication and division. Question ❷ looks at the use of a bar model. Children will need to realise that the length of the bars is the same – hence the total number has not changed. The way each bar is split will affect the quantity each part represents. Question ❸ requires children to solve a 2-step question and characters prompt children to use bar models. Encourage children to draw the bars neatly one above the other. Ask: *How many parts will each bar have? Why? What does each part represent?* It is important for children to understand fully how the bar model can be used to solve the problem.

STRENGTHEN Encourage children to work through each problem line by line. Ask them to highlight the key information in the question (each question in this example is a multiplication first, followed by division). Ask children to draw or use a bar model to explain each of the steps.

DEEPEN For question ❸, probe the question further and ask children what would happen if the 5 toy cars cost the same as 2 planes. What about 6 planes?

ASSESSMENT CHECKPOINT Children can solve simple multiplication and division word problems and represent a 2-step problem as a bar model to help them to see whether they need to do a multiplication or division. They recall multiplication facts to work out the answers.

ANSWERS

Question ❶ a): $5 \times 8 = 40$
There are 40 pencils in total.

Question ❶ b): $40 \div 2 = 20$
Each class gets 20 pencils.

Question ❷: $4 \times 6 = 24$
$24 \div 3 = 8$
$? = 8$

Question ❸: $5 \times 6 = 30$
$30 \div 3 = 10$
A toy plane costs £10.

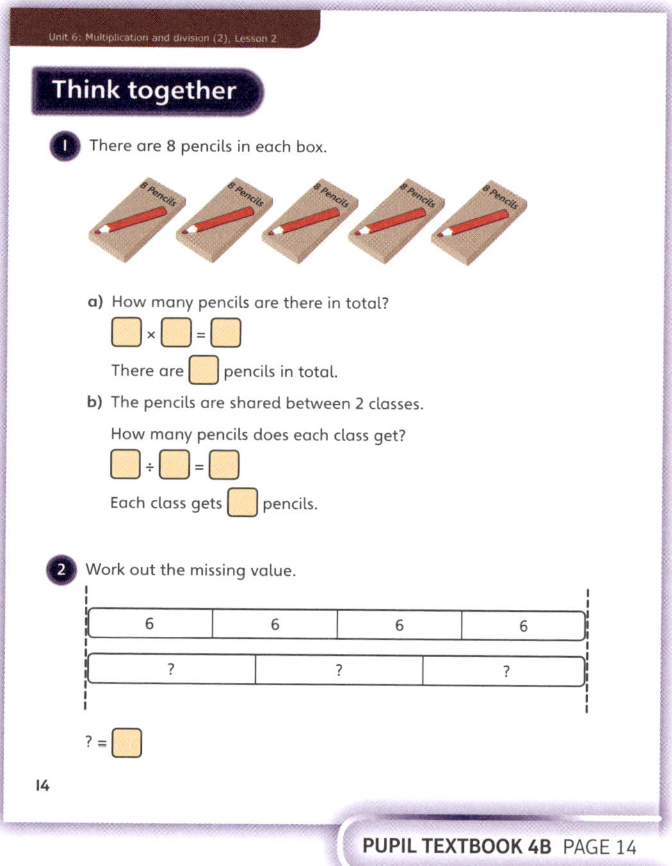

PUPIL TEXTBOOK 4B PAGE 14

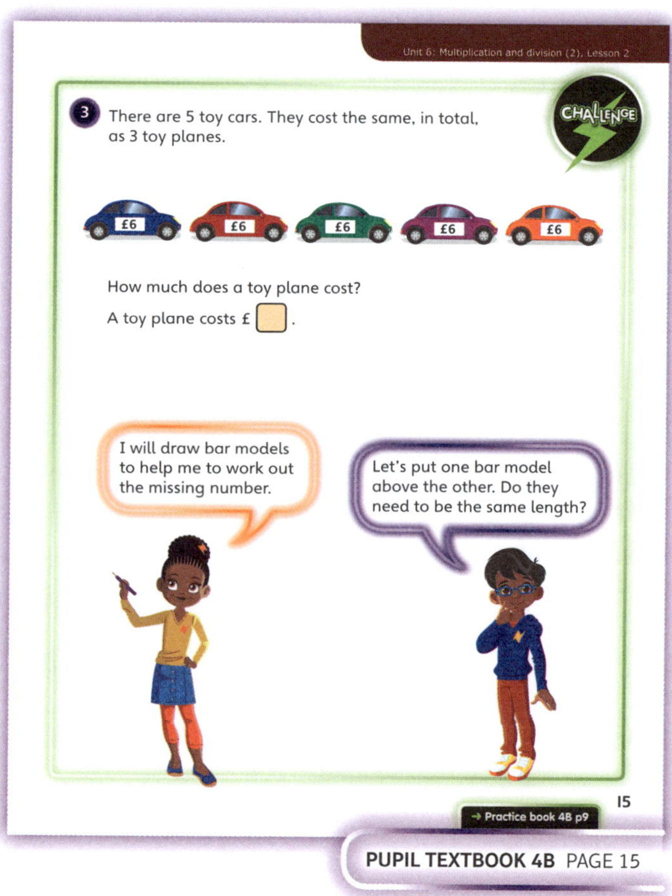

PUPIL TEXTBOOK 4B PAGE 15

Practice

WAYS OF WORKING Independent thinking

IN FOCUS Questions ❶ and ❷ provide children with an example structure for the answers: they will need to use this structured approach when problem solving later on in the unit. Using a single bar model may help children to understand the steps they need to take, so they are more confident when solving questions later. The use of a bar model also reminds children that bar models can be used in problem-solving questions to find an unknown.

STRENGTHEN For 2-step problems, it is helpful for children to see a visualisation of why they have to work out one part before the other. To support understanding, use counters and cubes to represent the different groups of objects; this will help children to see that, in questions ❶ and ❷, they first need to do a multiplication and then division. Encourage children to draw a bar model as this may help them to see whether they need to multiply or divide. For question ❸, provide children with cubes so that they can build the towers for themselves and visualise the question better; this may help them to see whether they need to multiply or divide.

DEEPEN In question ❻, children can make up their own multi-step problems based on the information given. For example: *How much do 5 small cones and 2 large cones cost? How much change will you get from £30?*

THINK DIFFERENTLY Question ❺ uses the context of weight and asks children to work in hundreds, incorporating their learning from the previous term.

ASSESSMENT CHECKPOINT Can children convert a word problem into a bar model? Can they use the bar model to work out whether they need to do a multiplication or division in order to find the answer? Can children solve multi-step problems involving multiplication and division?

ANSWERS Answers for the **Practice** part of the lesson appear in the separate **Practice and Reflect answer guide**.

Reflect

WAYS OF WORKING Pair work

IN FOCUS Drawing and using bar models allows children to understand and visualise the question. If children are unsure how to draw a bar model, or what the bars represent, they will not be able to use the bar model structure when problem solving. Children need to understand that if the bars are equal, they represent the same quantity.

Using the bar model correctly will help children to move from concrete to abstract calculations and to decide which operation they will use.

ASSESSMENT CHECKPOINT Check whether children have correctly created bar models that demonstrate the problem and that they can explain them.

ANSWERS Answers for the **Reflect** part of the lesson appear in the separate **Practice and Reflect answer guide**.

After the lesson ⏸

- Can children solve a range of multiplication and division multi-step word problems?
- Can children identify key information that is important to a problem and realise what type of calculation they have to do?

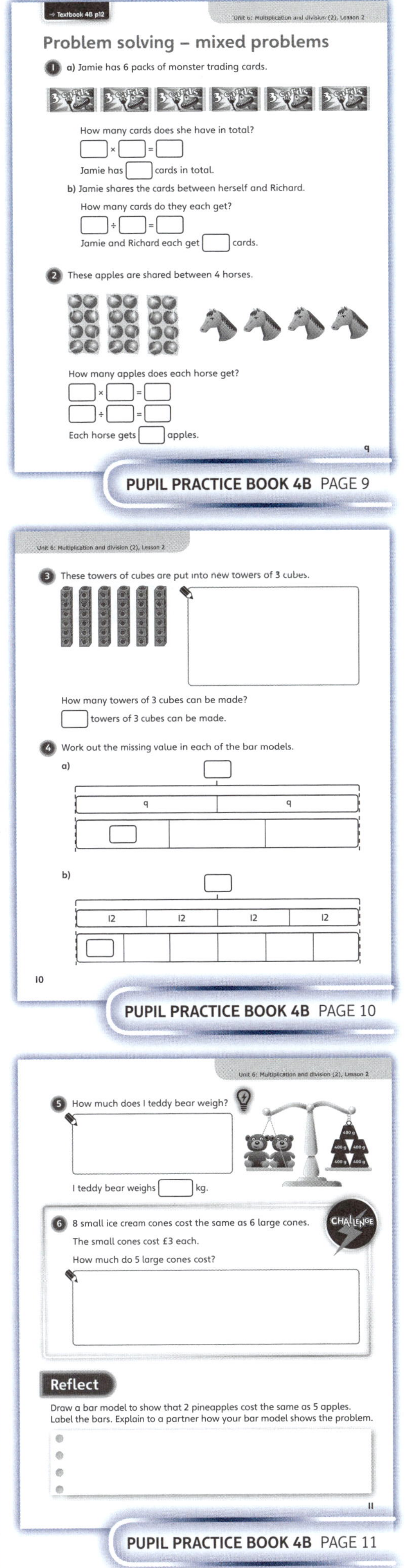

PUPIL PRACTICE BOOK 4B PAGE 9

PUPIL PRACTICE BOOK 4B PAGE 10

PUPIL PRACTICE BOOK 4B PAGE 11

Using written methods to multiply

Learning focus

In this lesson, children will build on their understanding of using expanded methods to multiply a 2-digit number by a 1-digit number, in preparation for using a formal written layout (column method).

Small steps

→ Previous step: Problem solving – mixed problems
→ **This step: Using written methods to multiply**
→ Next step: Multiplying a 2-digit number by a 1-digit number

NATIONAL CURRICULUM LINKS

Year 3 Number – Multiplication and Division

Multiply two-digit and three-digit numbers by a one-digit number using a formal written layout.

ASSESSING MASTERY

Children can use an expanded method to multiply a 2-digit number by a 1-digit number. They can solve problems involving multiplying, adding and subtracting.

COMMON MISCONCEPTIONS

Children may know the multiplying facts without understanding the properties of multiplication, so they are unable to partition the numbers correctly. For example, children calculate 6 × 13 by multiplying 6 × 1 and 6 × 3, or they can calculate 2 × 13 with ease but find it difficult to calculate 13 × 2. Ask:

- *Which numbers are you multiplying? How many groups of 13 are there? Show this using blocks. How many 10s and 1s are there? Count the 10s and the 1s.*

Children may follow the instructions without being able to explain why this method works. Show counters in an array of 13 × 2, with a gap between the 10th and 11th in each row, splitting them into two groups. Ask:

- *How can you work out how many counters there are, without counting them all? What multiplications can you do? If you find 10 × 2, how many are left? What happens if you add them together?*

STRENGTHENING UNDERSTANDING

Ask quick questions to practise partitioning numbers and using number bonds. *Which numbers can you add to make 14? Which numbers can you subtract to get 9? Which numbers can you add to make 23?* To support understanding of the concept of related facts, ensure that children have access to concrete objects such as counters and base 10 equipment.

Focus children's attention on the idea that the 1s and the 10s are both multiplied by 3. Provide children with base 10 equipment so that they can visualise the multiplication.

GOING DEEPER

Ask children to use counters to show 12 × 4 and 4 × 12. Ask: *What is the same between them? How do they differ? Will the result be different if you count in groups of 4 or in groups of 12?*

KEY LANGUAGE

In lesson: multiplication fact, part-whole model

Other language to be used by the teacher: related fact, commutativity, division fact, array

STRUCTURES AND REPRESENTATIONS

arrays, number line, part-whole model

RESOURCES

Mandatory: counters or cubes

 In the eTextbook of this lesson, you will find interactive links to a selection of teaching tools.

Before you teach 

- Do children know all the multiplication facts?
- Do children know how to use 6 × 5 = 30 to work out 7 × 5?
- Do children know that multiplication is commutative and distributive?

Discover

WAYS OF WORKING Pair work

ASK

- Question **1** a): *How many trees are there in Andy's orchard? How many apples are on each tree? How many trees are there in Jamilla's orchard? How many apples are on each tree? What multiplication facts can you see? How many can you find?*
- Question **1** b): *How many apples are there in each orchard? How can you find how many apples there are in total? Can you think of a different way to solve it?*

IN FOCUS In question **1** b), children use their previous knowledge of problem solving using the distributive law of multiplying to work out that 10 groups of 6 and 8 groups of 6 make 18 groups of 6 altogether. Children should use what they know so far to multiply a 2-digit number by a 1-digit number; this is a good opportunity to clarify any potential misconceptions. Children should realise that they need to consider the number of groups they have. When they partition, they need to consider the number of 10s and 1s, rather than single digits.

PRACTICAL TIPS Ensure there are enough counters or cubes for children to be able to make arrays to fit the problems.

ANSWERS

Question **1** a): $10 \times 6 = 60$ and $8 \times 6 = 48$
Andy's orchard has 60 apples. Jamilla's orchard has 48 apples.

Question **1** b): $60 + 48 = 108$
There are 108 apples in total.

Share

WAYS OF WORKING Whole class teacher led

ASK

- Question **1** a): *What does each number line show?*
- Question **1** b): *Why do you think Dexter added together both answers?*
- Question **1** b): *Why did Ash multiply 18 by 6?*

IN FOCUS Question **1** b) provides an opportunity to use the part-whole model and the distributive property of multiplying. The model allows children to see the relationship between a number and its component parts. This will be useful in later questions (see the **Practice Book** for this lesson). Developing part-whole reasoning will help children not only to manipulate calculations so that they are easier, but also to visualise, interpret and solve word problems. For example, 18 groups of 6 is made from 10 groups of 6 and 8 groups of 6, but 18 groups of 6 can also be made by subtracting 2 groups of 6 from 20 groups of 6.

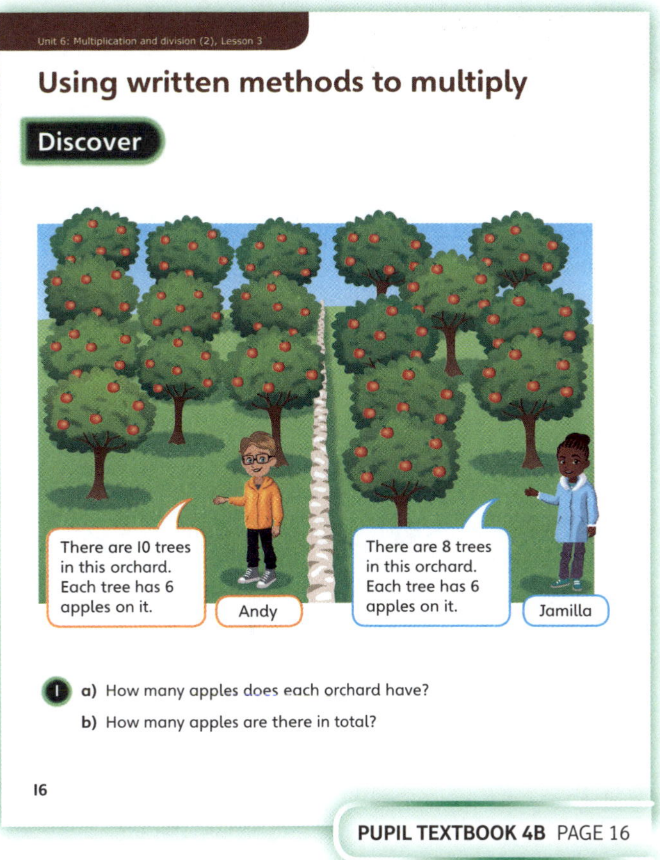

PUPIL TEXTBOOK 4B PAGE 16

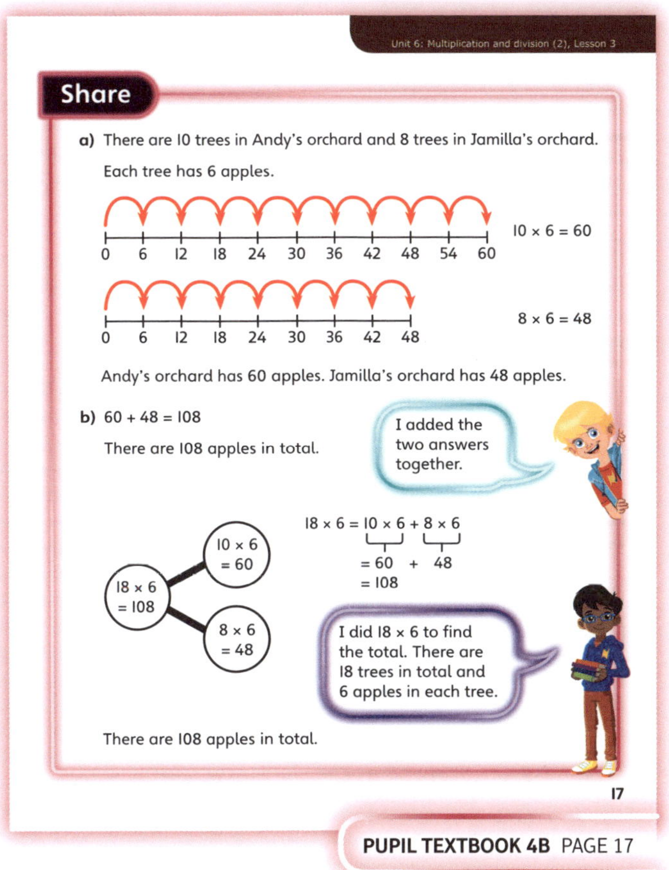

PUPIL TEXTBOOK 4B PAGE 17

Think together

 Whole class teacher led (I do, We do, You do)

ASK

- Question **1**: *What multiplication sentences will you write?*
- Question **2** a): *What is the question asking? What numbers will you multiply?*
- Question **3**: *What is the connection between the three groups that Richard uses? What is the connection between the groups that Kate uses? Will Richard's method and Kate's method achieve the same answer? What is the reason for grouping the counters? Why does Dexter suggest drawing around the different groups? Is that helpful?*

IN FOCUS Questions **1** and **2** provide an opportunity to link the abstract calculations with the problems that children are working out. Children need to use the way the answers are structured in questions **1** and **2**, in the rest of the questions in this lesson. Rather than simply writing numbers, they need to explain what these numbers mean: for example, 'There are 84 oranges in total'.

In question **3**, children need to look for different ways to multiply a 2-digit number by a 1-digit number. Children should start to see that, when multiplying, the number facts can be used to make the calculations easier. In order to explain their methods, Kate and Richard are drawing on the distributive law of multiplication.

STRENGTHEN To support understanding of the related facts, provide children with concrete experiences to help them to see the calculation in terms of 10s and 1s. For example, in question **2**, provide them with an array of 10 by 6 and an array of 4 by 6. Ask: *How many groups of 6 do you have? Write each calculation clearly.*

DEEPEN After working through question **3**, ask children to make up their own question for other children to solve. Give them a 2-digit number multiplied by a 1-digit number and ask: *What is the question? How would you solve it? Can you solve it in a different way?* Ask children to write all their calculations. Do they all lead to the same answer?

ASSESSMENT CHECKPOINT Children use different written methods to multiply a 2-digit number by a 1-digit number. They use the properties of multiplication and apply their knowledge to problem solve in more than one way.

Children can explain their methods of multiplying.

ANSWERS

Question **1**: Lexi: 10 × 5 = 50   Jamilla: 6 × 5 = 30
50 + 30 = 80   They have 80 pears in total.

Question **2** a): 10 × 6 = 60 and 4 × 6 = 24
60 + 24 = 84   There are 84 oranges in total.

Question **2** b): 14 × 6 = 84

Question **3** a): Richard has calculated each array separately and then added them together. Kate has worked out the number of counters in the first two arrays as one calculation and then added on the answer to the third array.

Question **3** b): An alternative calculation would be 5 × 23 = 115 or 23 × 5 = 115.

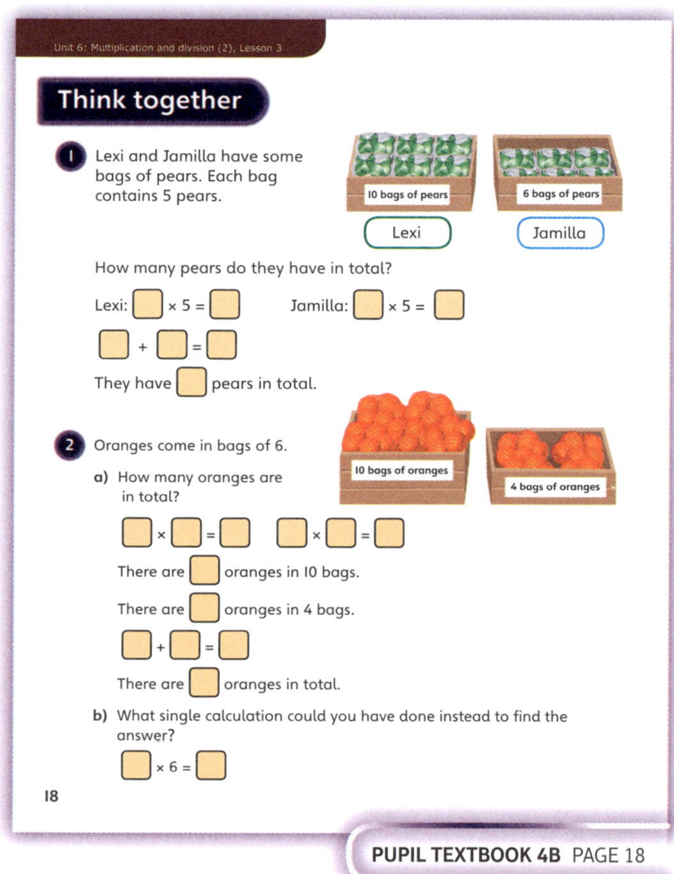

PUPIL TEXTBOOK 4B PAGE 18

PUPIL TEXTBOOK 4B PAGE 19

Practice

WAYS OF WORKING Independent thinking

IN FOCUS Questions ① and ② develop fluency of using written methods to multiply. Question ① a) provides scaffolding by splitting a 2-digit number into 10s and 1s. Children multiply the 10s and the 1s, and then add the results together. Question ③ requires children to use the properties of multiplying (distributive law). Question ④ asks children to use part-whole models and to link multiplication properties (that they are commutative and distributive) to finding different ways to multiply. Question ⑤ f) is an example of procedural variation, where children need to be flexible in how they think about the concept of multiplying: they use the commutative law of multiplication that $2 \times 10 = 10 \times 2$ and then the distributive law to derive the answer.

STRENGTHEN To support understanding, provide children with two arrays of counters. Both arrays should have the same number of rows. Ask children to write down the multiplication each array represents. Connect the number of rows and columns with the numbers used in each question. Ask: *Can you find two ways of working out the total number of counters?*

DEEPEN Expand on question ④ by giving children an answer and asking them to work out what three numbers they have to multiply and add to reach that answer.

ASSESSMENT CHECKPOINT Children should show that they know how to multiply a 2-digit number by a 1-digit number using a written method. They should use the properties of multiplying to make the calculations simpler and be able to multiply in two or more ways.

ANSWERS Answers for the **Practice** part of the lesson appear in the separate **Practice and Reflect answer guide**.

Reflect

WAYS OF WORKING Independent thinking

IN FOCUS Children use what they have learnt in the lesson to coherently explain the potential methods. Encourage children to show their working for each method. Do children also mention the commutative law to explain that 10×5 is the same as 5×10? Children may use a part-whole model to show the two parts of the whole calculation.

ASSESSMENT CHECKPOINT Children can describe two ways to solve the multiplication problem, by either grouping then adding their two answers or counting all the pencil boxes first.

ANSWERS Answers for the **Reflect** part of the lesson appear in the separate **Practice and Reflect answer guide**.

After the lesson ❙❙

- Can children use the distributive law to partition numbers in different ways to create equivalent calculations?
- Can children use arrays confidently?
- Are they able to find the answer to a calculation such as 14×6, if they know the related facts ($4 \times 6 = 24$ and $10 \times 6 = 60$)?

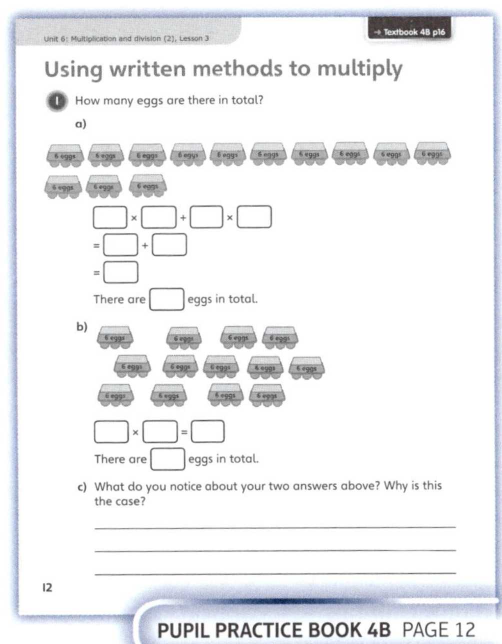

PUPIL PRACTICE BOOK 4B PAGE 12

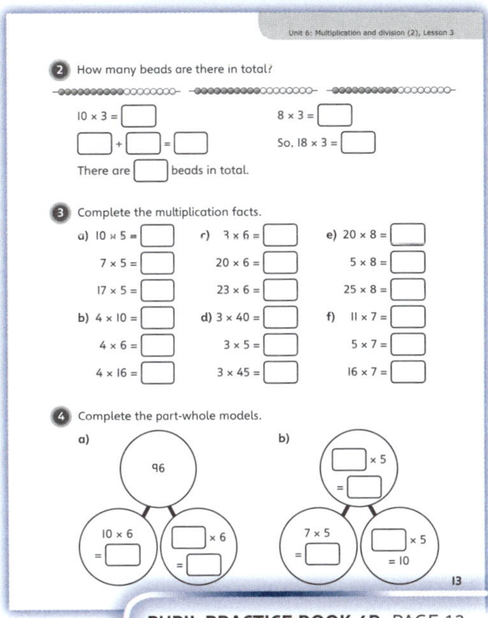

PUPIL PRACTICE BOOK 4B PAGE 13

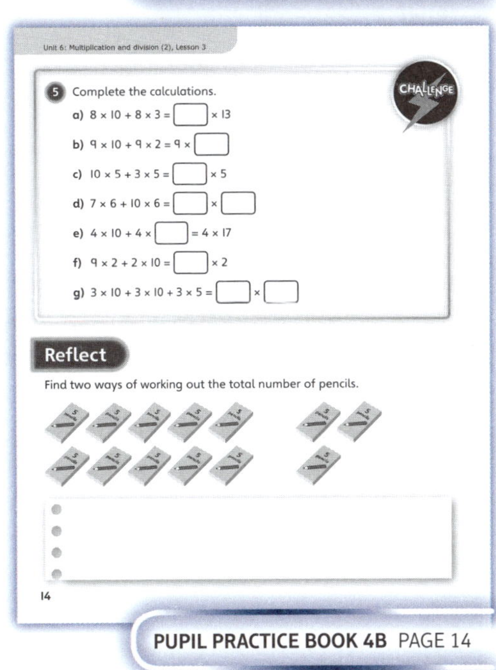

PUPIL PRACTICE BOOK 4B PAGE 14

Multiplying a 2-digit number by a 1-digit number

Learning focus

In this lesson, children will start using the compressed, single-line (short) formal multiplication. They will progress to examples that require exchange of 1 ten, and then of more than 1 ten.

Small steps

→ Previous step: Using written methods to multiply
→ **This step: Multiplying a 2-digit number by a 1-digit number**
→ Next step: Multiplying a 3-digit number by a 1-digit number

NATIONAL CURRICULUM LINKS

Year 4 Number – Multiplication and Division

Multiply two-digit and three-digit numbers by a one-digit number using formal written layout.

ASSESSING MASTERY

Children can use a formal written method to multiply a 2-digit number by a 1-digit number. They are able to use the expanded method or the standard method of multiplication.

COMMON MISCONCEPTIONS

Children may misapply the procedure of the formal written method and place the 10s and 1s in the wrong place. For example, when multiplying 23 by 5, they multiply 3 × 5 = 15, but put the 10s and 1s in the wrong columns. Ask children to estimate the answer:

• *What can 23 round to? Is 23 closer to 20 or 30? What is 20 × 5? 30 × 5? What would you expect the answer to be?*

When using the expanded written method, children may multiply each digit as if it represents the number of 1s. Ask:
• *Does your answer make sense? How many 10s are there in 23? How many 1s?*

STRENGTHENING UNDERSTANDING

To support children in understanding what they need to do, explain step by step how the multiplication is done using place value language. Start with simpler questions where no exchange is required, and then move to exchange of 1 ten. When children are ready, move to the exchange of more than 1 ten. Emphasise writing the digits of the product in the appropriate place value column.

GOING DEEPER

Ask children to complete questions where exchange is required, such as 23 × 8, or 34 × 6. It is important to remind children that the exchanged 1 ten is placed in the tens column. Explain that the 1 ten is added to the tens column after the multiplication of 10s has been done.

KEY LANGUAGE

In lesson: place value, ones, tens, expanded multiplication, single line multiplication, exchange, calculation

Other language to be used by the teacher: times-table, recall, grouping, digit

STRUCTURES AND REPRESENTATIONS

formal written short multiplication

RESOURCES

Mandatory: place value counters

Optional: cubes, counters

 In the eTextbook of this lesson, you will find interactive links to a selection of teaching tools.

Before you teach 

• Do children know 2, 3, 4, 5, 8 and 10 multiplication facts?
• Can they multiply 2-digit numbers by 10 and 100?
• Can they partition by multiples of the divisor and by place value?

Discover

 Pair work

ASK

- Question ❶ a): *What numbers does Bella multiply first? Where does she put the answer? Where did 100 come from in the second row? What does the small 1 represent in Danny's calculation? What do you have to do with this number?*
- Question ❶ b): *How many tens and unit counters do you need to represent 23? How many tens and unit counters do you need to represent 5 groups of 23?*

IN FOCUS This is the first time that children have come across the standard (short) and expanded (long) method of multiplying in formal layout. Both methods are very important, as children will be using these methods not only in this unit but also in their future learning. The use of concrete materials will help children to visualise and to understand both methods. It is important that children are provided time and opportunities to practise and explore both the methods before deciding which one they prefer. Show children 5 rows of 23. Remind them that 3 × 5 = 15 and that 10 'one' counters can be exchanged for 1 ten counter. Show children how 10 ten counters can be exchanged for 1 hundred counter.

PRACTICAL TIPS Provide children with whiteboards and pens. Ask children to work out the calculations on their whiteboard before writing them in their books.

ANSWERS

Question ❶ a): Both methods use columns. Both methods give the same answer. Bella has used long (expanded) multiplication, but Danny has used short (single-line) multiplication.

Question ❶ b): 11 tens and 5 ones = 1 hundred, 1 ten and 5 ones. So 23 × 5 = 115.

Share

 Whole class teacher led

ASK

- Question ❶ a): *Why does Dexter think Danny has gone wrong? Is Danny's method quicker? Does it give the same answer as Bella's method?*
- Question ❶ b): *How many groups of 23 are there? How has this been shown using place value counters? How many ones can you exchange for one ten? How many tens can you exchange for one hundred?*

IN FOCUS This section introduces children to the formal written, short method of multiplication, which they have not used before. Explain clearly how the calculations are made in each method. Emphasise writing the digits of the product in the appropriate place value column. Show children how the place value counters can be used to support each calculation. Emphasise the exchanging when there are 10 or more ones or 10 or more tens. Show children how the exchanged 1 ten is placed in the tens column. Explain that the 1 ten is added to the tens column after the multiplication of tens has been done.

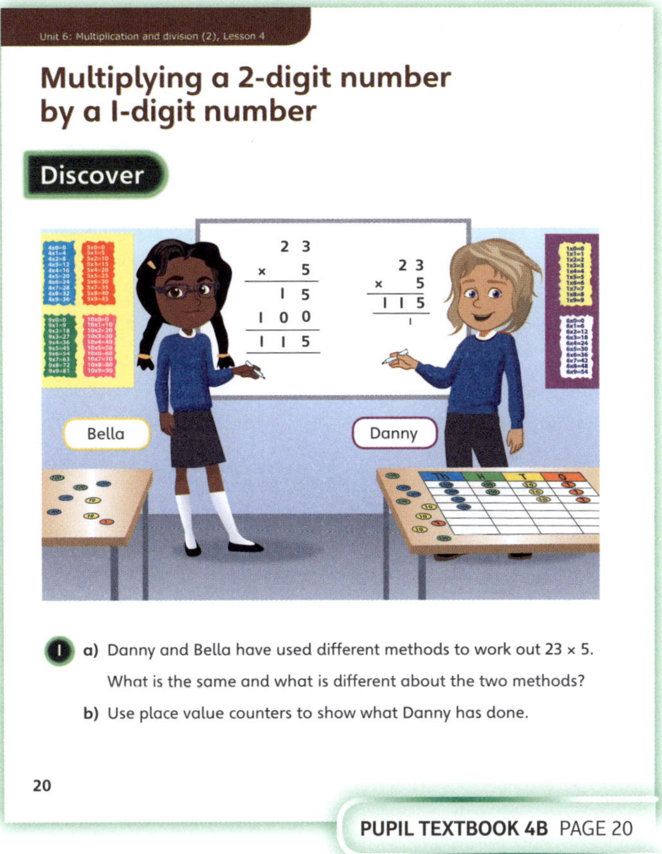

PUPIL TEXTBOOK 4B PAGE 20

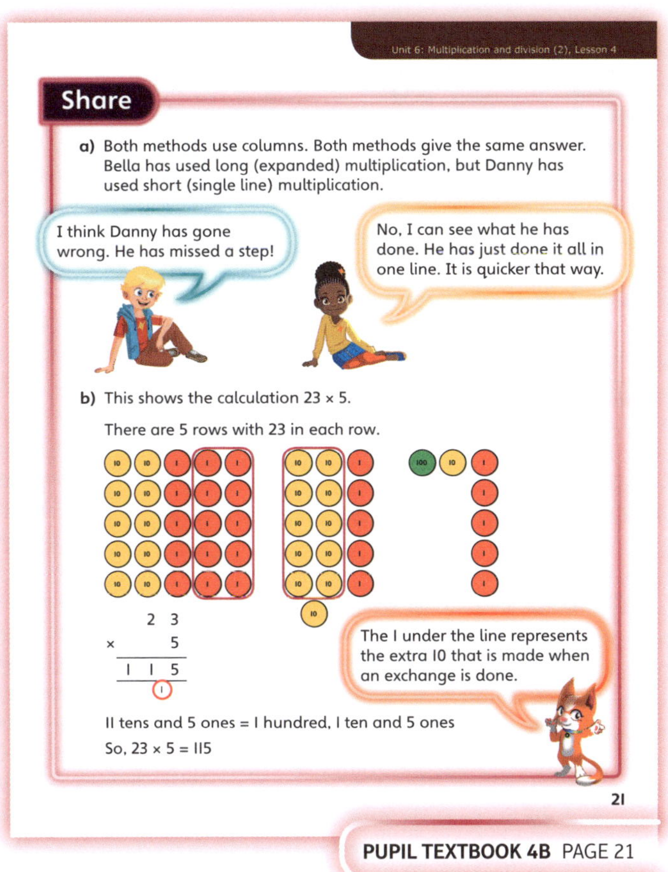

PUPIL TEXTBOOK 4B PAGE 21

Think together

WAYS OF WORKING Whole class teacher led (I do, We do, You do)

ASK

- Question **1**: *How many place value counters will you need? Which numbers are you multiplying? Which numbers should go in each box?*
- Question **2**: *How many ones are you exchanging? How many tens do you have? What do you do with these tens? Do you add them before or after the tens have been multiplied?*
- Question **3**: *What pattern is Ash noticing? How many ones have been exchanged for tens each time? Which multiplication facts is Flo going to use?*

IN FOCUS This section takes children through a series of multiplication questions. The key aim is for children to learn how to use the formal method of multiplying for 2-digit numbers. In question **1**, children are asked to multiply a 2-digit number by a 1-digit number. They exchange 10 ones for 1 ten. Making sure that children understand this concept is very important for the next questions, where children change 20, 30, 40 and 50 ones for tens. It is important to clarify any potential misconceptions children may have and to support the written calculations by using place value counters and other equipment before moving on in the lesson.

STRENGTHEN Encourage children to work through each problem line by line. Provide children with place value counters to support their learning. Help them to set the calculations correctly and as neatly as they can. Ask: *What does each column represent? Where should each number go?*

DEEPEN For question **3**, ask children to use both methods introduced in the **Discover** section. Ask: *Which one is the quickest? Which one do you prefer? Why?* Challenge children to make up their own questions. Can they make up a multiplication with the answer 12?

ASSESSMENT CHECKPOINT Children are learning to multiply 2-digit numbers by 1-digit numbers and exchange 1s for 10s. They know how to use resources such as place value counters if they are unsure or want to check their answers.

ANSWERS

Question **1**: $22 \times 6 = 132$

Question **2** a): $26 \times 5 = 130$

Question **2** b): Children exchanged 30 ones for 3 tens, instead of 10 ones for 1 ten.

Question **3** a): $32 \times 4 = 128$
$23 \times 6 = 138$
$43 \times 7 = 301$
$38 \times 4 = 152$

Question **3** b): The number of tens exchanged for ones increases by 1 each time.

Question **3** c): To continue the pattern, 4 tens would be exchanged in the fifth question, for example: $27 \times 6 = 162$.

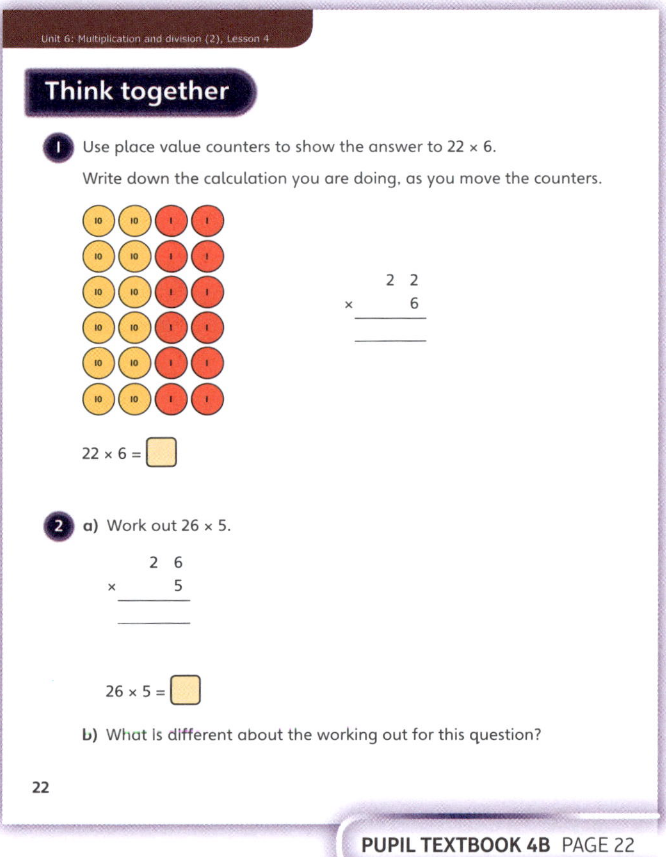

PUPIL TEXTBOOK 4B PAGE 22

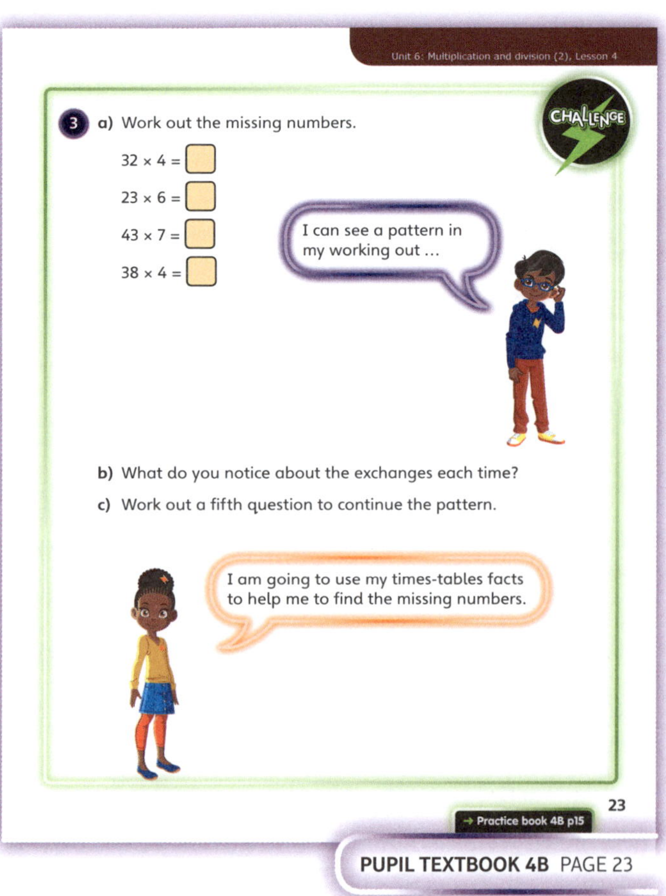

PUPIL TEXTBOOK 4B PAGE 23

Practice

WAYS OF WORKING Independent thinking

IN FOCUS Clarifying misconceptions is very important for future learning, otherwise children carry on making the same mistakes in their work. Make a note of any mistakes children make throughout the lesson and create a 'learning area'; keep adding any multiplying mistakes that children make. Ask children to correct each other's and their own mistakes using a different coloured pen.

All the multiplying questions here feature answers from times-tables that children should know, so encourage recall before using other methods to work out an answer.

STRENGTHEN To support children in understanding what they need to do, read the questions line by line and ask children to identify which numbers they are multiplying. Are they fluent with their multiplication facts? Have they partitioned the 2-digit numbers correctly? Do they know how to exchange 1s for 10s? Using Astrid's advice to think about place value columns is important. Encourage children to use the correct mathematical vocabulary when explaining their reasoning.

DEEPEN In response to question ❺, write on the board all the misconceptions and mistakes that children may have made during the lesson. Allow children time to look at each multiplication. Where is the mistake? What should the answer have been? Are children thinking about place value? Are they exchanging correctly? Are they fluent with their multiplication facts?

THINK DIFFERENTLY Question ❺ requires children to identify the mistake in an answer to a multiplication. This addresses the common mistake where children carry the 10 straight over to the answer space, rather than writing it below and, thus, create an extra place value.

ASSESSMENT CHECKPOINT Can children use the standard method of multiplication? Can they use the column format and write each digit of the product in the appropriate place value column?

ANSWERS Answers for the **Practice** part of the lesson appear in the separate **Practice and Reflect answer guide**.

Reflect

WAYS OF WORKING Independent thinking

IN FOCUS This section gives children the opportunity to demonstrate their understanding of the transition from concrete to pictorial to abstract. This deepens children's understanding of what multiplication means and why the written calculations work the way they do. Understanding why the 2 tens are carried and not the 4 ones is very helpful as this is one of the potential misconceptions that children may encounter.

ASSESSMENT CHECKPOINT Children explain how place value counters can be used to multiply a 2-digit number by a 1-digit number. They are able to exchange 20 ones for 2 tens, and 10 tens for 1 hundred, and place each counter in the correct position to complete the calculation.

ANSWERS Answers for the **Reflect** part of the lesson appear in the separate **Practice and Reflect answer guide**.

After the lesson ⏸

- Can children solve a range of multiplication word problems?
- Can they identify key information that is important to the problem?

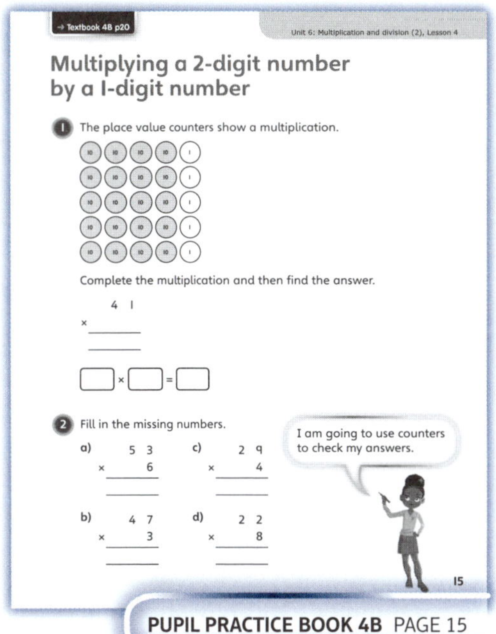

PUPIL PRACTICE BOOK 4B PAGE 15

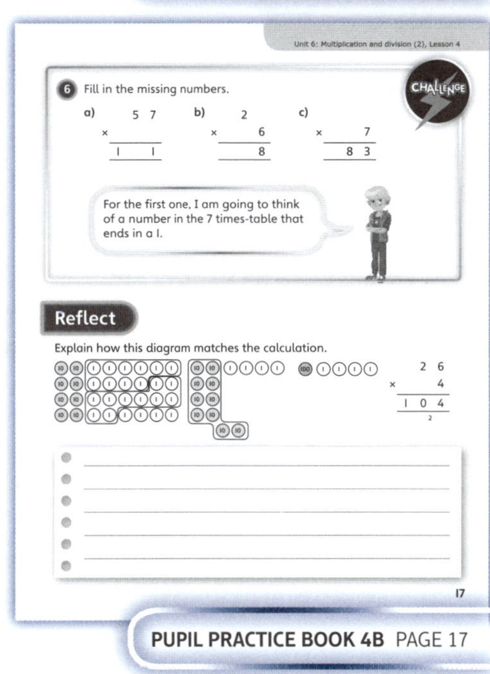

PUPIL PRACTICE BOOK 4B PAGE 16

PUPIL PRACTICE BOOK 4B PAGE 17

Multiplying a 3-digit number by a 1-digit number

Learning focus

In this lesson, children will start multiplying 3-digit numbers by a 1-digit number. They will progress from no exchange to examples that require exchange of 1, then of more than 1.

Small steps

→ Previous step: Multiplying a 2-digit number by a 1-digit number

→ **This step: Multiplying a 3-digit number by a 1-digit number**

→ Next step: Problem solving – multiplication

NATIONAL CURRICULUM LINKS

Year 4 Number – Multiplication and Division

Multiply two-digit and three-digit numbers by a one-digit number using formal written layout.

ASSESSING MASTERY

Children can use a formal written method to multiply a 3-digit number by a 1-digit number. They are able to use the expanded column method or the standard column method of multiplication.

COMMON MISCONCEPTIONS

Children may misapply the procedure of the standard written method of multiplication. They may firstly add the amount that is exchanged to the amount in the multiplication and then multiply (instead of multiplying first and then adding the amount that is exchanged). For example, when multiplying 206 by 8, they find $6 \times 8 = 48$ and they carry 4 tens, which they add to 0 tens, before multiplying by 8 again. Ask:

• *Does the answer make sense? What do you expect the answer to be?*

STRENGTHENING UNDERSTANDING

Ask children to think of the way they multiplied a 2-digit number by a 1-digit number. Ask: *How do you start multiplying? What things do you have to watch out for?* When moving to multiplying 3-digit numbers, start with simpler questions, where no exchange is required, and then move to exchange of 1, until children are ready to move to the exchange of more than 1 ten or 1 hundred. Emphasise writing the digits of the product in the appropriate place value column, using the correct place value language.

GOING DEEPER

Ask children to complete questions where the exchange of 100s and 10s is required. It is important to remind children that the exchanged 1 ten is placed in the tens column and the exchanged 1 hundred is placed in the hundreds column. Include questions with 0 ones or 0 tens.

KEY LANGUAGE

In lesson: multiplication, place value, digit

Other language to be used by the teacher: times-table, recall, grouping, ones, tens, hundred, column

STRUCTURES AND REPRESENTATIONS

formal written short multiplication

RESOURCES

Optional: cubes, counters

 In the eTextbook of this lesson, you will find interactive links to a selection of teaching tools.

Before you teach 

• Do children understand 0 as a placeholder?
• Can they multiply 2-digit numbers by 10 and 100?
• Can they partition by place value?

Discover

 Pair work

ASK

- Question ① a): *How many rows are there in each section? How many seats are there in each row? How can you write this as a multiplication statement? How can you work out the total?*
- Question ① b): *How many sections are there in the stadium? How many seats are there in a section? What multiplication statement can you write? How is this similar to, or different from, multiplying 3-digit numbers?*

IN FOCUS Children recap multiplying a 2-digit number by a 1-digit number in question ① a). This may be a good opportunity to clarify any mistakes that children are still making or to answer any questions they have, before moving on to multiplying a 3-digit number by a 1-digit number for the first time in part b). Ensure that children have enough place value counters and allow them time to try things for themselves, rather than simply copying calculations from the book or the board. Show children 3 rows of 312. Demonstrate how they can multiply 312 by 3, making links with their previous knowledge of multiplying 2-digit numbers.

PRACTICAL TIPS Use place value counters to work out each multiplication.

ANSWERS

Question ① a): $6 \times 52 = 312$
There are 312 seats in a section.

Question ① b): $312 \times 3 = 936$
There are 936 seats in total.

Share

 Whole class teacher led

ASK

- Question ① a): *Would it be the same if you multiplied 52 by 6? Does the order of multiplying matter? Which way is easier?*
- Question ① b): *What method was used in part a)? Can you use this method to multiply a 3-digit number by a 1-digit number? What place value counters do you need? How can you use counters to show the answers?*

IN FOCUS Children have not used the written standard method for 3-digit numbers before, so explain clearly how the calculations are made. Emphasise writing the digits of the product in the appropriate place value column. Show children how the place value counters can be used to support each calculation. Showing multiplication using place value counters reinforces the fact that multiplication is repeated addition. Discuss the method in question ① a). Provide children with different examples and ask them to discuss when it is easier to partition and multiply and when it is easier to multiply using a column method. For example, 121×4 can be easily multiplied, but the column method might be more appropriate to work out 368×7.

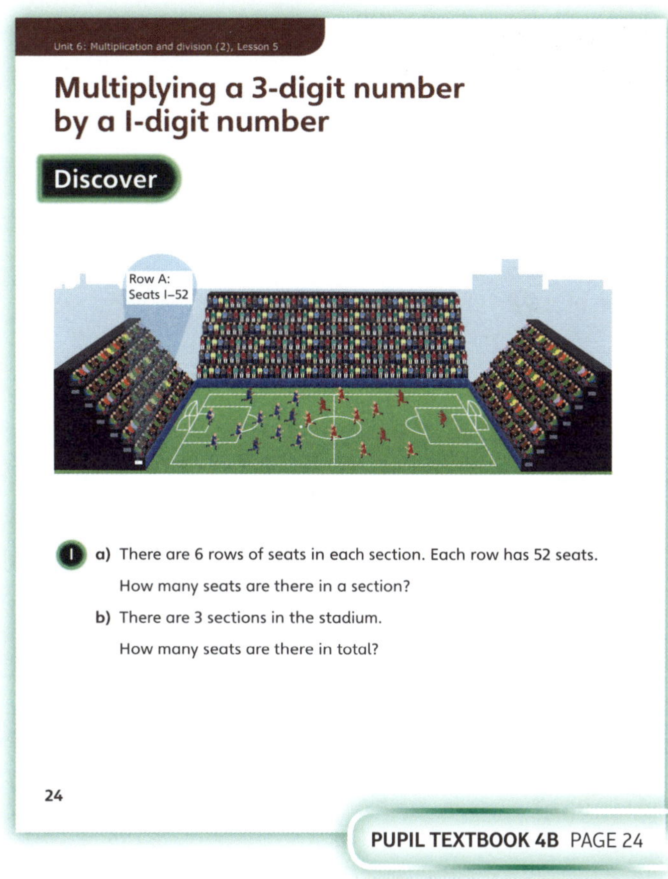

PUPIL TEXTBOOK 4B PAGE 24

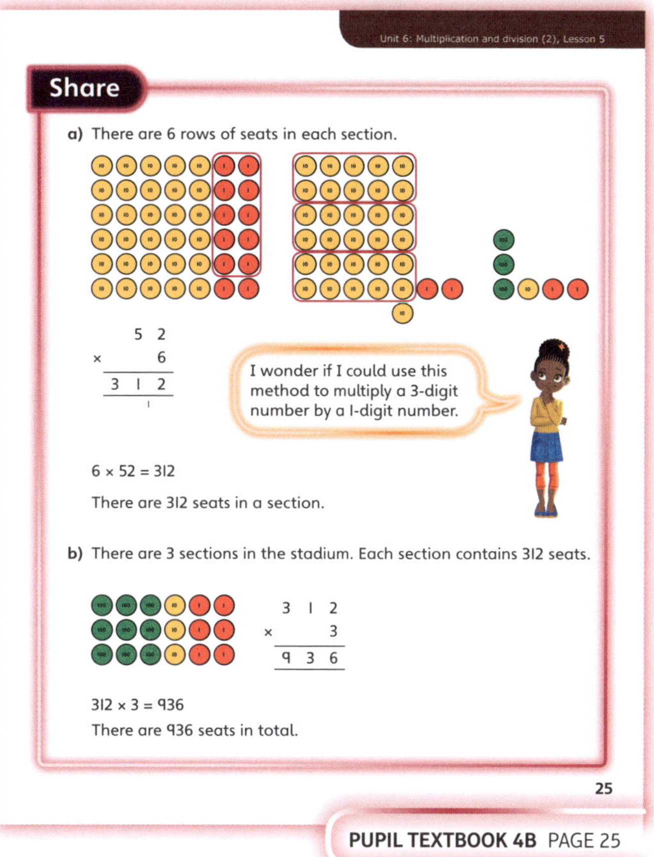

PUPIL TEXTBOOK 4B PAGE 25

Think together

WAYS OF WORKING Whole class teacher led (I do, We do, You do)

ASK

- Question **1**: *Which numbers are you multiplying? What place value counters can you use?*
- Question **2**: *How many 1s are you exchanging? How many 10s do you have now?*
- Question **3**: *How did Astrid use the last digit to work out the missing numbers? How can you check the answer? What multiplication facts do you need to know? What number should there be in the tens column so that when you multiply by 6 the answer is 8? Why?*

IN FOCUS Questions **1** and **2** provide a pictorial and abstract representation of multiplying 3-digit numbers by 1-digit numbers. Using the concrete, pictorial, abstract approach is very effective for progressing children to abstract concepts such as multiplying 3-digit numbers. To be able to solve questions and problems that require multiplying big numbers, children need to be secure in their written methods of multiplying.

STRENGTHEN Encourage children to work through each problem line by line. To help them complete the single-line multiplication, use the expanded method until they become more confident. Provide children with place value counters to support their learning. Help children to set the calculations correctly. Ask: *What does each column represent? Where should each number go?*

DEEPEN For question **3**, ask children to use their knowledge of times-tables to work out the answer. Challenge children to check their answers by working out each multiplication.

ASSESSMENT CHECKPOINT Can children multiply a 3-digit number by a 1-digit number? Can they use their multiplication facts to predict what the last digit of the answer will be? Are children able to use the facts they know to derive new ones? For example, if they know that $3 \times 6 = 18$ do they know what 30×6 would be? What about 300×6?

ANSWERS

Question **1**: $146 \times 2 = 292$; there are 292 sweets in 2 jars.

Question **2**: $136 \times 4 = 544$; there are 544 passengers in 4 full planes.

Question **3**: $123 \times 6 = 738$, $231 \times 6 = 1,386$, $312 \times 6 = 1,872$

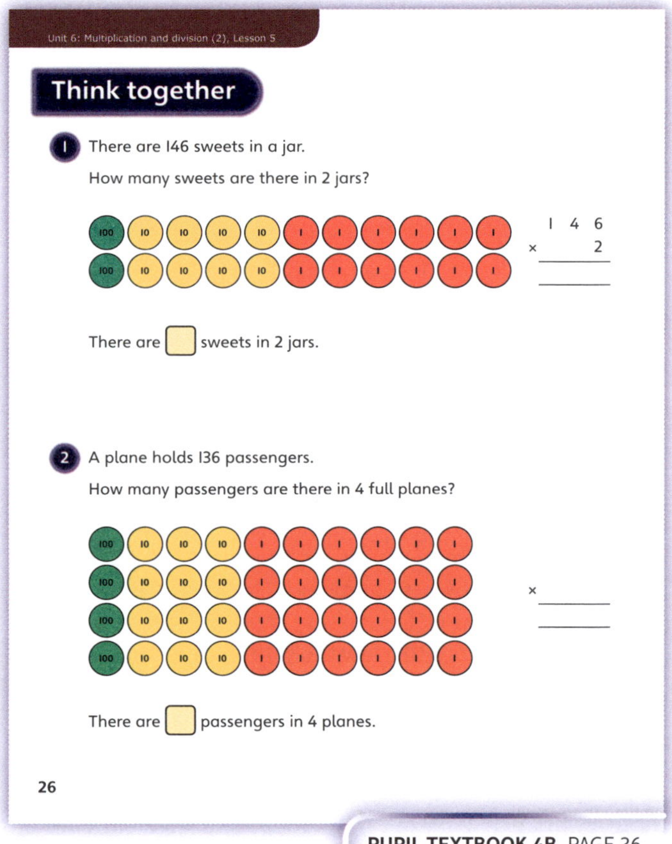

PUPIL TEXTBOOK 4B PAGE 26

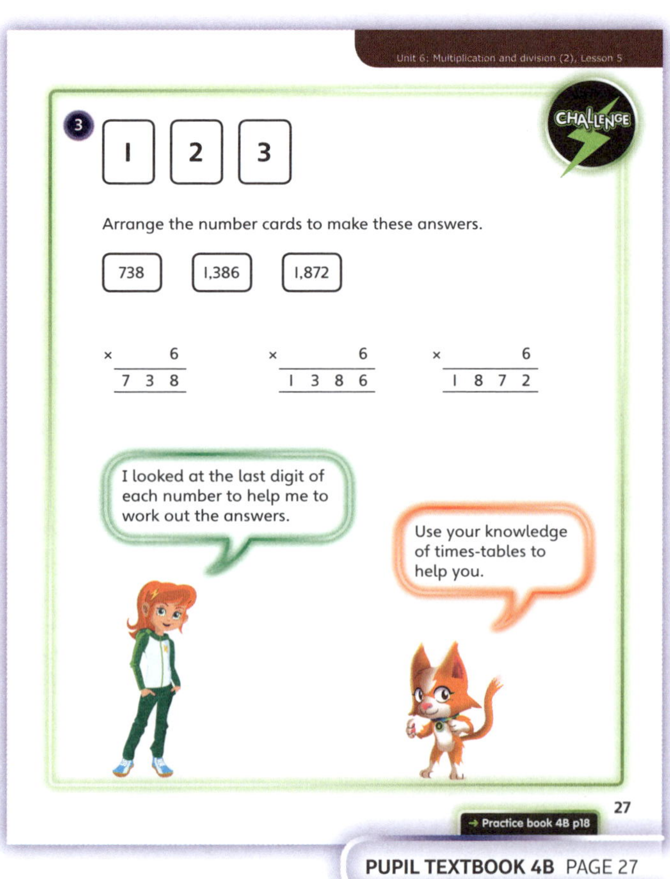

PUPIL TEXTBOOK 4B PAGE 27

Practice

WAYS OF WORKING Independent thinking

IN FOCUS Question **3** requires children to set up the column method themselves and then multiply.

Question **4** extends their problem-solving skills by asking them to find the missing numbers in each calculation. In question **5** children are asked to multiply to solve a word problem. Question **7** requires children to think of numbers from the times-tables that multiply to give a certain result. All the questions feature answers from times-tables that they should know, so encourage recall before using other methods to work out an answer.

STRENGTHEN To support children's work, ensure that they are fluent with their multiplication facts. For example, before starting question **7**, recap the 7 times-table. Some children may find it useful to have the 7 times-table written in front of them so that they can check which numbers they can use. Encourage children to write the digits of the product in the appropriate place value column and to use the correct mathematical vocabulary when explaining their reasoning.

DEEPEN In response to question **4**, write on the board all the misconceptions and mistakes that children may have made during the lesson. You may want to expand on the misconceptions and errors from last lesson and have a table of 'learning points'. Allow children time to look at the multiplication. Can they find the mistake? What should the answer have been? Are children thinking about place value? Are they exchanging correctly? Are they fluent with their multiplication facts?

THINK DIFFERENTLY In question **6** children use their knowledge of multiplying to find and explain the mistake in the calculation, consolidating their own understanding. This question addresses the common mistake of writing the digit carried over in the answer space, instead of below the answer line.

ASSESSMENT CHECKPOINT Children use the written column method of multiplication to multiply 3-digit numbers by 1-digit numbers. They are able to use their multiplicative reasoning to solve missing number questions, and word problems. Can they estimate what the answer could be?

ANSWERS Answers for the **Practice** part of the lesson appear in the separate **Practice and Reflect answer guide**.

Reflect

WAYS OF WORKING Pair work

IN FOCUS Children explain the different methods they can use to multiply a 3-digit number by a 1-digit number.

ASSESSMENT CHECKPOINT Check if children can use place value counters to multiply 3-digit numbers by 1-digit numbers. Listen for children's clear structure and reasoning. Which method did they use? Can they use a different one? Did they use the correct mathematical language?

ANSWERS Answers for the **Reflect** part of the lesson appear in the separate **Practice and Reflect answer guide**.

After the lesson ⏸

- Are children confident in multiplying?
- Are they able to use place value counters in their multiplication?
- Are they confident in exchanging 1s and 10s? 10s and 100s?

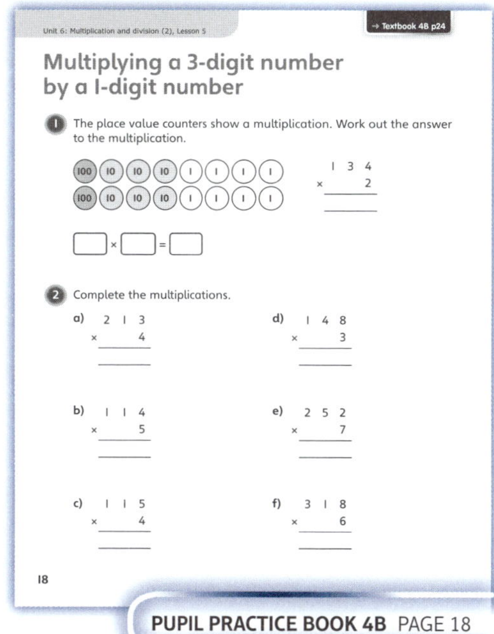

PUPIL PRACTICE BOOK 4B PAGE 18

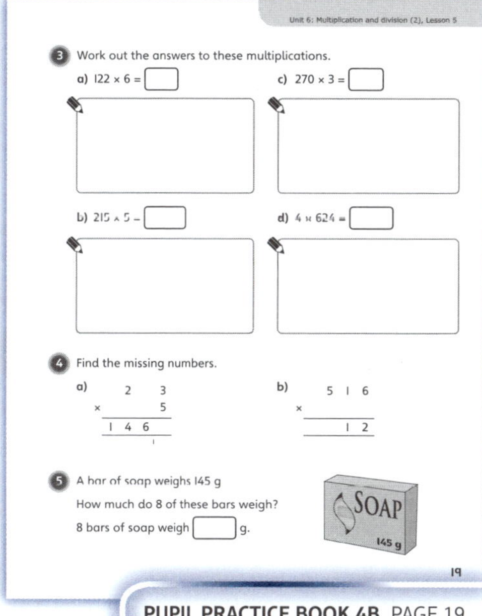

PUPIL PRACTICE BOOK 4B PAGE 19

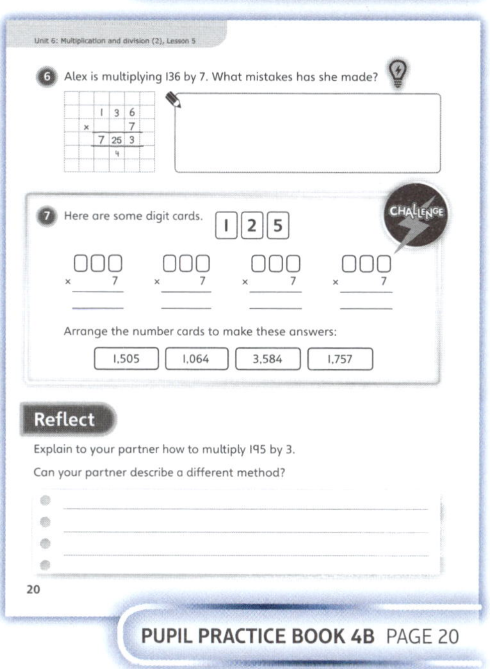

PUPIL PRACTICE BOOK 4B PAGE 20

Problem solving – multiplication

Learning focus

In this lesson, children will solve a mixture of problems by using the formal written method. Bar models are used to reveal the structure of more complex problems.

Small steps

→ Previous step: Multiplying a 3-digit number by a 1-digit number
→ **This step: Problem solving – multiplication**
→ Next step: Multiplying more than two numbers (1)

NATIONAL CURRICULUM LINKS

Year 4 Number – Multiplication and Division
• Solve problems involving multiplying and adding, including using the distributive law to multiply two digit numbers by one digit, integer scaling problems and harder correspondence problems such as *n* objects are connected to *m* objects.
• Multiply two-digit and three-digit numbers by a one-digit number using formal written layout.

ASSESSING MASTERY

Children can solve a mixture of problems by using the formal written method. They are able to use key information in a question to draw a bar model and to work out what operation to use to solve the problem.

COMMON MISCONCEPTIONS

In a multi-step problem-solving question, children may not be sure what operation to use and which numbers to add, subtract, multiply or divide. Provide scaffolding by helping children to think about the unit in each problem. Drawing a bar model to represent the situation may help them to see this. Encourage children to look at the words in the question for clues. Ask:
• *What will your bar model look like? Do you know how many parts there should be, or the value of each part?*

STRENGTHENING UNDERSTANDING

Children may not understand the question and what they need to do. To support understanding, read each question with children line by line. Encourage them to question the information given. Ask: *Why does the question say this? What do these numbers represent? What can I use them for?*

Try to encourage children to draw a bar model, which may help them to see whether they need to multiply or divide. Some children may look at the numbers and worry about the actual operations that they need to use rather than solving the problem. Simplify the question by using smaller numbers. Ask: *What would the answer be if there were 2 buses and 5 children on each bus?*

GOING DEEPER

Ask children to create their own problems based on a given bar model, or diagram. Ask: *What operations are needed to solve the problem? Can it be solved in two ways? Which way is easier?* Provide children with open-ended problems. Ask: *Can you find another answer? And another?*

KEY LANGUAGE

In lesson: multiply, bar model, total

Other language to be used by the teacher: times-table, recall, grouping, repeated addition, open-ended, divide, equal, sharing, word problem, multi-step

STRUCTURES AND REPRESENTATIONS

bar model, formal written short multiplication

RESOURCES

Optional: cubes, counters, number lines

 In the eTextbook of this lesson, you will find interactive links to a selection of teaching tools.

Before you teach 

• Do children know the multiplication facts?
• Can children draw a bar model to solve the questions?
• Do they know how to calculate a multiplication or division if they do not know the relevant multiplication fact?

Discover

WAYS OF WORKING Pair work

ASK

- Question ➊ a): *How many coaches are there? What does 'per' mean? How many people are in each coach? Can you represent this as a bar model? What type of problem are you solving? Is it division or multiplication?*
- Question ➊ b): *How many coaches are going to Adventure Park? How many people are there in each coach? Can you show this on a bar model? How do you work out the total? What calculation do you need to do?*

IN FOCUS This question is about using bar models to solve a problem. It is very important that children identify what one part of the bar model will represent and the number of parts they need to draw in order to answer the question. Children should be encouraged to draw the bar models neatly and label them accordingly. This will help children to identify what they already know and to think of the method they need to use to answer the question. Question ➊ b) is a two-step question. It is important to read the question carefully and answer it fully.

PRACTICAL TIPS Use counters or cubes of two different colours to represent the number of people in each coach.

ANSWERS

Question ➊ a): There are 168 people going to Bolton Towers.

Question ➊ b): There are 408 people going on a trip today.

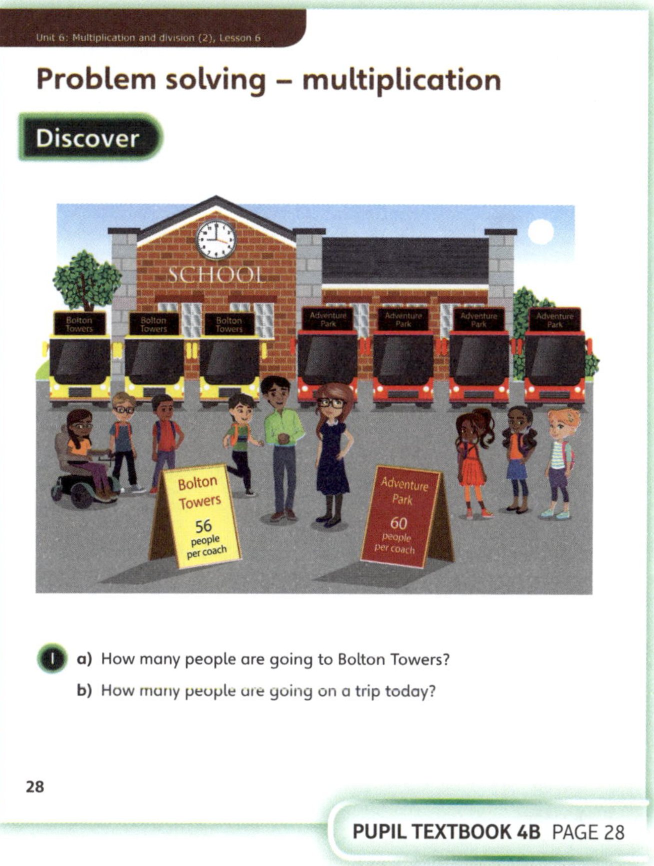

PUPIL TEXTBOOK 4B PAGE 28

Share

WAYS OF WORKING Whole class teacher led

ASK

- Question ➊ a): *How many groups of 56 children are there? How has this been shown in the bar model?*
- Question ➊ b): *What does the bar model look like? Why is there a separate part with 168 and four parts each with 60?*

IN FOCUS Ensure that children understand how to use bar models when solving a multiplication and addition question. In question ➊ a), showing multiplication as a bar model reinforces the fact that multiplication is repeated addition. In question ➊ b), the bar model is used to add two different quantities. It is important for children to link the numbers in the question with the bar model representation. Some children may have used other models, such as part-whole models or drawings/diagrams to find the answer. Discuss the importance of drawing diagrams in helping children to visualise the information given in order to answer the question.

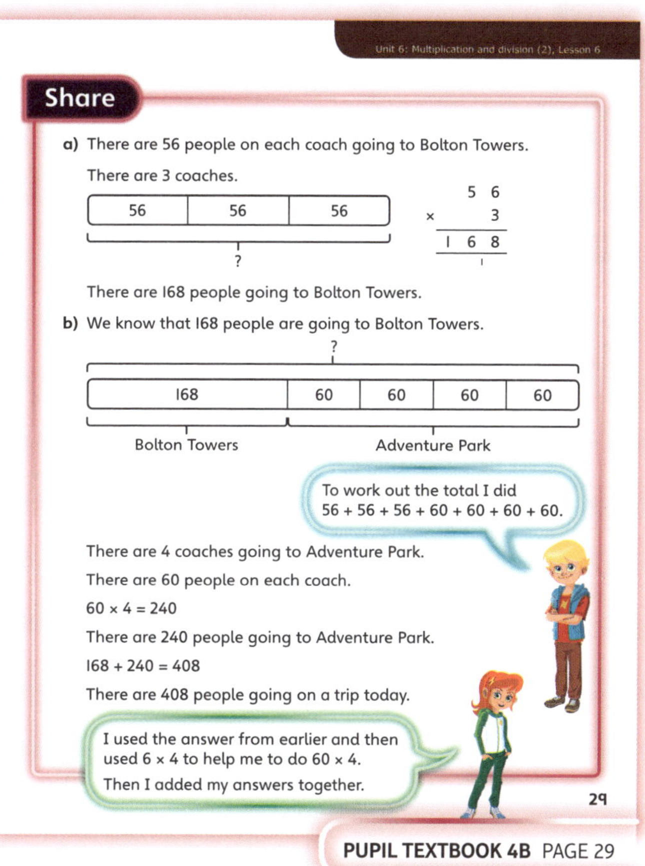

PUPIL TEXTBOOK 4B PAGE 29

Think together

WAYS OF WORKING Whole class teacher led (I do, We do, You do)

ASK

- Question **1**: *How many coaches are going to Bolton Towers? How many people are there per coach? How many coaches are going to Adventure Park? How many people are there per coach? Can you draw a bar model to show this?*
- Question **2**: *What does 'difference' mean? How many children are going ice skating? What is the cost per child? How can you present this on a bar model? How many children are going bowling? What is the cost per child?*
- Question **3**: *What information do you need to find the answer? Can you represent each calculation on a bar model?*

IN FOCUS Question **1** is a multiplication and addition problem, although children must first understand the question and work out that they need to do a multiplication. Question **2** is a multiplication and difference question where children find the difference between the total costs of two trips. Children may find it difficult in question **3** a) to decide which operation they need to use. There are no 'clue words' such as 'difference' or 'total'.

STRENGTHEN Encourage children to work through each problem line by line. Have they been given the information needed to find the answer to the question directly or is it a multi-step question? Support children in drawing a bar model to help them. Can they see the difference between when to add and when to subtract?

DEEPEN For question **3** a), ask: *Do you need the price of the ticket in the first part? Why not?* Discuss Ash's question in part b). Ask: *Could Ash find the answer this way? What does 385 represent? What is the question asking for? What is Flo suggesting? What information do you need to use to find the total cost of sending everyone to see the play?* Challenge children to discuss whether Flo or Ash is correct.

ASSESSMENT CHECKPOINT Children can use bar models to represent the problems, and written methods to multiply, add or subtract to solve word problems.

ANSWERS

Question **1**: 48 × 4 = 192 (Bolton Towers)
45 × 5 = 225 (Adventure Park)
225 + 192 = 417 (going on a trip)

Question **2**: Ice skating: 117 × 9 = £1,053
Bowling: 136 × 7 = £952
Difference: £1,053 − £952 = £101

Question **3** a): There are 336 people in total. (7 × 48 = 336)
There are 385 tickets available.
There will be 49 tickets available after the school has their tickets. (385 − 336 = 49)

Question **3** b): The total cost of sending everyone to the play is £2,016. (336 × 6 = 2,016)

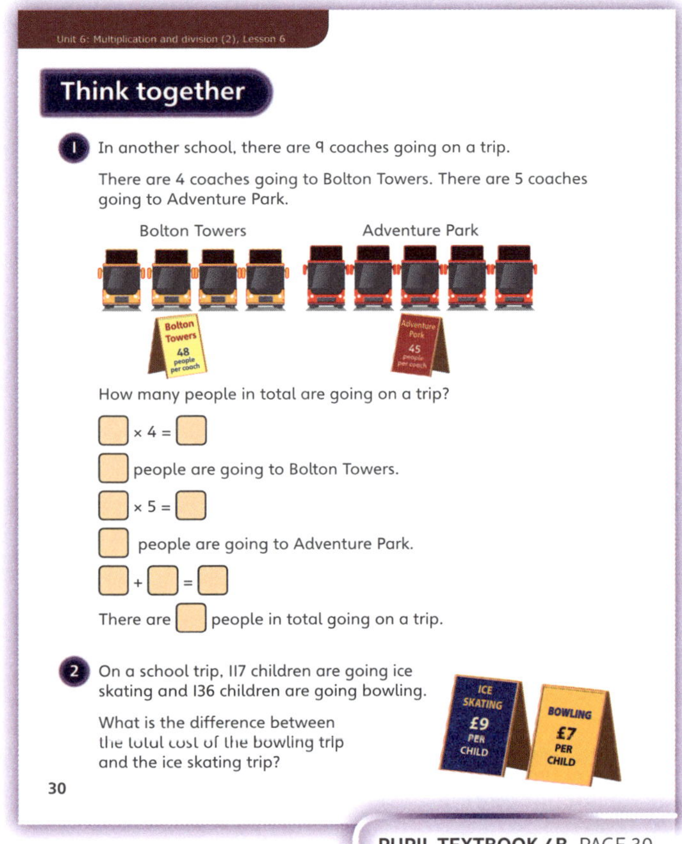

PUPIL TEXTBOOK 4B PAGE 30

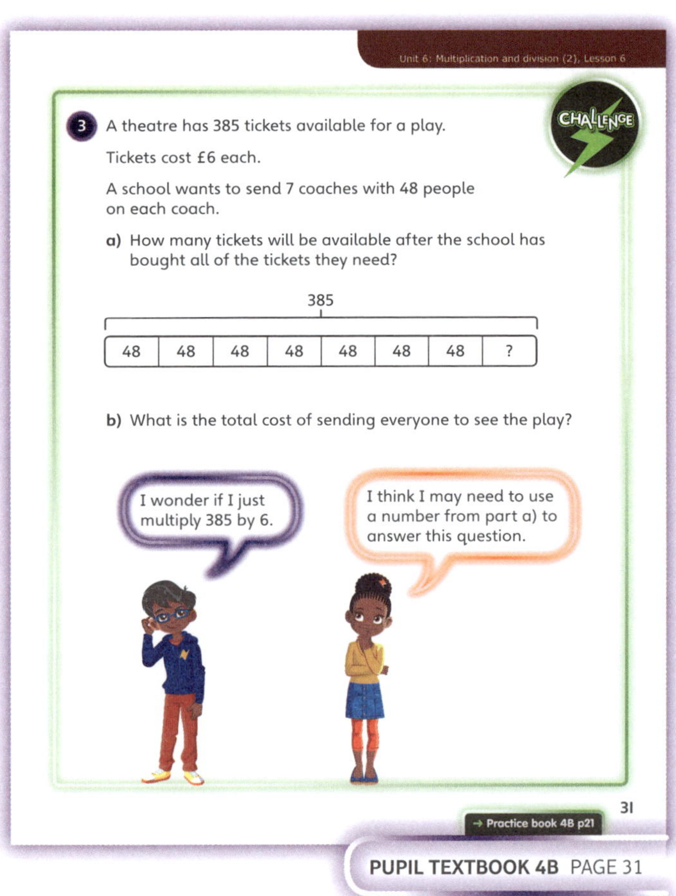

PUPIL TEXTBOOK 4B PAGE 31

Practice

IN FOCUS Question **4** is a 2-step word problem which has had the scaffolding removed, so children may find it difficult to work out which operation they will need to use. Encourage children to highlight any key information in the question, including numbers and keywords.

STRENGTHEN To support children's understanding, encourage them to use a bar model to translate the question in context into a mathematical problem. Drawing the bar model is an important step in consolidating their understanding. Challenge children to solve the problem in a different way, still using the bar model, or to explain why they are using the method they have chosen. Read the questions line by line and ask children to identify key information. If necessary, ask children to represent the elements of the problem using counters or cubes, which may help them to understand the more abstract questions. Provide opportunities for children to solve mathematical problems as often as possible in order to develop their thinking and reasoning skills while practising the operations and methods they have learnt.

DEEPEN In question **5**, ask children to change the height of the second tower to centimetres only. To further deepen understanding, ask children to draw two bars one above the other. The first bar is split into 7 equal parts. Write '86' in each of the parts. The second bar is split into 4 parts; each part is 1 m 42 cm high. Ask: *What multiplication statements can you write? Which of the towers is taller? How much taller is it?*

ASSESSMENT CHECKPOINT Can children convert a word problem into a bar model? Can they use the bar model to work out what calculations they need to do?

ANSWERS Answers for the **Practice** part of the lesson appear in the separate **Practice and Reflect answer guide**.

Reflect

IN FOCUS This section links children's multiplicative understanding with their understanding of using bar models to solve multi-step problems. They should clarify how the bar models can be used to support or check their answer.

ASSESSMENT CHECKPOINT Check children's understanding and reasoning when using bar models to compare two values.

ANSWERS Answers for the **Reflect** part of the lesson appear in the separate **Practice and Reflect answer guide**.

After the lesson ▮▮

- Can children use bar models confidently?
- Can they solve a range of multiplication word problems?
- Can they identify key information that is important to a problem?

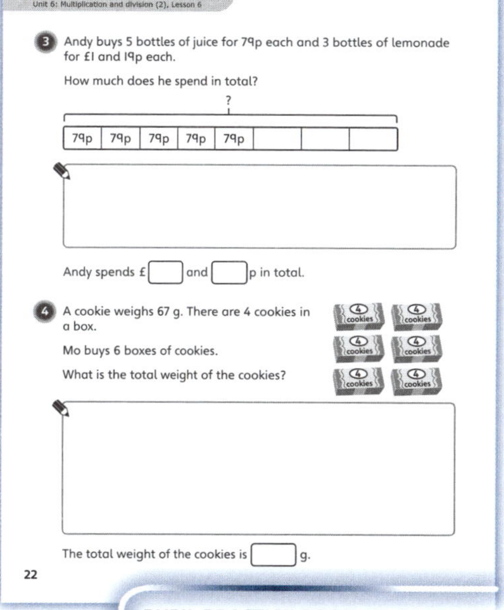

PUPIL PRACTICE BOOK 4B PAGE 21

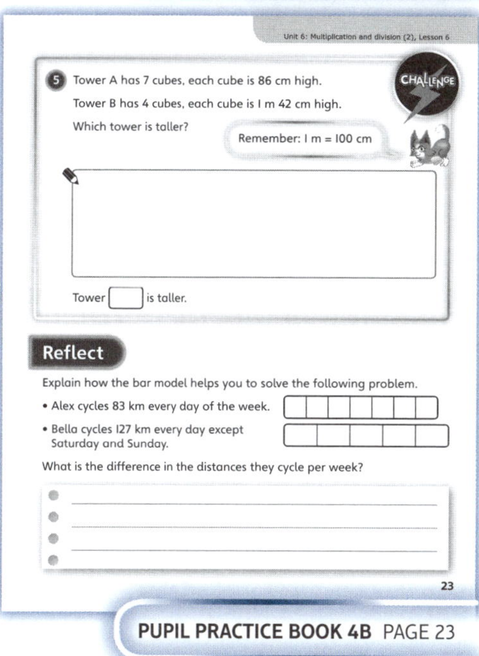

PUPIL PRACTICE BOOK 4B PAGE 22

PUPIL PRACTICE BOOK 4B PAGE 23

Multiplying more than two numbers 

Learning focus

In this lesson, children will learn to find more efficient ways to multiply. They will use the commutative properties of multiplication to calculate 'in a different order', such as $2 \times 7 \times 5 = 7 \times 10$, to increase their ability to calculate mentally.

Small steps

→ Previous step: Problem solving – multiplication
→ **This step: Multiplying more than two numbers (1)**
→ Next step: Multiplying more than two numbers (2)

NATIONAL CURRICULUM LINKS

Year 4 Number – Multiplication and Division

Solve problems involving multiplying and adding, including using the distributive law to multiply two digit numbers by one digit, integer scaling problems and harder correspondence problems such as *n* objects are connected to *m* objects.

ASSESSING MASTERY

Children use the properties of multiplication (that it is associative and distributive) and are learning to recognise the most efficient way to multiply three numbers. Children are able to recall multiplication facts rapidly.

COMMON MISCONCEPTIONS

Children may know the associative property of multiplication but fail to apply it to simplify when multiplying. For example, when multiplying $4 \times 7 \times 5$, they first multiply 4 and 7, and then use a written method to multiply the answer by 5. Ask:
• *How else could you work it out? Which numbers do you multiply first?*

Children may find it challenging to complete the second part of the multiplication, getting 'lost' when having three numbers to multiply. Increased fluency in multiplication facts will help. Ask:
• *What multiplication facts can you see in this multiplication? Are there any more?*

STRENGTHENING UNDERSTANDING

To strengthen understanding of multiplying three numbers, show them as a visual or concrete representation, for example, two sets of arrays, both with 3 columns and 5 rows. Ask children to compare this with three sets of arrays, each with 5 rows and 2 columns. Ask:
• *What multiplication can you see? Which multiplication is easier? Why? What would happen if, instead of 3, you had a different number, such as 9? What would be easier to multiply: $5 \times 9 \times 2$ or $5 \times 2 \times 9$?*

GOING DEEPER

Ask children to reason why both sets of arrays described above show the same answer, without actually working out the multiplications. Ask: *If you joined the columns together so that there was no gap between the columns, what multiplication would you see?*

KEY LANGUAGE

In lesson: multiplication, **commutative**

Other language to be used by the teacher: multiplicand, recall, multiply, divide, product, equal, grouping, array

STRUCTURES AND REPRESENTATIONS

number line, arrays

RESOURCES

Mandatory: cubes, counters, number lines

Optional: 4 times-table flashcards

 In the eTextbook of this lesson, you will find interactive links to a selection of teaching tools.

Before you teach 

• Can children recall multiplication facts quickly?
• Do children know that multiplication is commutative?
• Do they know how to use arrays to multiply?

Discover

 Pair work

ASK

- Question ① a): *How many columns of stickers are there? How many stickers are there in each column? What multiplication is shown here?*
- Question ① b): *How many sheets of stickers are there? What multiplication can you use to work out the number of stickers? Why?*

IN FOCUS Children are required to use their multiplicative reasoning in order to solve this problem. This is an opportunity to explore the commutative property of multiplying. To ensure children's concrete understanding of multiplying three numbers, it is important to link multiplying three numbers with their experience of using counters and arrays. Encourage children to rearrange the objects in different ways and to demonstrate why the answer does not change, regardless of the order of multiplying.

PRACTICAL TIPS Provide children with counters and ask them to practically solve the problem posed in the picture. Adapt the challenge by varying the numbers used in the question.

ANSWERS

Question ① a): 2 × 5 = 10
There are 10 stickers on one sheet.

Question ① b): 2 × 5 × 3 = 30, 2 × 3 × 5 = 30
There are 30 stickers, in total, on the teacher's desk.

Share

 Whole class teacher led

ASK

- Question ① a): *What does each group show?*
- Question ① b): *Can you explain Dexter's method? Can you explain Flo's method? What do you notice about your answers? Which multiplication facts did you use? Why do you think it is important to know your times-tables off by heart?*

IN FOCUS Use this section as an opportunity to clarify any misconceptions that children may have. Encourage children to talk about the number of groups and the number of objects in each group. Question ① b) provides an opportunity for children to explore the commutative property of multiplication and the variety of ways in which they can find the correct answer by multiplying three numbers in a different order. Ask children to use counters or arrays to show each multiplication and discuss how resources can be used in this way.

PUPIL TEXTBOOK 4B PAGE 32

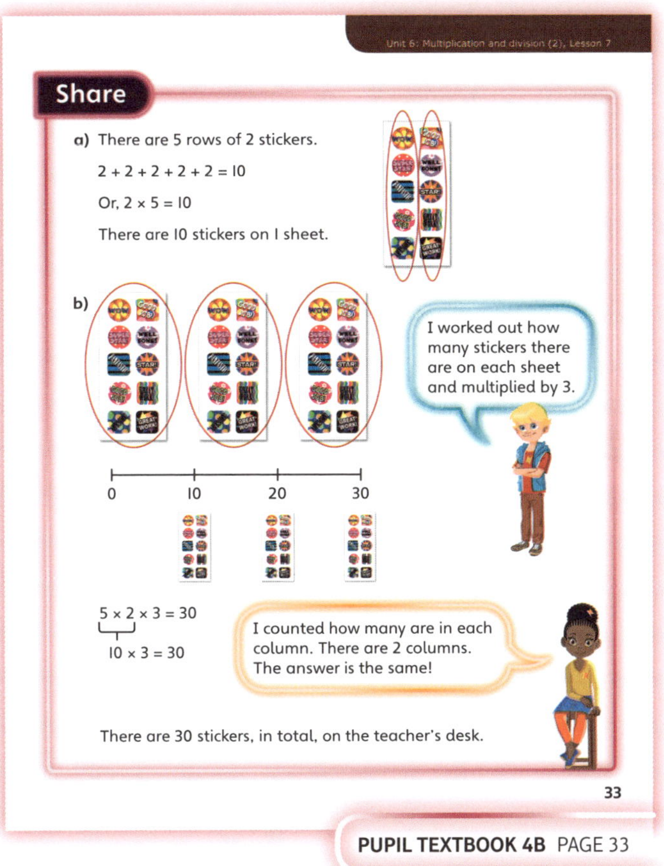

PUPIL TEXTBOOK 4B PAGE 33

Think together

WAYS OF WORKING Whole class teacher led (I do, We do, You do)

ASK

- Question ❶: *What facts can you see? How can you use these facts to work out the number of stickers? Can you multiply in a different way?*
- Question ❷: *How many doughnuts are there in each row? How many rows are there in each box? How did you multiply the numbers? Can you use a different method?*
- Question ❸: *Which multiplication facts do you find easier? Why? Look at Luis's and Isla's calculations. Which one is easier to do in your head? Which do you prefer?*

IN FOCUS In question ❸, children start to explore the commutative property even further and discuss the different methods they can use to multiply three numbers. They should notice that it is easier to multiply by 10, or to double a number, than to multiply by 3 or 9. Ask children to discuss what Astrid says and link the way children can add three numbers to the way they can multiply three numbers. Write '9 + 2 + 8' and '9 × 2 × 8'.

STRENGTHEN To strengthen understanding of multiplying three numbers, use visual or concrete representations, for example with towers of cubes, or arrays. It is important that children can visualise and understand how to use their multiplication facts to multiply three numbers. Reinforce that they need to develop rapid recall of the multiplication and division facts, but also need to understand what they are. In question ❷, discuss Sparks's advice about drawing a diagram. Ask: *What diagram can you draw? How many groups will there be?*

DEEPEN To extend question ❸ ask children to calculate 45 × 8. Encourage them to write 45 as '5 × 9'. Ask: *What is the easiest way to calculate this?* Write '5 × 9 × 8' on the board and ask: *Which numbers can you multiply first?* Provide other examples and encourage children to generalise that multiplication is commutative.

ASSESSMENT CHECKPOINT Children use their multiplication multiplication facts to multiply three numbers. They can work out which fact will help them to work out the answer in the easiest and quickest way.

ANSWERS

Question ❶: 5 × 2 × 6 = 60 or 6 × 2 × 5 = 60
10 × 6 = 60 or 12 × 5 = 60
There are 60 stickers on 6 sheets.

Question ❷ a): 3 × 6 × 2 = 36; there are 36 doughnuts in 2 boxes.

Question ❷ b): 3 × 6 × 5 = 90; there are 90 doughnuts in 5 boxes.

Question ❸ a): Example answer: Isla's method is better as it is easier to multiply by 10 than by 5 or by 18.

Question ❸ b): 7 × 6 = 42, 42 × 2 = 84
4 × 5 = 20, 20 × 3 = 60
9 × 8 = 72, 72 × 2 = 144

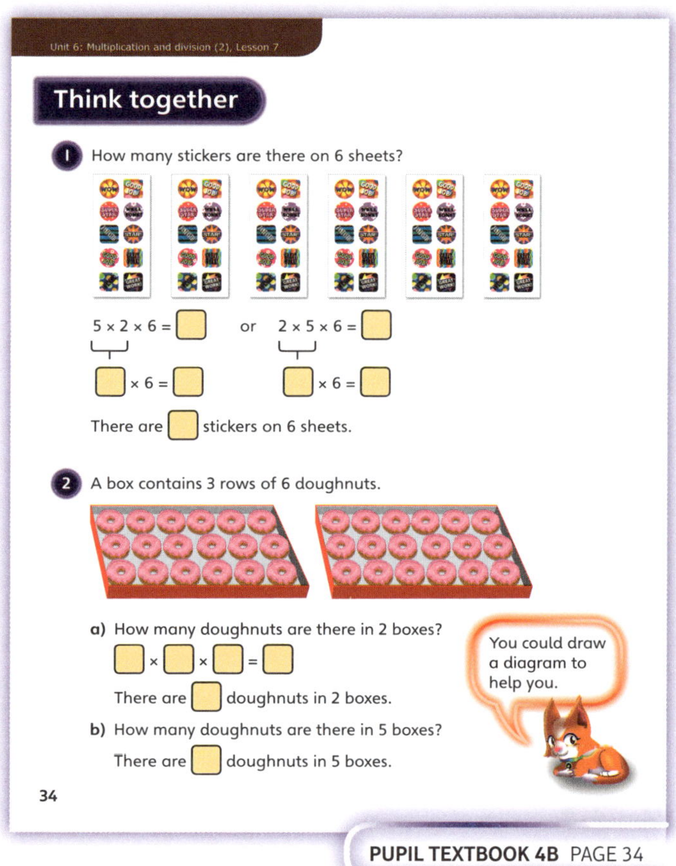

PUPIL TEXTBOOK 4B PAGE 34

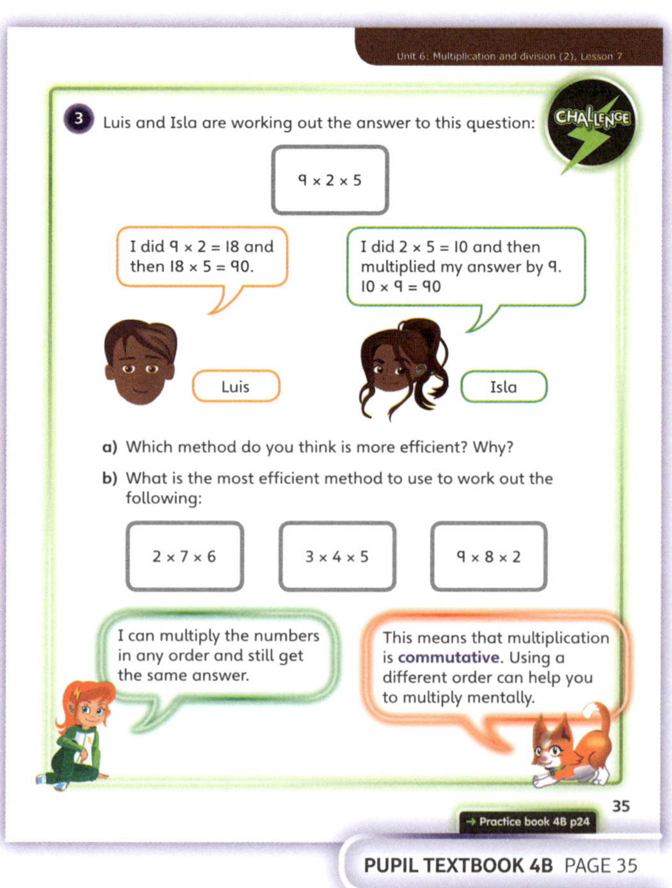

PUPIL TEXTBOOK 4B PAGE 35

Practice

WAYS OF WORKING Independent thinking

IN FOCUS Question ❶ reinforces visual images of multiplying three numbers. It is important to cement this understanding. Question ❸ asks children to discuss the method of multiplying three numbers and to consider efficiency when choosing which numbers to multiply first.

STRENGTHEN To strengthen understanding of multiplying three numbers, show the numbers using visual or concrete representations. Question ❶ will help with this. As children move through the rest of the exercise, the questions become more abstract, but children may still need to use cubes or counters to represent multiplications. It is important that children recognise the importance of multiplication facts. Reinforce that they need to use their mathematical knowledge, not only to find a way to solve a question, but also to find the most efficient way to do so.

DEEPEN In question ❽, ask children to explain how they can find the missing numbers. They need to focus on what is given in the question and ask themselves how to use the information. Ask: *What is special about 216? Is it even or odd? What numbers can it be divided by? How do you know?*

THINK DIFFERENTLY In question ❼, children need to notice that the numbers are multiplied by 0, hence they do not need to multiply them, as the answer will be 0. Children need to explain their reasoning and mathematical thinking rather than guess the answer. This may be an opportunity to explain the misconception that a number does not change when multiplied by 0.

ASSESSMENT CHECKPOINT Can children find the most efficient way to multiply three numbers? They should have a secure recall of their multiplication facts (both multiplication and associated division facts) and use these to find the most efficient way to multiply three numbers.

ANSWERS Answers for the **Practice** part of the lesson appear in the separate **Practice and Reflect answer guide**.

Reflect

WAYS OF WORKING Independent thinking

IN FOCUS This brings together the commutative property of multiplying and the knowledge of multiplication facts. Children should notice that it is easier to multiply by 2 last. Encourage them to take a strategic approach rather than just guessing. Ask them to explain their reasoning clearly.

ASSESSMENT CHECKPOINT Check whether children have instant recall of multiplication facts and are able to explain the properties of multiplication. They should be able to explain that multiplication can be done in any order because it is commutative. Check if there are any multiplication facts that they are confused about or any properties of multiplication that they do not understand.

ANSWERS Answers for the **Reflect** part of the lesson appear in the separate **Practice and Reflect answer guide**.

After the lesson 

- Can children recall division and multiplication facts from the 12 times-table?
- Are they able to manipulate the facts they know in order to solve problems in an efficient way?

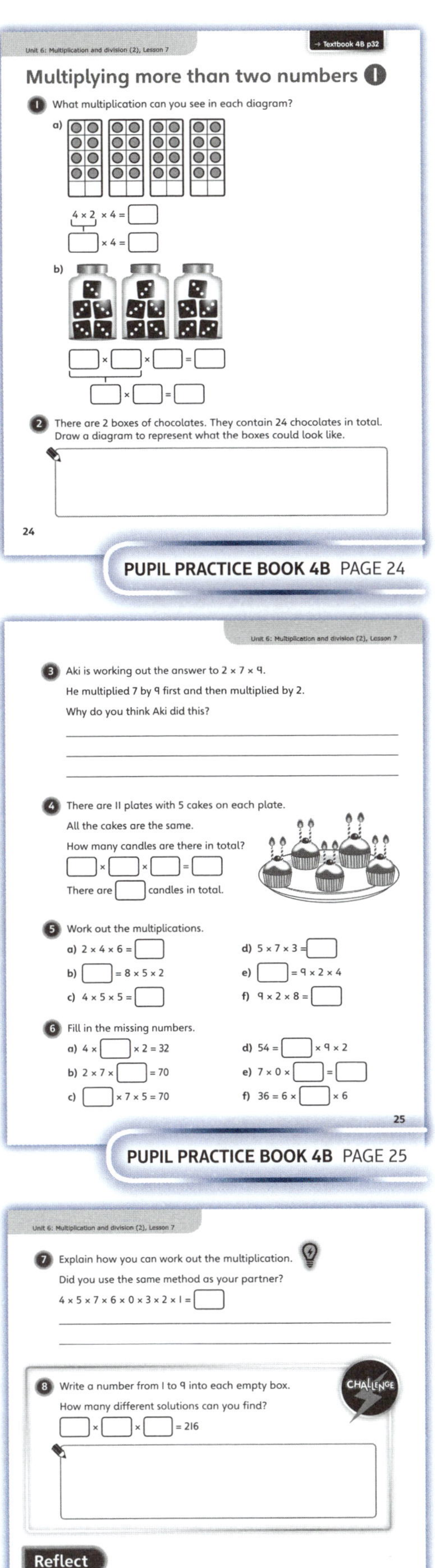

PUPIL PRACTICE BOOK 4B PAGE 24

PUPIL PRACTICE BOOK 4B PAGE 25

PUPIL PRACTICE BOOK 4B PAGE 26

Multiplying more than two numbers 2

Learning focus

In this lesson, children will focus on learning to simplify multiplications by finding factor pairs of 2-digit numbers and then using commutativity to help them to perform mental calculations.

Small steps

→ Previous step: Multiplying more than two numbers (1)

→ **This step: Multiplying more than two numbers (2)**

→ Next step: Problem solving – mixed correspondence problems

NATIONAL CURRICULUM LINKS

Year 4 Number – Multiplication and Division

Recognise and use factor pairs and commutativity in mental calculations.

ASSESSING MASTERY

Children are able to identify and use appropriate factor pairs to help them to multiply numbers in the most efficient way. For example, $5 \times 36 = 5 \times 6 \times 6 = 30 \times 6 = 180$.

COMMON MISCONCEPTIONS

Children may confuse factor pairs with partitioning. For example, when multiplying 5×22, they write $22 = 20 + 2$, $5 \times 20 = 100$, $100 \times 2 = 200$.

Ask children to estimate in order to check their answers:
- *What number is 22 close to? What is 5×20? Should the answer be closer to 100 or 200? Why?*

STRENGTHENING UNDERSTANDING

To strengthen understanding of multiplying numbers, show them as a visual or concrete representation, for example, 5 towers made of 22 cubes each. Ask: *What is the multiplication?*

Split the towers in half. Ask: *How many towers do you have now? How many cubes are there in each tower? What is the multiplication? Is 10×11 easier or harder to find than 5×22?* Children should see for themselves that $5 \times 22 = 5 \times 2 \times 11 = 10 \times 11 = 110$ cubes.

GOING DEEPER

Ask children to use the factors of 25 and 22 to calculate 25×22.

Children could be encouraged to set challenges for each other. Ask: *How can you calculate 25×12 mentally? What would the answer be? Can you use a different way to calculate? Would the answer change?*

KEY LANGUAGE

In lesson: multiply, equal to, groups, factor, factor pair

Other language to be used by the teacher: efficient, divide, product

STRUCTURES AND REPRESENTATIONS

number line, arrays

RESOURCES

Mandatory: cubes, counters

 In the eTextbook of this lesson, you will find interactive links to a selection of teaching tools.

Before you teach ⏸

- Can children recall the multiplication facts quickly?
- Do children know how to use arrays to help them to multiply?

Discover

 Pair work

ASK

- Question **1** a): *Is 24 even or odd? What numbers multiply to make 24? Can you think of a different pair? How do you know you have all the answers?*
- Question **1** b): *List all the factors of 24 that you found. Which of these numbers can be easily multiplied by 5?*

IN FOCUS Question **1** a) asks children to find pairs of numbers that multiply to make 24. This is a good opportunity to revisit the 2, 3, 4, 6, 8 and 12 times-tables. Children could use 24 counters to find different ways to make arrays. Children also explore the commutative property of multiplying; for example, 2 × 12 = 12 × 2.

In question **1** b), children use the answer from part a) to multiply three numbers. They use a pair of numbers from the list from part a) and multiply it by 5. Their aim is to find an efficient way of multiplying the three numbers and make their calculation as easy as possible.

PRACTICAL TIPS Provide children with 5 groups of 24 counters. Encourage them to explore all the different arrays they can make.

ANSWERS

Question **1** a): 1 × 24 = 24, 2 × 12 = 24, 3 × 8 = 24, 4 × 6 = 24

Question **1** b): Richard can use the numbers from part a) to help him find the numbers that are easiest to multiply together: 24 × 5 = 12 × 2 × 5 = 12 × 10 = 120; so, 24 × 5 = 120.

Share

 Whole class teacher led

ASK

- Question **1** a): *How many pairs of numbers did you find? Can you find another one? And another? Can you make arrays to show each of the pairs? How can you make sure that you have all the answers?*
- Question **1** b): *Which of the factors from part a) will you use in part b) to make the calculation easier? Discuss Sparks's advice and remind children of their learning in the previous lesson. Does the order of multiplying matter? How do you choose which numbers to multiply first? Which multiplication facts did you use? Why do you think it is important to know your times-tables off by heart?*

IN FOCUS In question **1** a), the aim is for children to use their knowledge of multiplying to find all the factor pairs of a number. Draw out the importance of being systematic. In part b), children explore the commutative property of multiplication. Children should be encouraged to explore the multiple ways they may get to the correct answer. This is an opportunity for children to think of numbers that may be multiplied together to make a calculation easier.

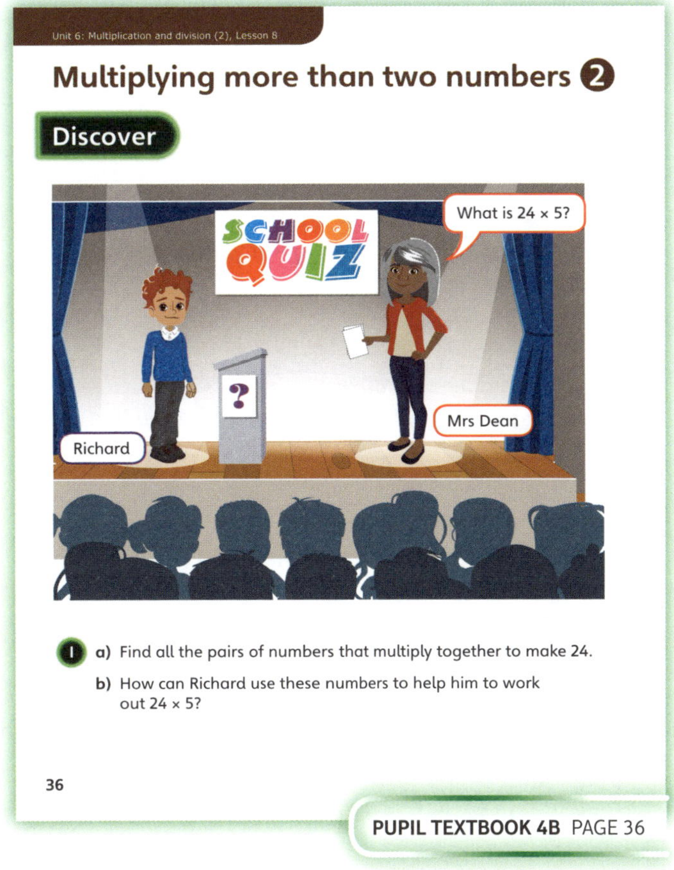

PUPIL TEXTBOOK 4B PAGE 36

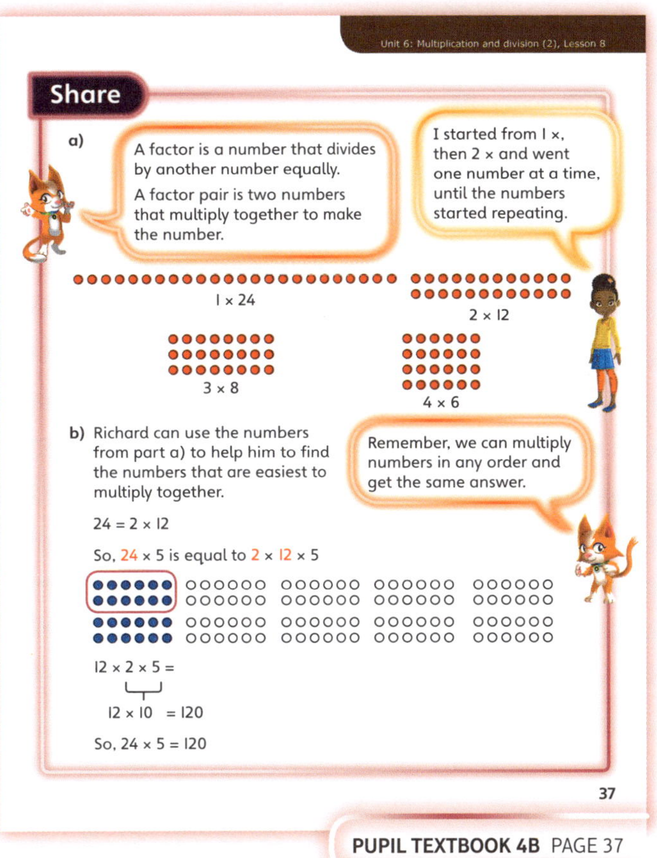

PUPIL TEXTBOOK 4B PAGE 37

Think together

WAYS OF WORKING Whole class teacher led (I do, We do, You do)

ASK

- Question **1**: *Which numbers is Richard multiplying? What are the factors of 55? How did Richard decide which numbers to multiply first?*
- Question **2**: *Write down the pairs of factors of 36. Choose a pair that will make 36 × 5 easy to work out in your head. Why did you choose these numbers?*
- Question **3**: *Which of Richard's methods do you find the easiest? Which methods can you do in your head? Has Richard found all the methods?*

IN FOCUS Question **1** requires children to link their multiplicative understanding to their new understanding of factors by choosing the numbers that are easier to multiply first. Use this opportunity to clarify any misconceptions from the previous part of the lesson. Encourage children to talk about the pair of factors of a number and how they can find them. Question **2** aims to develop fluency with using the properties of multiplying and factors of a number to multiply 1- and 2-digit numbers. Explain Ash's and Astrid's methods. Discuss how it is easier to double twice than to multiply by 4.

STRENGTHEN To strengthen understanding of multiplying three numbers, show them as a visual or concrete representation, for example with towers of cubes, or arrays. In question **1**, you could use an 11 × 5 array and then repeat this 6 times. Discuss with children the importance of knowing the 12 times-table in order to find factors and make the multiplications easier.

DEEPEN Allow children time in question **3** to explore all the different ways to make 48. Discuss Ash's and Astrid's comments and ask children to separate the different methods into two groups – according to whether they can calculate in their heads or write the calculations down. Children should discuss the methods. Ask: *What is the easiest way to multiply 6 × 48?* Listen for children who can justify their answer using the correct mathematical vocabulary.

ASSESSMENT CHECKPOINT Children use their multiplication multiplication facts to find pairs of factors and to simplify their workings. They can work out which fact will help them to work out the answer in the easiest and quickest way.

ANSWERS

Question **1** a): $5 \times 6 = 30$

Question **1** b): $11 \times 30 = 330$

Question **2** a): $36 \times 5 = 6 \times 6 \times 5 = 6 \times 3 \times 2 \times 5$
$= 18 \times 10 = 180$

Question **2** b): $18 \times 18 = 2 \times 9 \times 2 \times 9 = 81 \times 2 \times 2 = 324$

Question **3** a): Richard is correct.

Question **3** b): He could have used:
$6 \times 6 \times 8 = 6 \times 6 \times 2 \times 4 = 6 \times 6 \times 2 \times 2 \times 2$
$= 36 \times 2 \times 2 \times 2 = 288$
Or, $2 \times 3 \times 6 \times 8 = 2 \times 3 \times 2 \times 3 \times 4 \times 2$
$= 9 \times 2 \times 2 \times 2 \times 2 \times 2 = 288$

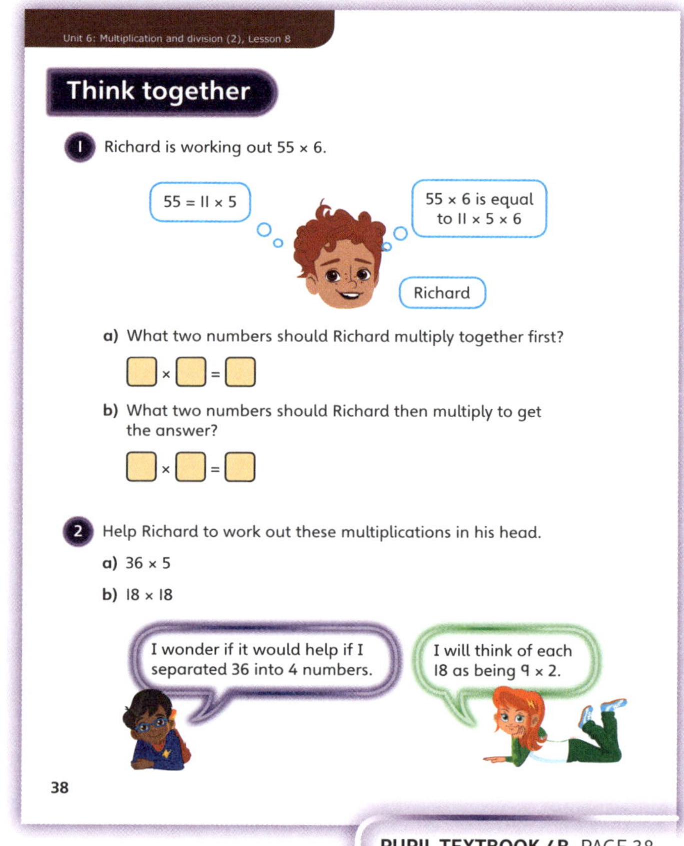

PUPIL TEXTBOOK 4B PAGE 38

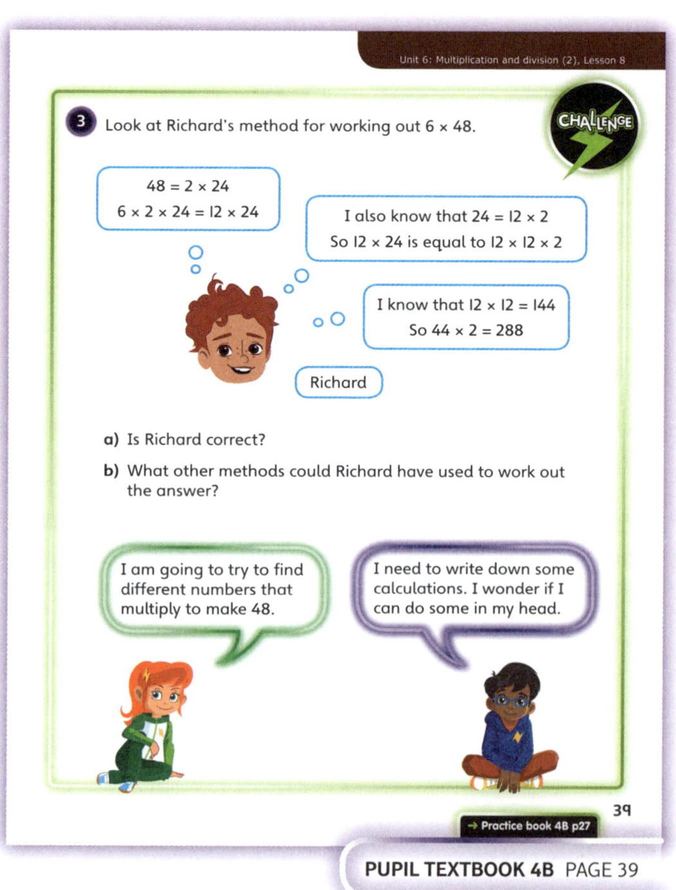

PUPIL TEXTBOOK 4B PAGE 39

Practice

WAYS OF WORKING Independent thinking

IN FOCUS Questions ❶ and ❷ reinforce the method used throughout the lesson of finding factor pairs when multiplying numbers. Question ❸ requires children to reason and compare the different methods that can be used to multiply two numbers depending on the pair of factors used.

STRENGTHEN Questions ❶ and ❷ require children to find factors that they can use to make multiplications easier. As children move through the rest of the exercise, the questions become more abstract; children may still need to use cubes or counters to represent multiplications. It is important that children recognise the multiplication facts. Reinforce that children need to use their maths knowledge not only to find a way to solve a question, but also to find the most efficient way to do so.

DEEPEN While solving question ❺, deepen children's thinking about the link between numbers. Ask: *Does knowing 35 × 16 help you to find 70 × 16? Is the answer to 70 × 16 bigger or smaller than 35 × 32? What about 70 × 8 and 35 × 16? Can you convince me?*

THINK DIFFERENTLY Question ❹ requires children to look at two different methods of multiplying two numbers and explain why they are correct. This will help to consolidate their learning.

ASSESSMENT CHECKPOINT Can children find the most efficient way to multiply numbers? Are they confident in finding the factor pairs of a number? They should have a secure recall of their multiplication facts (both multiplication and associated division facts) and use these to find the most efficient way to multiply.

ANSWERS Answers for the **Practice** part of the lesson appear in the separate **Practice and Reflect answer guide**.

Reflect

WAYS OF WORKING Pair work

IN FOCUS This question exposes children's depth of understanding of the commutative property of multiplying. Children can use concrete materials, or a diagram such as an array, to support their answer. Ask children to explain their reasoning clearly.

ASSESSMENT CHECKPOINT Check whether children identify the factor pairs of 3 × 4 and 4 × 3, and whether they can explain why these give the same answer.

ANSWERS Answers for the **Reflect** part of the lesson appear in the separate **Practice and Reflect answer guide**.

After the lesson 

- Can children recall division and multiplication facts from the 12 times-table?
- Are children able to list the factors pairs of a number?
- Are they able to manipulate the facts they know in order to solve problems in an efficient way?

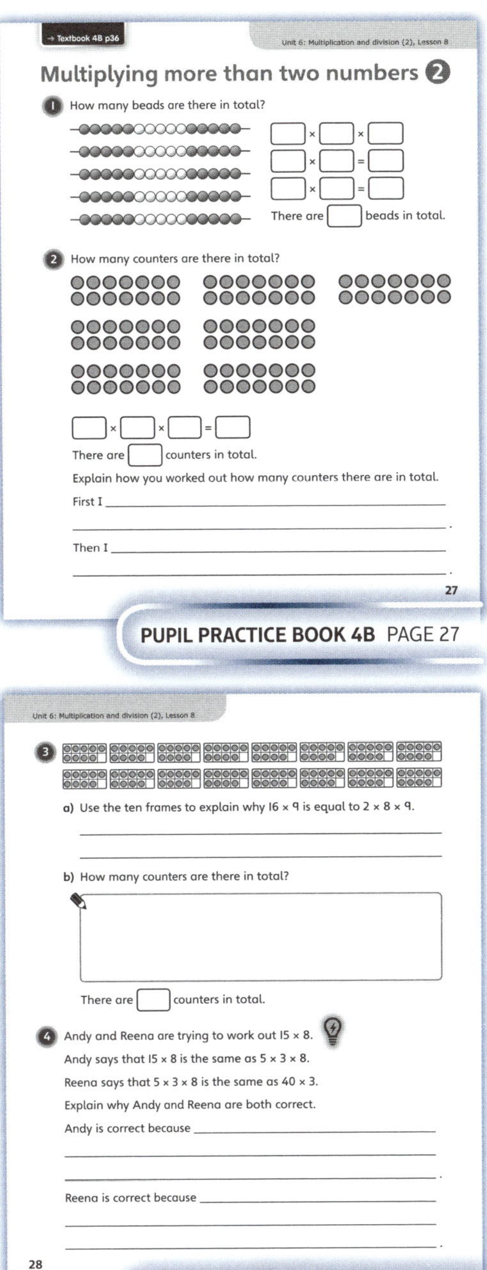

PUPIL PRACTICE BOOK 4B PAGE 27

PUPIL PRACTICE BOOK 4B PAGE 28

PUPIL PRACTICE BOOK 4B PAGE 29

Problem solving – mixed correspondence problems

Learning focus

In this lesson, children will solve more complex correspondence problems, working out how *n* objects relate to *m* objects, finding all solutions and noticing how to use multiplication to solve these problems.

Small steps

→ Previous step: Multiplying more than two numbers (2)
→ **This step: Problem solving – mixed correspondence problems**
→ Next step: Dividing a 2-digit number by a 1-digit number

NATIONAL CURRICULUM LINKS

Year 4 Number – Multiplication and Division

Recognise and use factor pairs and commutativity in mental calculations.

ASSESSING MASTERY

Children should be able to work out how many possible combinations of two simple sets of objects there are, using correspondence, and be able to identify the multiplication they should use to work this out efficiently.

COMMON MISCONCEPTIONS

Children add the number of combinations rather than multiply them. For example, if they have 3 scarves and 2 hats, they think they have 5 combinations they can wear. Ask:
• *How many hats do you have? For each hat, how many scarves do you have to choose from?*

STRENGTHENING UNDERSTANDING

To support children in understanding correspondence problems, start by exploring the number of possibilities arising from two choices, before adding a third. Children need to have a lot of practice in combining different items, such as socks, shoes, hats, scarves, before they are able to make the link between the correspondence problems and multiplying to find all the possible combinations.

GOING DEEPER

Encourage children to think of real life examples of correspondence problems: for example, ask children to look around the classroom and find correspondence between table, children, pens and so on – there might be one table for every two children, or each child might have three different-coloured pens.

KEY LANGUAGE

In lesson: multiplication, correspondence

Other language to be used by the teacher: multiply, divide, grouping, sharing, multiplication statement, division statement, recall, combinations

RESOURCES

Optional: picture cards of buckets and spades, pictures of shorts, shirts and different-coloured socks

 In the eTextbook of this lesson, you will find interactive links to a selection of teaching tools.

Before you teach 

• Do children know multiplication facts?
• Can children show multiplication using a diagram?
• Can children work out the answer to simple 1-step multiplication word problems?

Discover

WAYS OF WORKING Pair work

ASK

- Question **1** a): *What does the picture show? How many different buckets are there? How many different spades are there? Look at the first bucket. Match it with a spade. Are there other spades you could have matched this bucket with? How many different matches might there be?*
- Question **1** b): *How can you match the buckets with spades? What type of diagram can you make so that you do not miss any of the matches? Will you record the possible options by using a table, or by drawing arrows to show all the possible matches? Will you draw the pictures of each item, or represent each item by a colour or by a letter or number?*

IN FOCUS Question **1** a) introduces the idea of matching two items. It is important that children fully understand the context and use pictures and diagrams when moving from concrete to pictorial representations of the objects in the question. For both parts of the question, children should be encouraged to be systematic in their approach. Discuss all the possible options of recording the combinations available and ask children to explain which one is their favourite and why.

PRACTICAL TIPS Using real objects, picture cards or counters will help children to visualise the question better.

ANSWERS

Question **1** a): There are 20 different ways to match up the buckets with the spades.

Question **1** b): This links to the multiplication 5 × 4 because there are 5 rows of 4 different matches. This is 5 lots of 4, which is the same as 5 × 4 = 20 matches.

Share

WAYS OF WORKING Whole class teacher led

ASK

- Question **1** a): *How has Astrid answered the question? How will the number of lines help her? What does each line represent? How can she check that she has not missed any matches?*
- Question **1** b): *How has Dexter answered the question? How did Flo link 5 × 4 with the information in the table? How many boxes does the table have? What does each box represent?*

IN FOCUS This question provides an opportunity to focus on the different methods that can be used to answer correspondence problems. Astrid shows how you can draw a line from each item to the corresponding item. Dexter and Flo use a table to write their answers. Discuss both methods that are used, and encourage children to follow a systematic approach so they do not miss any of the possible matches.

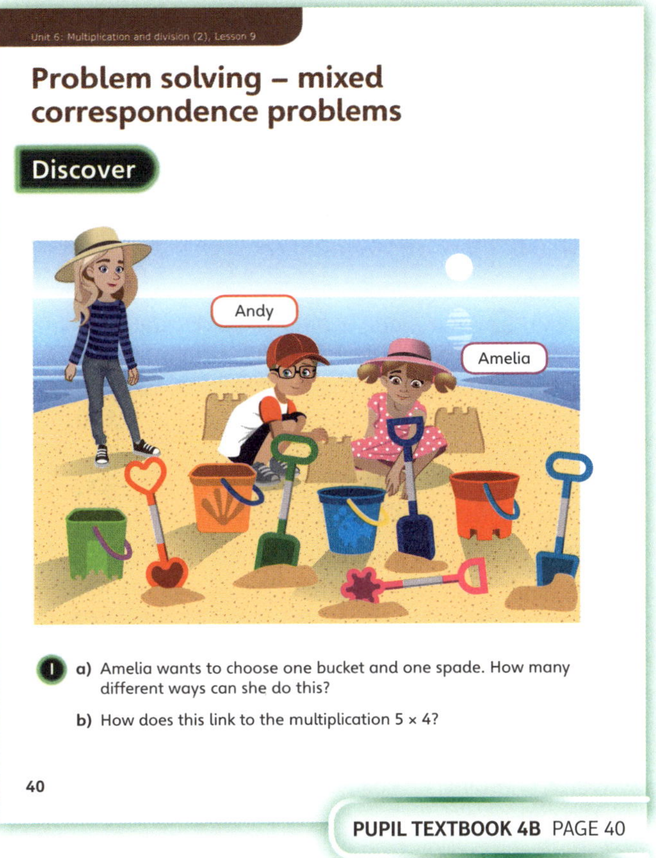

PUPIL TEXTBOOK 4B PAGE 40

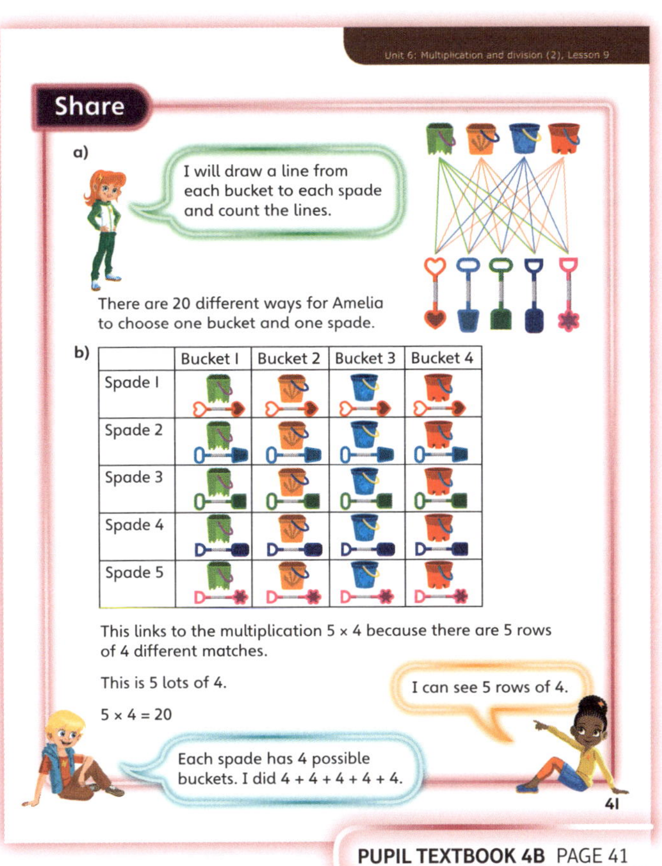

	Bucket 1	Bucket 2	Bucket 3	Bucket 4
Spade 1				
Spade 2				
Spade 3				
Spade 4				
Spade 5				

PUPIL TEXTBOOK 4B PAGE 41

Think together

WAYS OF WORKING Whole class teacher led (I do, We do, You do)

ASK

- Question **1**: *How many buckets are there? How many spades are there? How does the diagram help you to understand it is a multiplication?*
- Question **2**: *How many T-shirts does Zac have? What number multiplied by 6 gives you 30? What would you do to find the answer?*
- Question **3**: *How many socks are needed to make a pair? How can you make sure that you have found all the possible pairs? Are any pairs in your grid twice? Take one sock first: how many other socks could it match with?*

IN FOCUS Question **1** builds on the **Discover** task by asking children to find the total number of different combinations of 6 buckets and 4 spades. Question **2** gives children an opportunity to find the missing number of objects if the total of the combinations and one of the objects used in the combinations is known. Children reverse the multiplication to find the missing number. Ensure they understand what is given in the question and are able to link the concrete objects (buckets and spades, or T-shirts and shorts) with their pictorial representation (drawing, diagrams) and to the abstract concept (multiplying or dividing) when solving similar questions.

STRENGTHEN Encourage children to draw lines joining the objects, similar to the method that Astrid used; or they can draw a table like Dexter and Flo did. Can children see where the multiplication is in the diagram? For question **3**, provide children with 8 different socks or 8 different picture cards of socks. Ask them to start with one sock and see how many others it could be matched with. Ask children to be systematic in their approach.

DEEPEN In question **3**, children could be encouraged to investigate the relationship between the number of socks in each row and the number of matches for each of the socks. Children need to realise that each sock can only be matched with 7 others (a sock cannot be matched with itself), so they should start by multiplying 8 × 7. However, a pink sock paired with a purple sock is the same as a purple sock paired with a pink sock, so they then need to divide by 2 to find the number of unique combinations. Ask: *What would the total number of matches be if you had 10 socks, 15 socks or 100 socks? How many rows and columns would your table have?*

ASSESSMENT CHECKPOINT Children can solve correspondence problems by drawing diagrams and using multiplication to find the total number of possibilities. Children should make a table to write their findings systematically.

ANSWERS

Question **1**: 6 × 4 = 24; there are 24 different ways to match the buckets with the spades.

Question **2**: 6 × 5 = 30; Zac has 5 pairs of shorts.

Question **3** a): There are 28 unique pairs.

Question **3** b): 8 × 7 will give the total number of options; divide by 2 to see the unique combinations.

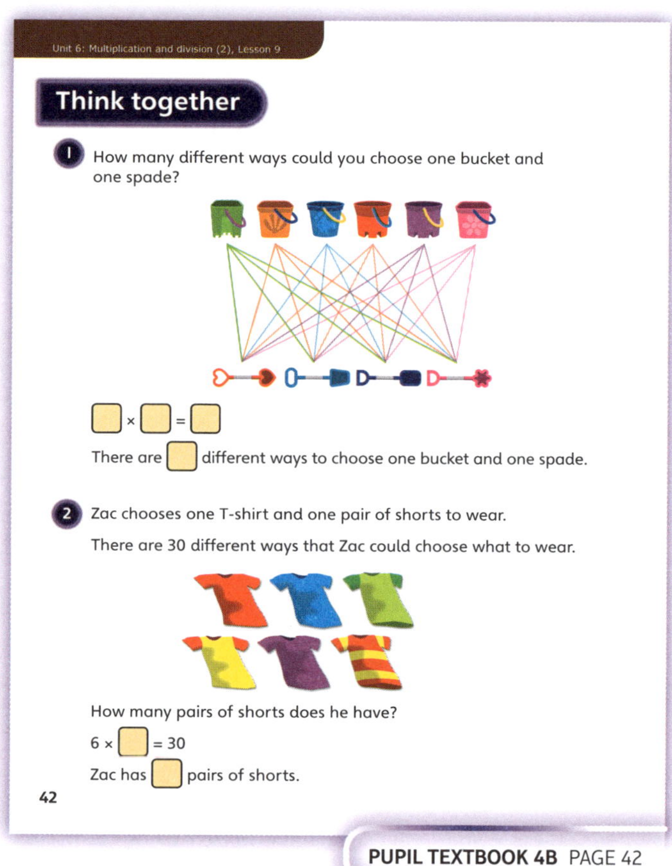

PUPIL TEXTBOOK 4B PAGE 42

PUPIL TEXTBOOK 4B PAGE 43

Practice

WAYS OF WORKING Independent thinking

IN FOCUS Question **5** offers children an opportunity to find a systematic way to work out all possible combinations, including the possibility that the same object (a snack) may be chosen twice. To ensure children's understanding of the concept of correspondence, it is important to make links with their everyday experiences and choices/matches they may make.

STRENGTHEN To support understanding, use counters and cubes to represent the different types of object used in each question. Encourage children to draw a sample space diagram (see **Share** activity in the **Textbook**) and write all the possible answers. Ask them to be systematic in their approach. For question **5**, list all the possible snacks Reena can buy. Encourage children to draw a diagram and to list the options to check their answer. If necessary, point out that Reena could buy two of the same snack. Provide opportunities for discussion by asking children to work in pairs. Ensure that children look at one variable at a time. Ask: *How do you know that you have found all the possible answers?*

DEEPEN Ask children to make up their own correspondence problems: for example, choosing ice-cream toppings and a cone or a cup. How many variables will they choose in the question? What diagram will they use to find the answer?

ASSESSMENT CHECKPOINT Children can describe their method of solving correspondence problems by using a diagram and are able to find the total number of possibilities by multiplying the variables involved in the question.

ANSWERS Answers for the **Practice** part of the lesson appear in the separate **Practice and Reflect answer guide**.

Reflect

WAYS OF WORKING Independent thinking

IN FOCUS This question provides a final opportunity for children to link the total number of matches that they can make, with multiplication. Encourage them to support their answers by using diagrams and observe whether they can generalise the rule of finding the total number of matches when one group of objects is connected with another one.

ASSESSMENT CHECKPOINT Children's understanding of multiplicative relationships is deepened by exploring correspondence problems. They can generalise the rule of finding the total number of combinations when one group of objects is connected with another. To find the total number of combinations that *n* objects make when connected with *m* objects, they multiply the number of each group of objects by each other.

ANSWERS Answers for the **Reflect** part of the lesson appear in the separate **Practice and Reflect answer guide**.

After the lesson ⏸

- Can children solve a range of correspondence problems?
- Can they identify correspondence problems in real life?
- Can they draw diagrams to represent and solve the problems?

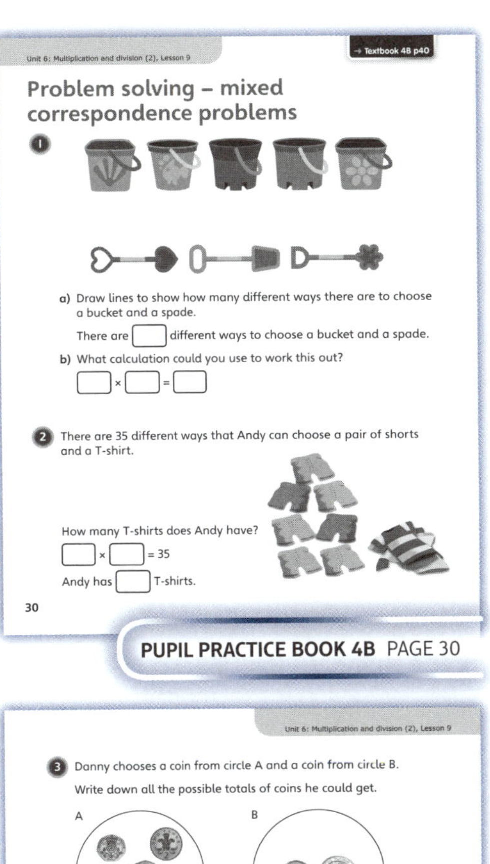

PUPIL PRACTICE BOOK 4B PAGE 30

PUPIL PRACTICE BOOK 4B PAGE 31

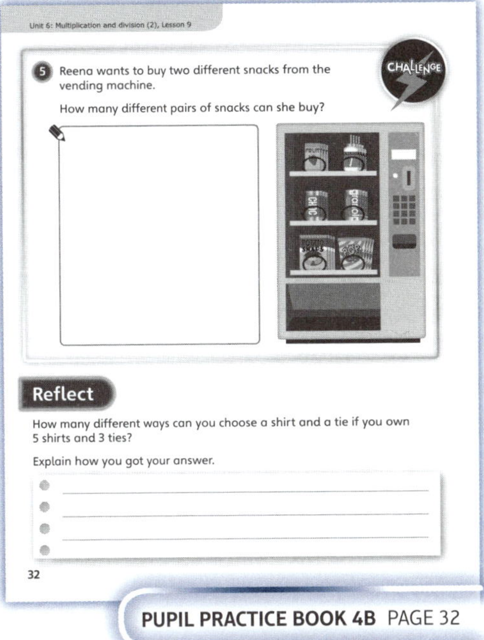

PUPIL PRACTICE BOOK 4B PAGE 32

Dividing a 2-digit number by a 1-digit number ①

Learning focus

In this lesson, children will learn how to divide 2-digit numbers by 1-digit numbers. They will focus on learning how to divide a 2-digit number where the tens digit and the ones are divisible by the divisor (for example, 96 divided by 3, 48 divided by 4, 55 divided by 5).

Small steps

→ Previous step: Problem solving – mixed correspondence problems
→ **This step: Dividing a 2-digit number by a 1-digit number (1)**
→ Next step: Division with remainders

NATIONAL CURRICULUM LINKS

Year 4 Number – Multiplication and Division
- Recognise and use factor pairs and commutativity in mental calculations.
- Solve problems involving multiplying and adding, including using the distributive law to multiply two digit numbers by one digit, integer scaling problems and harder correspondence problems such as *n* objects are connected to *m* objects.

ASSESSING MASTERY

Children develop a conceptual understanding of the written methods for division. They can explain and demonstrate how the written methods for multiplication and division work.

COMMON MISCONCEPTIONS

When dividing a 2-digit number, children may divide each of the digits and then add the results. For example, they may say that $68 ÷ 2 = 7$, calculating $6 ÷ 2 = 3$, $8 ÷ 2 = 4$, and then adding $3 + 4 = 7$. Alternatively, they may be unsure of what to do with the digits. To draw out these misconceptions, ask:
- *Why is 7 not a good estimation? What are 6 tens divided by 2? What are 8 ones divided by 2?*

STRENGTHENING UNDERSTANDING

Ensure children have access to concrete equipment that they can share into groups, or distribute. Encourage children to write out all their working, including drawing diagrams of what they have made. Being confident with their times-tables will help children to spot number patterns and factors of numbers, and this will support them when working through the division questions in this lesson.

GOING DEEPER

Draw out deeper thinking by asking children to use reverse operations to work out the number of objects in a group, or the number of groups. For example, ask: *What number multiplied by 2 is 68? How might you work this out?*

KEY LANGUAGE

In lesson: share equally, how many, equal groups

Other language to be used by the teacher: grouping, whole, division, factor, remainder

STRUCTURES AND REPRESENTATIONS

part-whole model

RESOURCES

Optional: counters, bricks

 In the eTextbook of this lesson, you will find interactive links to a selection of teaching tools.

Before you teach 

- Do children understand simple division by grouping?
- Do children know their times-tables?

Discover

WAYS OF WORKING Pair work

ASK

- Question **1** a): *How many pieces of pineapple do Richard and Toshi have in total? How many pieces of pineapple can fit on to each stick? Can you use counters to help you to represent the question? What operation do you need to do to solve this problem?*
- Question **1** b): *How many grapes are there in total? How many sticks are there in total? What does 'shared equally' mean?*

IN FOCUS Question **1** a) focuses on the core objective of the lesson, dividing a 2-digit number by a 1-digit number. Encourage children to physically represent the groups of 3 pineapple chunks by using concrete equipment such as counters. Question **1** b) reinforces learning by asking children to divide 48 grapes into groups of 4.

PRACTICAL TIPS To support children with this question, split them into groups and give each group 39 counters of one colour, to represent the pineapple chunks, and 48 counters of another colour, to represent the grapes. Work through each part of the question in turn. Can they group the 39 counters into equal groups of 3 to represent the pineapple sticks? Similarly, how many equal groups of 4 can they make with the 48 counters?

ANSWERS

Question **1** a): 13 full sticks can be made.

Question **1** b): There will be 12 grapes on each stick.

Share

WAYS OF WORKING Whole class teacher led

ASK

- Question **1** a): *How many sticks have been made? Are there any pineapple chunks left over? What calculation is represented here?*
- Question **1** b): *How has Ash organised his work? Is this a good way of doing it? Why? What method is Sparks reminding us of?*

IN FOCUS Question **1** b) offers an opportunity to discuss equal distribution, which is different from the equal grouping used in question **1** a). In equal distribution, the number of items in each group is not known and the items must be equally distributed to the given number of groups to work out the answer. Remind children about the importance of fully understanding the question. Ask: *What information do you know? What do you not know? What do you need to find out? How can you find it?*

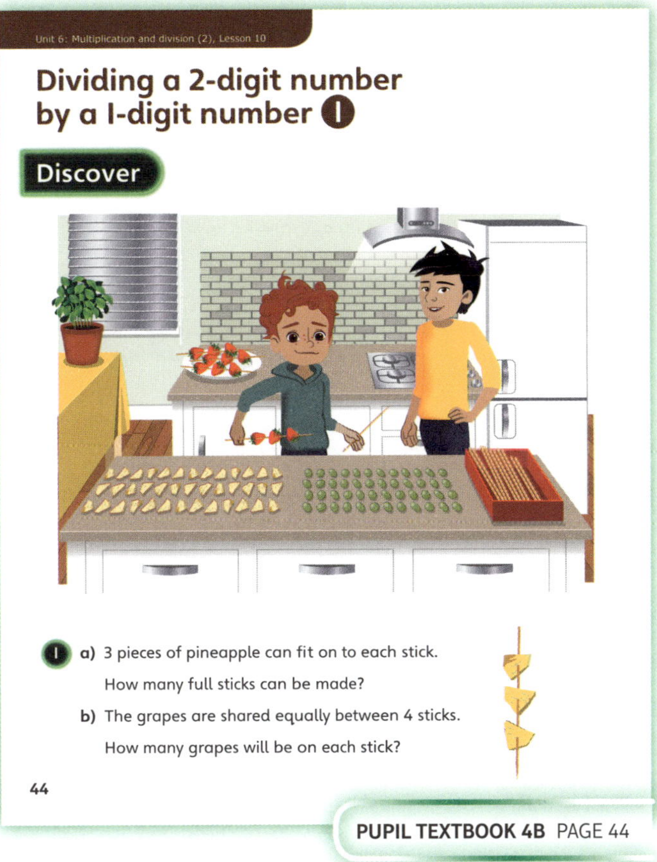

PUPIL TEXTBOOK 4B PAGE 44

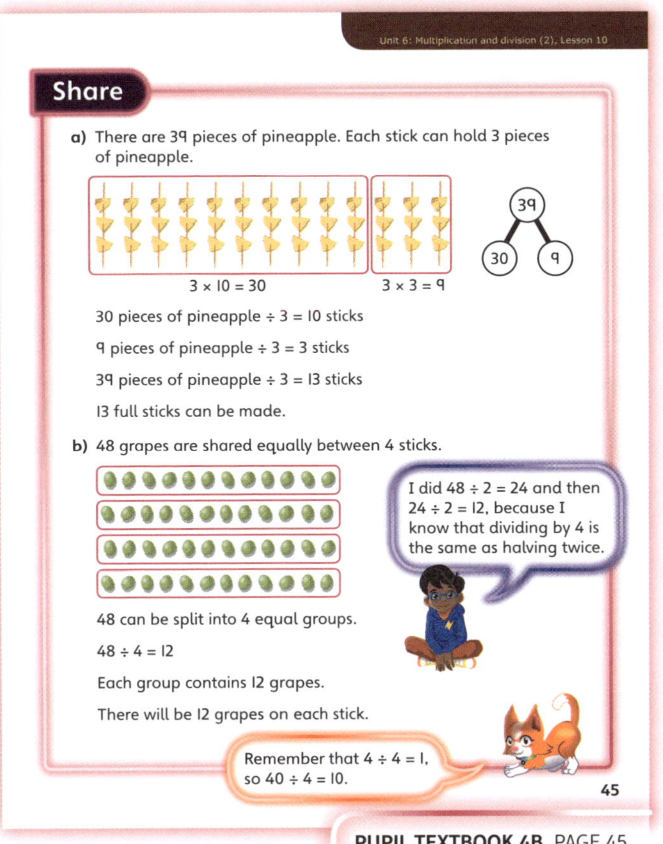

PUPIL TEXTBOOK 4B PAGE 45

Think together

WAYS OF WORKING Whole class teacher led (I do, We do, You do)

ASK

- Question **1**: *How many strawberries are there altogether? How many strawberries fit on to one stick? How can the part-whole model help you to work out the answer?*
- Question **2**: *How many limes are there altogether? How many boxes are used? Are the limes shared equally? What division can you see?*
- Question **3**: *What is Reena's method? What is Astrid's method? Which method do you prefer? Why?*

IN FOCUS In question **1**, children practise what they have learnt in the **Discover** and **Share** sections, working out the number of groups they can make with 69 strawberries if there are 3 strawberries on each stick. Question **2** consolidates understanding, asking children to distribute 55 limes equally between 5 boxes, referring children to a part-whole model to help them to structure their answer. Question **3** investigates the different ways in which a 2-digit number can be divided by a 1-digit number.

STRENGTHEN Ensure that all children have access to concrete equipment, such as arrays, and objects such as counters that they can divide into groups or share equally. Some children may see division and multiplication as separate operations. Model how multiplication can be used to check the accuracy of answers: for example, $11 \times 5 = 55$ and $55 \div 5 = 11$. Ask children to write the mathematical statements to represent each division and encourage them to do this for their answers in this section of the **Textbook**. Ensure children understand the concepts of equal grouping and equal distribution.

DEEPEN Ask children to make their own word problems. They could work in pairs or groups and make questions for their partner or another group. Give them a calculation, such as $68 \div 4$, or $81 \div 3$, and ask them to think of a context in which this question can be used. What would the answer be? What are the keywords in the question? How can the question be made easier/harder?

ASSESSMENT CHECKPOINT Children recognise that they have been grouping objects and sharing objects in equal groups. They are confident writing the correct mathematical statements to represent their divisions.

ANSWERS

Question **1**: 60 strawberries ÷ 3 = 20 full sticks
9 strawberries ÷ 3 = 3 full sticks
23 full sticks can be made from 69 strawberries.

Question **2**: 55 ÷ 5 = 11; there are 11 limes in each box.

Question **3** a): 68 ÷ 2 = 34 (3 groups of 10 × 2 and 1 group of 4 × 2) so Reena is correct.

Question **3** b): 44 ÷ 2 = 22 (2 groups of 10 × 2 and 1 group of 2 × 2)
84 ÷ 2 = 22 (4 groups of 10 × 2 and 1 group of 2 × 2)
104 ÷ 2 = 22 (5 groups of 10 × 2 and 1 group of 2 × 2)

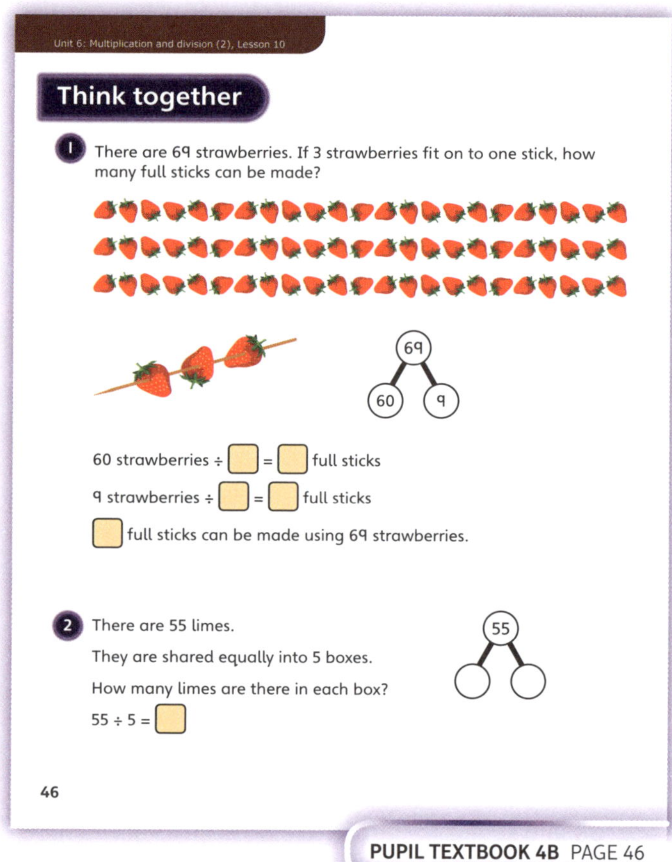

PUPIL TEXTBOOK 4B PAGE 46

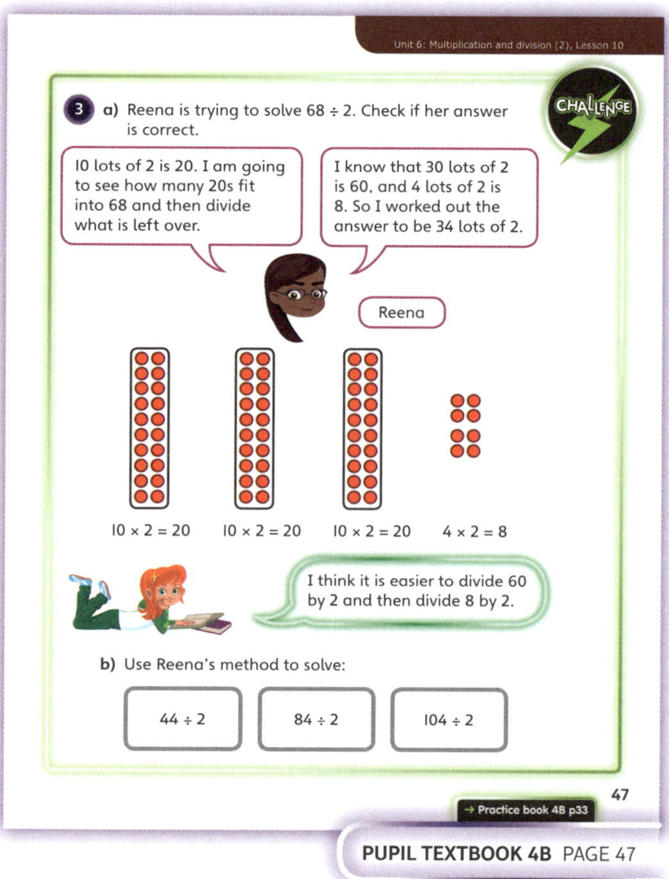

PUPIL TEXTBOOK 4B PAGE 47

Practice

WAYS OF WORKING Independent thinking

IN FOCUS Question **1** consolidates learning in the **Textbook** by practising grouping and sharing equally. Encourage children to discuss the differences between question **1** a) and question **1** b). How might they use their answer to **1** a) to check their answer to **1** b)? In question **2**, children extend their thinking by choosing their own method of division and working independently. Question **3** provides an opportunity to practise and expose any potential misconceptions. Question **4** deepens children's understanding by asking them to analyse a calculation with a mistake, encouraging them to recognise that this is not an appropriate method. Question **5** gives children an opportunity to spot a pattern in their answers. Can they link these calculations with the 4 and 3 times-tables?

STRENGTHEN Ensure children have access to concrete equipment such as blocks, counters and arrays, so they can physically represent the questions.

DEEPEN In question **7**, children are asked to think deeply and compare two division questions. Some children may be able to generalise what happens when the same amount is divided by two different numbers and that the bigger the number you divide by, the smaller the result. Ask children to explain their reasoning by using concrete equipment and real life examples.

ASSESSMENT CHECKPOINT Children understand the concept of division and the different scenarios in which division can be used. Children are able to explain their reasoning.

ANSWERS Answers for the **Practice** part of the lesson appear in the separate **Practice and Reflect answer guide**.

Reflect

WAYS OF WORKING Pair work

IN FOCUS To demonstrate their understanding of dividing a 2-digit number by a 1-digit number, children should feel confident communicating their methods through words or diagrams. They should be able to clearly explain how to solve $26 \div 2$ using an appropriate method. Ask them to compare their explanations with a partner. Have they both used the same method?

ASSESSMENT CHECKPOINT Check whether children are confident in explaining the different methods of division. Are they using the correct mathematical language? Is their diagram clear? Are they systematic in their approach?

ANSWERS Answers for the **Reflect** part of the lesson appear in the separate **Practice and Reflect answer guide**.

After the lesson ▐▐

- Do children know the difference between grouping and sharing equally?
- Can children explain when division is used?
- Can children explain how they can check that their answers are reasonable?

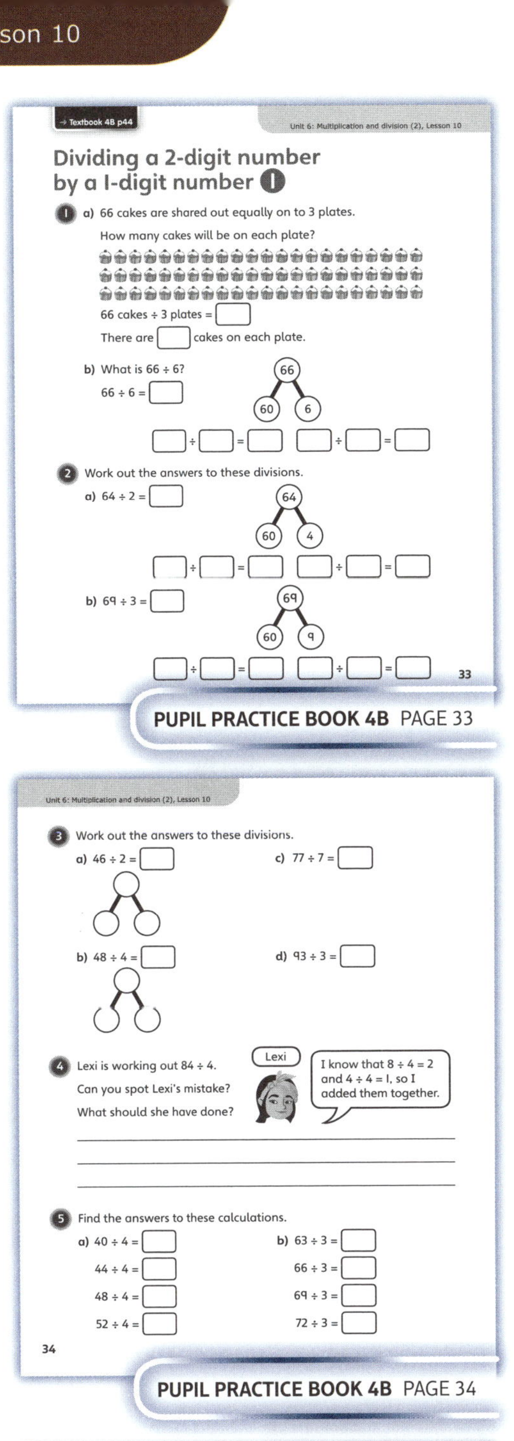

PUPIL PRACTICE BOOK 4B PAGE 33

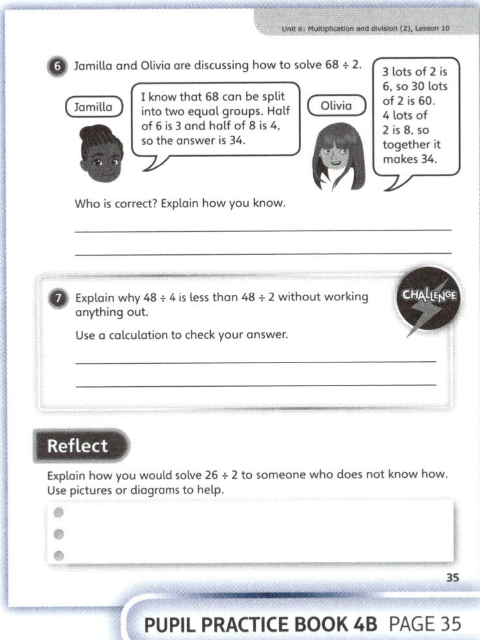

PUPIL PRACTICE BOOK 4B PAGE 34

PUPIL PRACTICE BOOK 4B PAGE 35

Division with remainders

Learning focus

In this lesson, children will recap the concept of remainders in division and be able to solve division problems that leave a remainder.

Small steps

→ Previous step: Dividing a 2-digit number by a 1-digit number (1)

→ **This step: Division with remainders**

→ Next step: Dividing a 2-digit number by a 1-digit number (2)

NATIONAL CURRICULUM LINKS

Year 4 Number – Multiplication and Division
- Multiply two-digit and three-digit numbers by a one-digit number using formal written layout.
- Use place value, known and derived facts to multiply and divide mentally, including: multiplying by 0 and 1; dividing by 1; multiplying together three numbers.

ASSESSING MASTERY

Children explore the rules of divisibility and can explain what a remainder is and how to express the remainder in a problem. Children can predict division calculations that will have remainders when dividing a 2-digit number by a 1-digit number.

COMMON MISCONCEPTIONS

Children may calculate the wrong remainder. For example, when dividing 39 by 4, they may say that the answer is 9 remainder 1, as they think they need 1 more to make 40, which is a multiple of 4. This indicates that children are not confident with the concept of remainders.

At this stage, children need as much concrete practical experience with division as possible. They should group and share a number of objects that have a remainder and see for themselves what the term 'remainder' means. Encourage children to describe what they see. Can they divide 39 blocks/counters/pens into 4 equal groups? Ask:
- *How many items are left unused? What is this called?*

STRENGTHENING UNDERSTANDING

Revisit the concept of remainders in division with children. Ask them to describe what a remainder is. Can children give examples of situations where there may be a remainder? Ask them to use objects such as counters or pencils and to divide them into equal groups. Focus on how many are left over. Encourage children to write out all their working, including diagrams of what they make, as a visual reference to support the abstract calculations.

GOING DEEPER

Ask children to write and solve word problems based on a number sentence showing a division, such as: there are 5 children in the playground and 41 hoops; How many hoops can each child have? $41 \div 5 = ?$ Ask: *What is the remainder? What could the numbers be so that there is no remainder?*

KEY LANGUAGE

In lesson: calculation, remainder, equally, divide

Other language to be used by the teacher: left over, pattern, grouping, whole, part

STRUCTURES AND REPRESENTATIONS

part-whole model

RESOURCES

Mandatory: base 10 equipment, place value counters
Optional: cubes

 In the eTextbook of this lesson, you will find interactive links to a selection of teaching tools.

Before you teach 

- Do children understand simple division by grouping?
- Do they understand simple division by sharing?
- Do they know what a remainder is?

Discover

 Pair work

- Question **1** a): *What calculations are shown on the board? How can you divide 39 by 3? What method are you using? What if you were dividing 40 by 3? Would there be a remainder then?*
- Question **1** b): *How many 10s are there in 85? How many 1s? Can you divide 8 by 4? What about 5? How can you check you are correct?*

  In question **1** a), children divide the 10s and the 1s by 3. Provide children with base 10 equipment and ask them to physically represent the calculation. Recap children's previous knowledge of remainders. In question **1** b), ask children to show the number 85 using base 10 equipment. Can they divide 85 by 4? How do they know if this is possible? Work through the other calculations in this way. Encourage children to draw diagrams of the groups they are making with the base 10 equipment to reinforce the concepts they are learning here.

  Provide children with base 10 equipment, counters and other objects that they can share out into groups to represent the division calculations.

Question **1** a): 39 ÷ 3 = 13; there is no remainder as it divides equally.

Question **1** b): The two calculations with a remainder are 85 ÷ 4 and 57 ÷ 5.

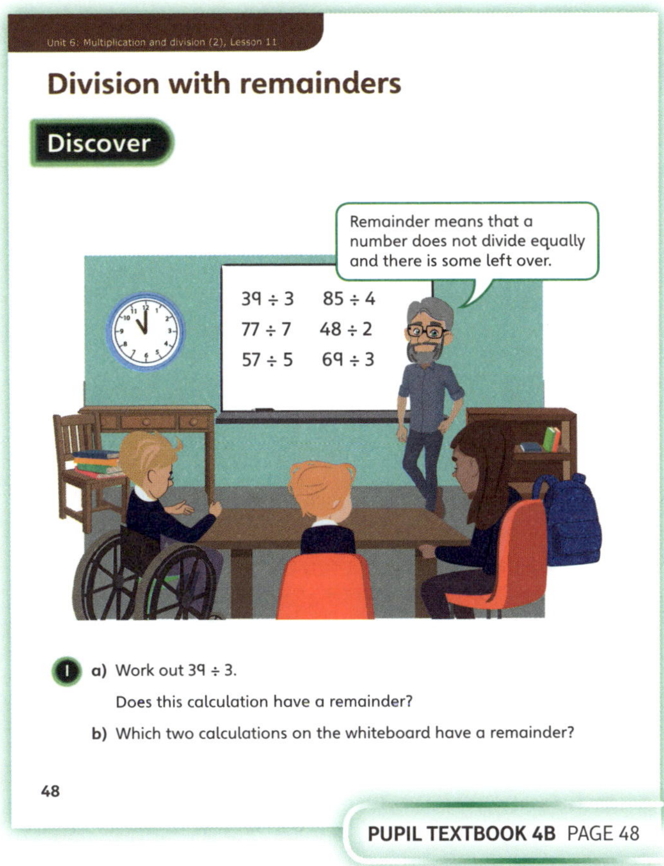

PUPIL TEXTBOOK 4B PAGE 48

Share

  Whole class teacher led

- Question **1** a): *How many groups of 10 have been made? How many groups of 1 have been made? How many are left over? What calculation is represented here?*
- Question **1** b): *How has Flo organised her work? Do you think this is a good way of doing it? Can you use blocks to show your answer? Can you draw a diagram to explain it?*

  Question **1** a) is an opportunity to look at wholes with no remainders and to explore the division of 10s and 1s, ensuring children are secure with this. Question **1** b) focuses on identifying numbers with remainders. Ask children how they can check their answers, prompting them with questions. For example, in the calculation 85 ÷ 4, model for children how 21 × 4 = 84. Discuss how, to make 85, they need to add the remainder of 1, so 85 ÷ 5 = 21 r 1. Ask children to use multiplication to check the other division calculations in this question.

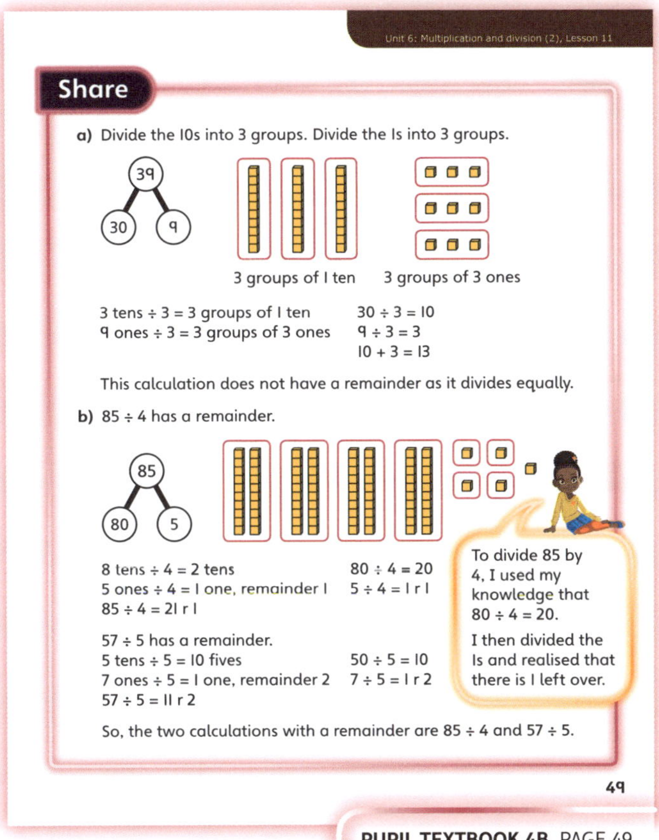

PUPIL TEXTBOOK 4B PAGE 49

Think together

 Whole class teacher led (I do, We do, You do)

ASK

- Question **1**: *What number should go in each of the parts? How did you find the remainder?*
- Question **2**: *What is the calculation here? What is the answer? What number can go in each box? Why? What is the remainder? Can you explain why?*
- Question **3** a): *What are you dividing by this time? How many 10s are in 64? How many 1s? Are they each divisible by 3? How do you know?*

IN FOCUS In question **1**, children use base 10 equipment to support them when partitioning the 10s and 1s. Question **3** encourages children to think about numbers that may have remainders, and also to think about how the number they are dividing by will affect this. Discuss Ash's advice. Ask: *Why might it be helpful to partition 64 in this way? What question could this part-whole model be showing?*

STRENGTHEN Ensure that children have access to base 10 equipment or counters. Ask them to predict the divisions before they complete them. For example, how do they know that the remainder of 45 ÷ 4 is 1? (It is 1 more than a multiple of 4.) Explore the rules of divisibility with the class. Children should know that any even number can be divided by 2; that any even number that is also a multiple of 4 can be divided by 4. They should notice that all the numbers that are multiples of 5 (numbers that are in the 5 times-table) end in 5 or 0 and all the numbers that are multiples of 10 (in the 10 times-table) end in 0.

DEEPEN Deepen understanding by asking children to predict the remainder of some division calculations. For example, ask: *How do you know that 43 ÷ 4 has a remainder of 3?* (It is 3 more than a multiple of 4.) Extend the rules of divisibility by writing multiples of 3 and asking children to notice what they have in common – the digit total is a multiple of 3. For multiples of 6, the digit total (up to 10 × 6) is a multiple of 3 and the numbers are also even. For multiples of 9 (up to 10 × 9), the digit total is always 9.

ASSESSMENT CHECKPOINT Can children predict the remainder that a division calculation could have? Can they explain the rules of divisibility by 2, 3, 4, 5, 6, 9 and 10?

ANSWERS

Question **1**: 37 = 30 + 7
30 ÷ 3 = 10; 7 ÷ 3 = 2 r 1
37 ÷ 3 = 12 r 1

Question **2**: 86 = 80 + 6
80 ÷ 4 = 20; 6 ÷ 4 = 1 r 2
86 ÷ 4 = 21 r 2

Question **3** a): Yes, Amelia is correct: 63 is a multiple of 3; 64 has a remainder of 1 when divided by 3.

Question **3** b): Accept any correct answers, such as: 35 ÷ 3, 45 ÷ 4, 67 ÷ 6, 89 ÷ 5.

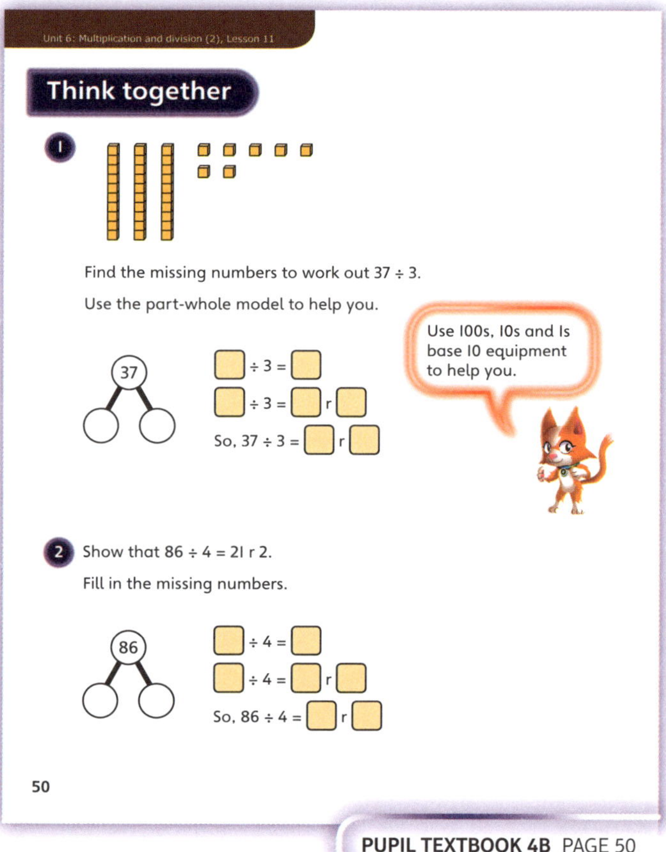

PUPIL TEXTBOOK 4B PAGE 50

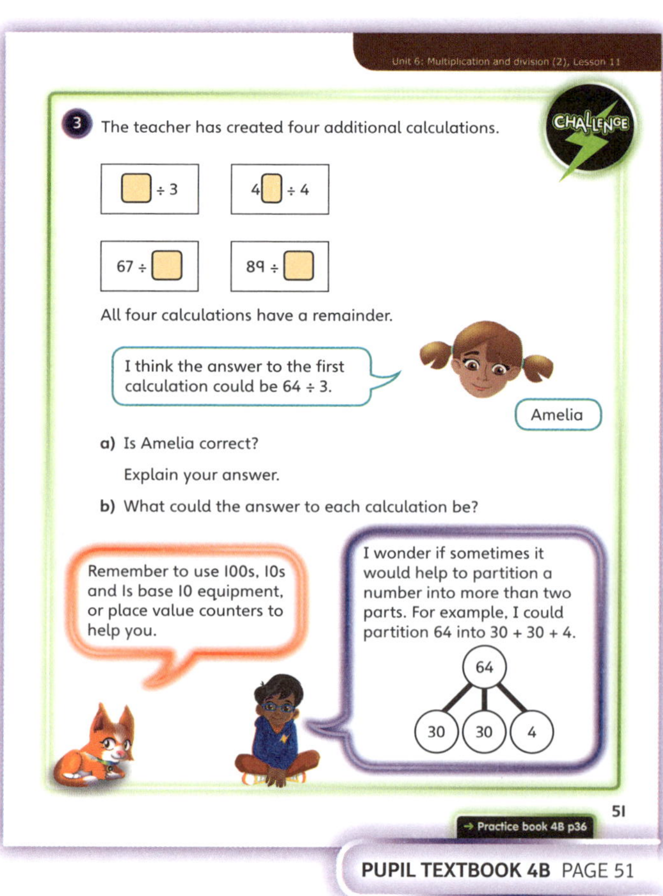

PUPIL TEXTBOOK 4B PAGE 51

Practice

WAYS OF WORKING Independent thinking

IN FOCUS Question ① builds confidence and supports independent learning by asking children to work with the concrete representation of blocks, the pictorial part-whole model and the abstract number statements for 29 ÷ 2 and 97 ÷ 3. Question ② asks children to identify the division shown in the picture, without being given the calculation. In question ③, the structure has been removed and children can choose the method with which they are most confident. Question ④ provides an opportunity for children to explore any misconceptions they may have. Check that their explanations of Luis's mistake are clear, and revisit key learning points from the lesson if further support is necessary.

STRENGTHEN Ensure children have access to the base 10 equipment and place value counters they were using in the main lesson. It may also be useful to recap all the times-tables prior to children working independently. To strengthen and support understanding in question ④, ask: *How could you change the picture so that it shows 63 ÷ 2? Where would you put the 1s? What would the remainder be?*

DEEPEN In question ⑤, ensure that children have access to digit cards 0 to 9 to make a series of 2-digit and 1-digit numbers. They must divide the numbers they make so the remainder is 1. To deepen thinking here, ask children to make a table of their findings. Ask them to think about what all the divisions have in common. They need to establish that the number they are dividing must be 1 more than a multiple of the divisor. Encourage children to discuss their findings with a partner.

ASSESSMENT CHECKPOINT Children can explain how a division calculation works and are able to find the remainder.

ANSWERS Answers for the **Practice** part of the lesson appear in the separate **Practice and Reflect answer guide**.

Reflect

WAYS OF WORKING Independent thinking

IN FOCUS This question requires children to give a clear and confident explanation, using pictures to support their answer. Look for evidence of reasoning, such as: '87 is not an even number', or '87 is not a multiple of 4'.

ASSESSMENT CHECKPOINT Check children know how to identify multiples of 4 and can explain why there are sometimes remainders when a number is divided by another one.

ANSWERS Answers for the **Reflect** part of the lesson appear in the separate **Practice and Reflect answer guide**.

After the lesson ⏸

- Can children recognise multiples of 2, 3, 4, 5, 6, 9 and 10?
- Can they explain why some division calculations have remainders?
- Can they use multiplication to check that their answer is correct?

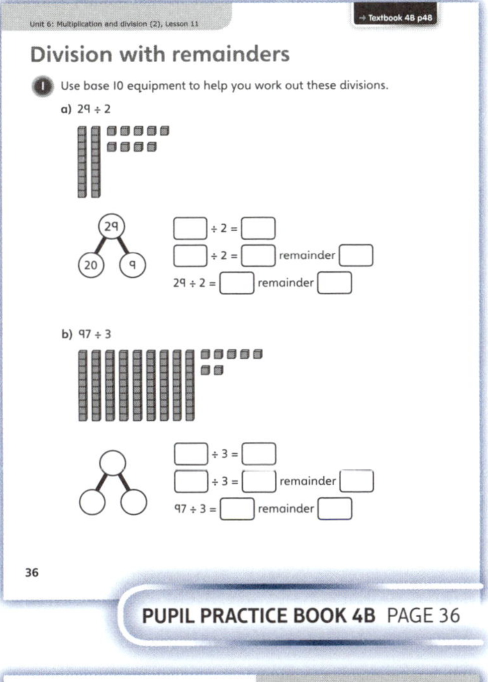

PUPIL PRACTICE BOOK 4B PAGE 36

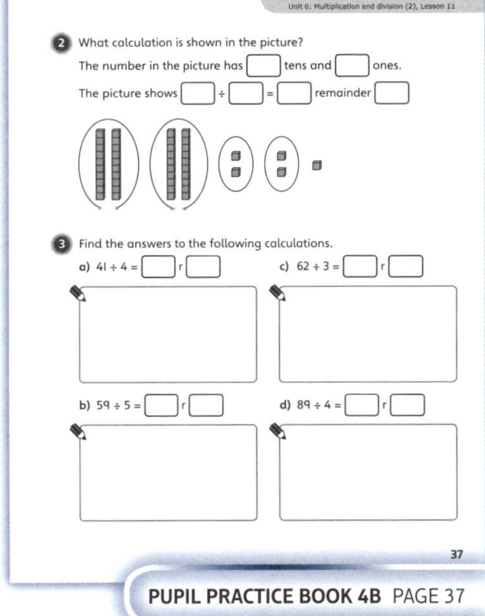

PUPIL PRACTICE BOOK 4B PAGE 37

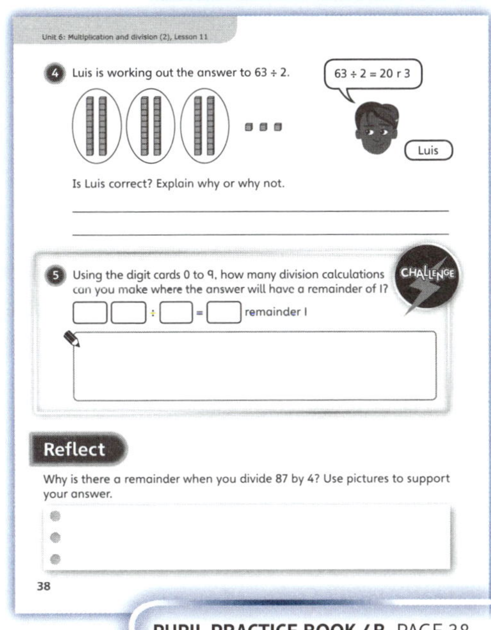

PUPIL PRACTICE BOOK 4B PAGE 38

Dividing a 2-digit number by a 1-digit number ❷

Learning focus

In this lesson, children will divide a 2-digit number by a 1-digit number using flexible partitioning and by focusing on mental methods.

Small steps

→ Previous step: Division with remainders

→ **This step: Dividing a 2-digit number by a 1-digit number (2)**

→ Next step: Dividing a 2-digit number by a 1-digit number (3)

NATIONAL CURRICULUM LINKS

Year 4 Number – Multiplication and Division

Use place value, known and derived facts to multiply and divide mentally, including: multiplying by 0 and 1; dividing by 1; multiplying together three numbers.

ASSESSING MASTERY

Children explore their knowledge of times-tables, known and derived facts, partitioning and place value and can explain and demonstrate calculation strategies for dividing.

COMMON MISCONCEPTIONS

Children may think that partitioning by 10s and 1s is the only way to divide a 2-digit number by a 1-digit number. For example, they may think that 45 cannot be divided by 3 as 45 = 40 + 5 and 40 and 5 are not multiples of 3. Instead, children need to partition 45 as $30 \div 3 = 10$ and $15 \div 3 = 5$; the answer is $10 + 5 = 15$. Encourage children to look at different ways of partitioning a number, choosing the partition that works best with the number they are asked to divide by. Ask:
• *How can the number be partitioned? Is there another way?*

STRENGTHENING UNDERSTANDING

Revisit the concept of partitioning with children. Encourage them to use base 10 equipment or counters and challenge them to find different ways of partitioning a given number. Reinforce the importance of knowing the times-tables and recap any that children find tricky.

GOING DEEPER

Ask children to partition numbers into two or three numbers so that each of the numbers is a multiple of the given number. Ask children to work in pairs or small groups. One child chooses a 'mystery number' and the rest of the group has to work out what the number is, based on information the first child gives about how it can be partitioned.

KEY LANGUAGE

In lesson: how many, furthest, altogether, divided equally, same, each, partition, division, greater, part-whole

Other language to be used by the teacher: number facts, place value, multiple, times-tables

STRUCTURES AND REPRESENTATIONS

part-whole model, number line

RESOURCES

Mandatory: counters, arrays

Optional: cubes, bean bags

 In the eTextbook of this lesson, you will find interactive links to a selection of teaching tools.

Before you teach 

• Do children understand what multiples are?
• Do they know how to partition?
• Are they fluent in their times-tables?

Discover

 Pair work

ASK

- Question **1** a): *How many lanes are there on the track? How many bean bags are there in total?*
- Question **1** b): *How far is one bean bag from the other? How many bean bags are in each lane? How can you check that you are correct?*

IN FOCUS Children need to understand the context of this **Discover** question before deciding how to solve the problem. Question **1** a) requires children to recognise that they need to use division when grouping. When dividing 56 by 4, children may halve and halve again; use concrete equipment, such as 56 blocks split into 5 groups of 10 and 6 ones, to encourage children to partition the number into 40 and 16 before dividing by 4. Question **1** b) provides a contrasting example, using multiplying to solve a problem. Children may expect to divide again and guess which numbers they should divide. At this stage, it is very important that children are encouraged to understand the question fully before deciding how to solve it.

PRACTICAL TIPS Support children by providing a number line and counters to represent each bean bag. Encourage children to represent each part of the question using concrete equipment. If you are able, take 56 bean bags outside to re-create the picture.

ANSWERS

Question **1** a): 56 ÷ 4 = 14; there are 14 bean bags in each lane.

Question **1** b): 14 × 10 = 140 metres; the furthest bag is 140 metres away from the start line.

Share

WAYS OF WORKING Whole class teacher led

ASK

- Question **1** a): *Why is 56 partitioned into 40 and 16? How many 4s make 40? How many 4s make 16? What calculation is represented here?*
- Question **1** b): *Can you draw a number line to show the bean bags? How can you check that you are correct? What multiplication would you need to do? Could Astrid be correct? Why not?*

IN FOCUS Question **1** a) reinforces earlier learning by partitioning a number so that it can be divided efficiently. Draw out from children that both 40 and 16 are multiples of 4. Question **1** b) is an opportunity to use knowledge gained so far in multiplying and dividing to problem solve. Look out for any misconceptions, using Astrid's comment as a prompt.

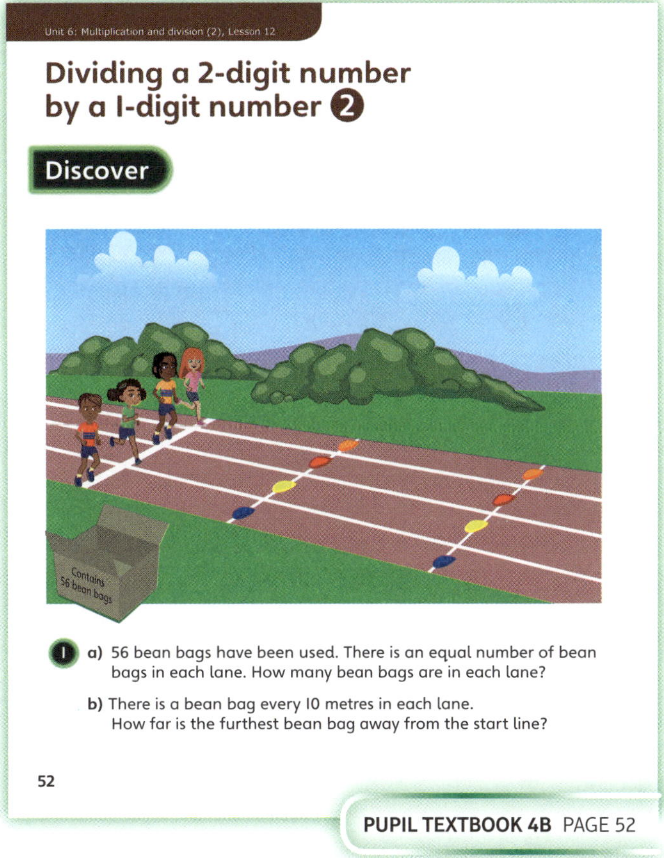

PUPIL TEXTBOOK 4B PAGE 52

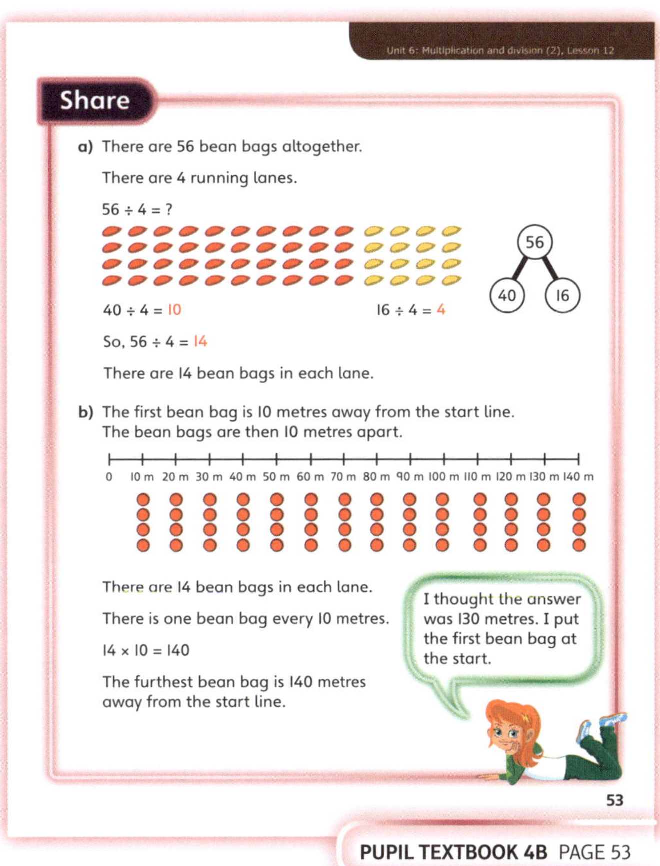

PUPIL TEXTBOOK 4B PAGE 53

Think together

WAYS OF WORKING Whole class teacher led (I do, We do, You do)

ASK

- Question ❶: *How many lanes are there on the racetrack? How many cones are there in total? How would you partition 78? What is special about both numbers that you chose?*
- Question ❷: *How can you work this out? What number should go in each box? Why do you think this is the case? What is the answer? What is the remainder? Can you explain why?*
- Question ❸: *What are you dividing by this time? How many 10s are there in 72? How many 1s? Can you divide them by 3? How do you know? How many cakes are there? How can you find the number of boxes needed to carry the cakes? How many cakes are there in one box?*

IN FOCUS In question ❶, children practise partitioning 78 in a way that helps them with the division. Investigate Ash's idea.

Question ❷ provides less scaffolding, so children need to partition the number and work out what would help them most with the division, encouraging them to develop procedural fluency.

Question ❸ is a multi-step question which requires children to think deeply and logically about how they can work out the answer. Firstly, children need to work out the number of cakes based on the total number of cherries and how many cherries go on each cake. The next step is to work out how many groups of 4 the total number of cakes can be broken down into.

STRENGTHEN In question ❸, prompt children to follow Dexter's advice. Read the question aloud and complete one division at a time. Pay attention to how the numbers are partitioned to make the calculations easier: children should notice that, when they have divided 72 by 3, the answer (24) is divisible by 4. After the second division, they find that they need 6 boxes. Use multiplication and knowledge from the previous lessons in the unit to check the answer. Write on the board: $3 \times 4 \times \_ = 72$. Ask: *What could the number be?*

DEEPEN Deepen understanding by asking children to explore what Flo is saying in question ❸. Instead of dividing by 3 and 4, what other number could 72 be divided by? Work with children to find all the possible ways of partitioning 72 and model how to use multiplication to check the answers: $\_ \times 6 + \_ \times 6 = 72$. Encourage children to notice that the sum of the pair of numbers is always equal to 12, which is the answer to $72 \div 6$.

ASSESSMENT CHECKPOINT Children can divide a 2-digit number by a 1-digit number mentally, by exploring partitioning and known and derived facts.

ANSWERS

Question ❶: There are 13 cones in each lane.

Question ❷: $76 \div 4 = 19$. There are 19 children in each team.

Question ❸: $72 \div 3 = 24$ cakes; $24 \div 4 = 6$ boxes of cakes
Or, $72 \div 12 = 6$ boxes of cakes.

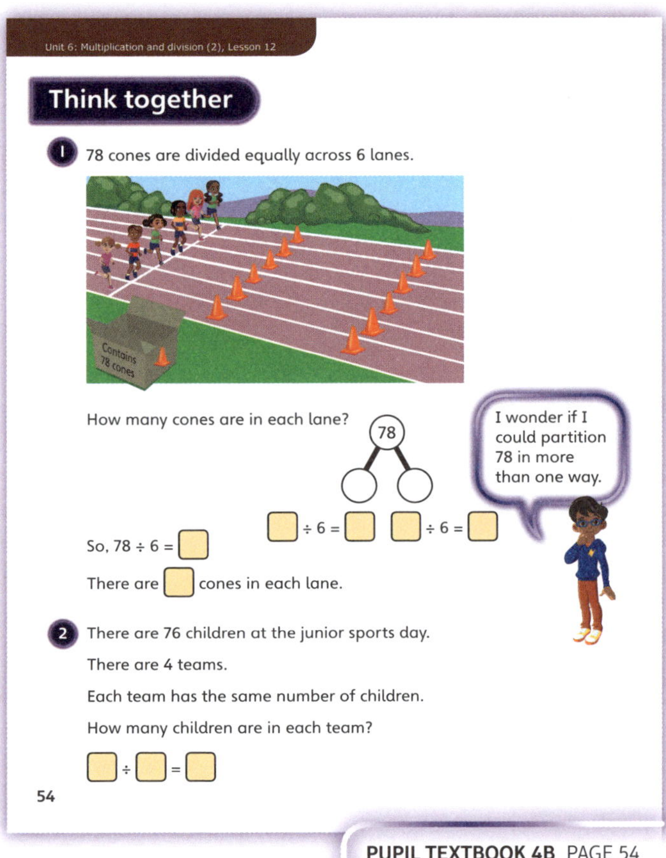

PUPIL TEXTBOOK 4B PAGE 54

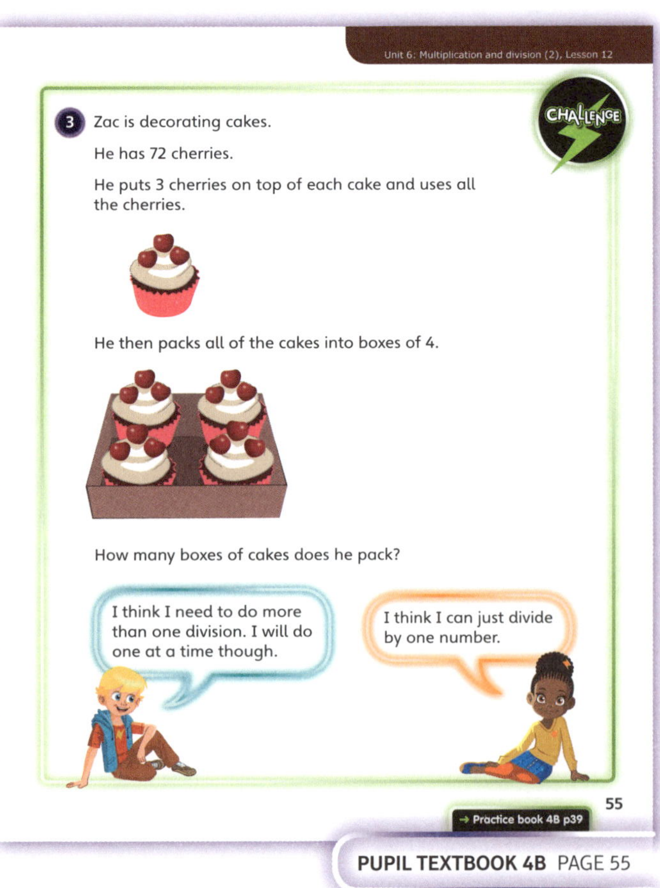

PUPIL TEXTBOOK 4B PAGE 55

Practice

WAYS OF WORKING Independent thinking

IN FOCUS Questions ➊, ➋ and ➌ require children to partition numbers using the part-whole model and to complete the calculations given, with decreasing scaffolding. In question ➍, the part-whole model is removed and children answer the question using their own method. Question ➎ requires children to use division to solve a word problem using their own method. In question ➏, children divide two 2-digit numbers by a 1-digit number and compare the calculations to find out which answer is greater.

STRENGTHEN Ensure children understand that there are different ways to partition a number and that there is not a strict right or wrong answer: how a number should be partitioned usually depends on what question is being asked. Discuss with children how they can tell whether a number is divisible by 2, 3, 5, 6, 9 or 10. Provide simple calculations and ask children to use counters or cubes to demonstrate how the numbers can be partitioned, before dividing each part.

DEEPEN Question ➐ challenges children to think deeply. Ask children to explain the way in which they found answers to the divisions. Did they think of numbers that can divide into 48? Or did they think of numbers that can divide into 30 and 18? Children need to be able to explain their reasoning clearly. To extend children's thinking, ask them to think of a division that cannot happen using the numbers in the part-whole model and ask them to explain why this is the case.

ASSESSMENT CHECKPOINT Children can explain how to partition numbers so that they can divide them using a mental strategy.

ANSWERS Answers for the **Practice** part of the lesson appear in the separate **Practice and Reflect answer guide**.

Reflect

WAYS OF WORKING Independent thinking

IN FOCUS Children must explain why partitioning 57 into 40 + 17 does not help them to find the answer to 57 ÷ 3 (because 40 and 17 are not multiples of 3). Children should then partition the number correctly so that it is easy to divide by 3. Encourage them to draw a part-whole model or draw base 10 equipment to explain their reasoning.

ASSESSMENT CHECKPOINT Can children partition a 2-digit number appropriately to help them work out divisions efficiently?

ANSWERS Answers for the **Reflect** part of the lesson appear in the separate **Practice and Reflect answer guide**.

After the lesson ⏸

- Can children partition 2-digit numbers effectively to help them to solve divisions of 2-digit numbers by a 1-digit number?
- Can children use multiplication calculations accurately to check that their division is correct?

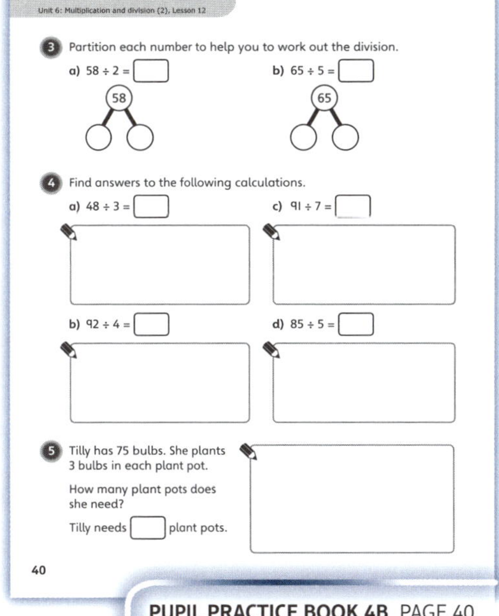

PUPIL PRACTICE BOOK 4B PAGE 39

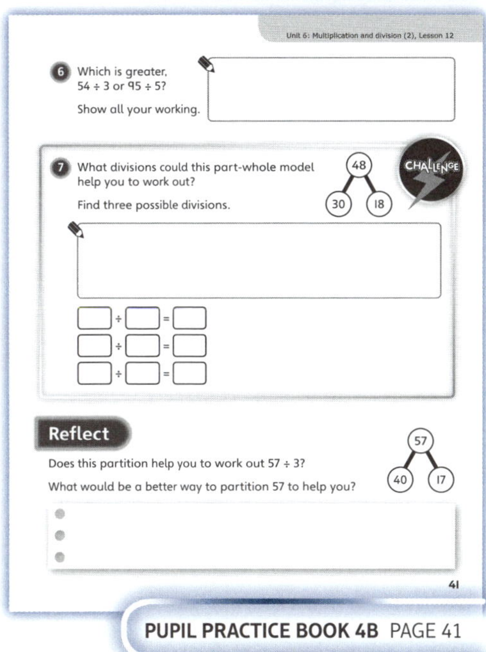

PUPIL PRACTICE BOOK 4B PAGE 40

PUPIL PRACTICE BOOK 4B PAGE 41

Dividing a 2-digit number by a 1-digit number ❸

Learning focus

In this lesson, children will focus on flexible partitioning and mental approaches, including division problems that leave a remainder.

Small steps

→ Previous step: Dividing a 2-digit number by a 1-digit number (2)

→ **This step: Dividing a 2-digit number by a 1-digit number (3)**

→ Next step: Dividing a 3-digit number by a 1-digit number

NATIONAL CURRICULUM LINKS

Year 4 Number – Multiplication and Division
- Use place value, known and derived facts to multiply and divide mentally, including: multiplying by 0 and 1; dividing by 1; multiplying together three numbers.
- Multiply two-digit and three-digit numbers by a one-digit number using formal written layout.

ASSESSING MASTERY

Children understand that some divisions are not exact and can leave a remainder. Children can partition and use multiplying facts to predict what the remainder could be.

COMMON MISCONCEPTIONS

When solving problems, if children divide and the answer has a remainder, they may assume that the answer is wrong. This may lead to manipulation of the question so that they get a 'nice answer' with no remainders. They cannot explain why they are dividing the numbers in the question unless directly told so. Ask:
- *What is the question asking? Can you describe it? What does this number represent? What does the answer show? If we put the answer back into the question, will it make sense?*

STRENGTHENING UNDERSTANDING

Use counters to represent the objects in the groups and use a bar model to represent the situation. To start with, use counters to represent the parts. To prompt thinking, ask: *What does each number show? What does each part represent?*

GOING DEEPER

To deepen understanding in this lesson, ask children to make their own division problems using concrete equipment such as counters or base 10 equipment to help them. Children should have experience of making one-step and multistep problems. They should try to solve each problem in more than one way. Encourage them to look at their word problems in a number of different ways. Ask: *Is there a different way to partition this number? Is there an easier way?*

KEY LANGUAGE

In lesson: remainder, triangles, hexagons, partition, calculation, shares, left over, division

Other language to be used by the teacher: grouping, whole, part, divide

STRUCTURES AND REPRESENTATIONS

part-whole model, bar model

RESOURCES

Mandatory: 2D shapes such as hexagons and triangles, calendar

Optional: counters, base 10 equipment

 In the eTextbook of this lesson, you will find interactive links to a selection of teaching tools.

Before you teach 

- Do children understand simple division by grouping?
- Do children know the names of basic shapes and the number of sides they have?

Discover

WAYS OF WORKING Pair work

ASK

- Question ❶ a): *What type of problem is this? What number should you divide by?*
- Question ❶ b): *What calculation did Lee make? Why did he divide these numbers? What are 'school days'? How many school days are there in a week? What should the calculation have been?*

IN FOCUS In question ❶ a), children need to understand the question properly before they decide what operation they need to use and the numbers involved in their calculations. Lee divides the numbers in the question to find 84 ÷ 7 = 12. In question ❶ b), encourage children to read the question carefully and prompt their thinking if necessary to draw out any misconceptions. Ask them if there is any information from the text that Lee has missed. Ensure that children understand what the term 'school days' means and how many there are in one week.

PRACTICAL TIPS Provide children with a calendar so they can visualise the question better. Can they put a counter on each of the school days in one week? How about the school days in one month?

ANSWERS

Question ❶ a): 84 ÷ 7 = 12. Lee has divided 84 by 7.

Question ❶ b): 84 ÷ 5 = 16 remainder 4. The answer should be that they have 16 full school weeks and 4 days left.

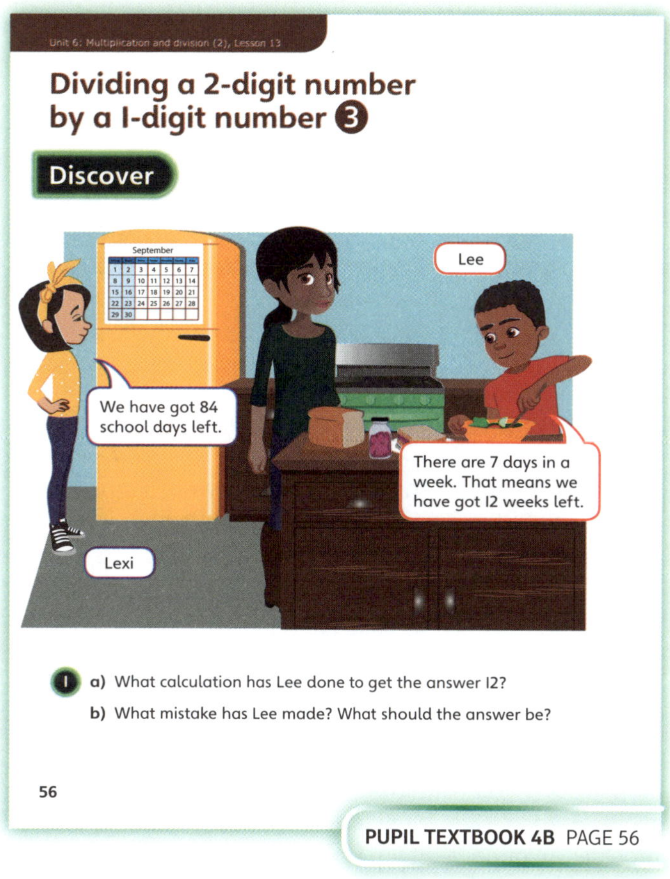

PUPIL TEXTBOOK 4B PAGE 56

Share

WAYS OF WORKING Whole class teacher led

ASK

- Question ❶ a): *What numbers did Lee divide? What do the numbers in the question represent? How did Lee partition 84?*
- Question ❶ b): *What numbers should Lee have divided? Why? Why is Dexter worried about the calculations? Does it matter if the division has a remainder?*

IN FOCUS Question ❶ a) requires children to look at the question carefully in order to explain the calculation that Lee has made. For this, they need to be confident in their division of 2-digit numbers. In question ❶ b), children need to explain the mistake that Lee has made. This is a good opportunity to clarify the potential misconception mentioned in the lesson introduction, that when dividing, if the answer has a remainder, it must be wrong. Discuss this misconception with children by asking them: *If a division has a reminder, does it mean that it is wrong? What do you have to do?* Reinforce the importance of reading the question carefully and answering the question that is asked.

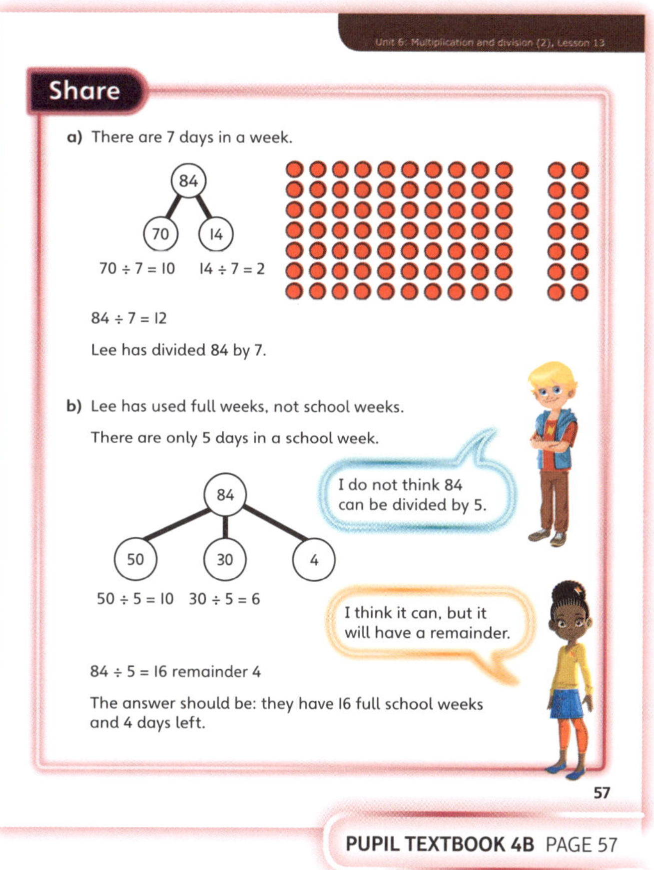

PUPIL TEXTBOOK 4B PAGE 57

Think together

WAYS OF WORKING Whole class teacher led (I do, We do, You do)

ASK

- Question ❶: *How many triangles are there altogether? Can Ambika use all the triangles? How do you know? What is the division in this question? Why?*
- Question ❷: *What does the part-whole model show? How is 79 partitioned? What does 'r' mean? How can you check your answer? What is the most the remainder can be?*
- Question ❸: *Why did Olivia partition 95 as 80, 12 and 3? What do the numbers 80 and 12 have in common? Can you predict the most you can have left over? Which method do you prefer? Why? Can you partition 95 differently?*

IN FOCUS Question ❶ is a good opportunity to recap on knowledge of shapes, by asking questions such as: *How many sides/vertices does a hexagon have? What does 'tessellation' mean? What is special about an equilateral triangle?* Children must organise the triangles into groups of 6. This leads into question ❷, where children are partitioning the number 79 and must write the number statement to complete the division. Discuss how the answer can be written when there is a remainder. Question ❸ provides an opportunity to look at different ways of working out the answer to the same problem. Encourage deeper thinking by referring children to Astrid's statement. Ask: *What is the difference between the methods used by Olivia and Lexi? Can you investigate Ash's idea of partitioning the number 95 in other ways?* Discuss with children that the numbers must be multiples of 4 so it is easy to divide them.

STRENGTHEN To strengthen understanding in question ❶, give children 18 paper or card triangles that they can move around. Can they predict how many regular hexagons they can make with the triangles? Creating a table to write all the information can be very helpful.

DEEPEN Deepen understanding of question ❶ by asking children to write some of the answers as divisions. Can children predict what the remainder will be if they have 73 triangles? How do they know that the remainder is 1? (It is 1 more than a multiple of 6.)

ASSESSMENT CHECKPOINT Children can use the part-whole model to partition numbers in different ways. They can divide 2-digit numbers by 1-digit numbers and predict what the remainder will be.

ANSWERS

Question ❶ a): Ambika can make 11 hexagons.

Question ❶ b): 2 triangles will be left over.

Question ❷: 79 ÷ 3 = 26 r 1

Question ❸ a): Olivia's method: Lexi's method:
    80 ÷ 4 = 20   80 ÷ 4 = 20
    12 ÷ 4 = 3   15 ÷ 4 = 3 r 3
    Remainder 3
    95 ÷ 4 = 23 r 3   95 ÷ 4 = 23 r 3

Question ❸ b): Accept any logical preference. Can children explain their reasoning to a partner?

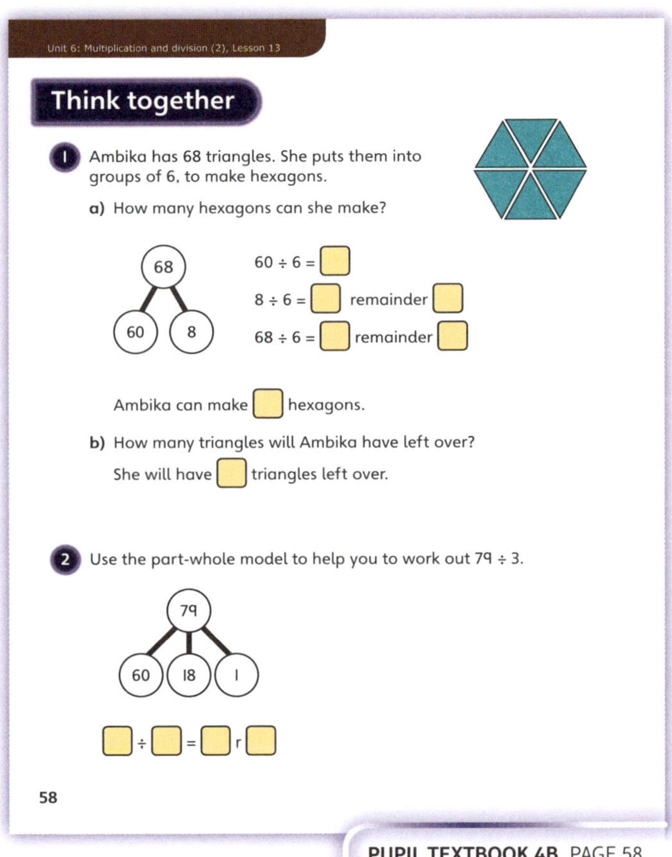

PUPIL TEXTBOOK 4B PAGE 58

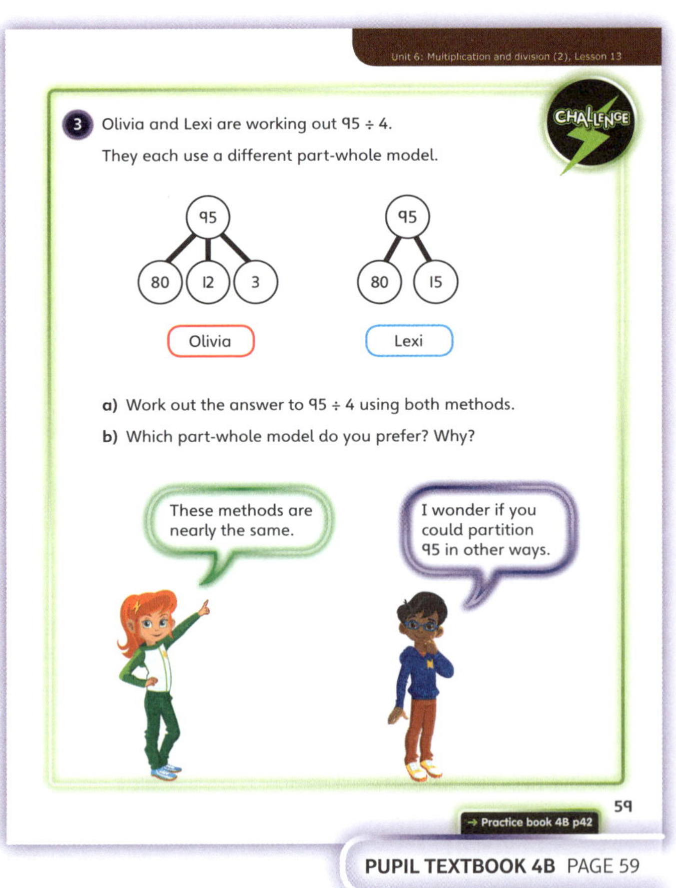

PUPIL TEXTBOOK 4B PAGE 59

Practice

WAYS OF WORKING Independent thinking

IN FOCUS Questions **1**, **2** and **3** build on the **Textbook** by recapping partitioning numbers in multiples of the divisors, with decreasing scaffolding, to build children's confidence in the method. Children complete number statements for each calculation and write the remainder. Encourage children to use correct mathematical language: for example, say 'remainder' rather than 'left over'. In question **4**, the scaffolding is reduced and children need to choose their own method of division. They should consider the number they are dividing by and how to partition their number. In each part, the total stays the same, so children can start to see the pattern emerging. To consolidate learning, ask: *Can you predict what the remainder would be for 67 ÷ 7?*

STRENGTHEN For children to confidently spot multiples they need to be confident with their times-tables. It may be helpful to practise the 2, 3, 4, 5 and 6 times-tables prior to starting this independent practice session. Ask children to use base 10 equipment to make the first 12 multiples of 5. Then ask children to use the cubes they already have to make numbers that have a remainder of 1. Allow children time to explore and to realise that they need to add 1 to all the multiples of 5. Repeat this exercise for remainders of 2, 3 and 4.

DEEPEN In question **6**, children must think deeply, working from the divisor and the remainder to work out which number Danny started with. Children may need to think of the missing number in terms of multiples. The number has to be 1 more than a multiple of 2 (so an odd number), a multiple of 3, and 1 more than a multiple of 5. Ask children to explain their reasoning. Ensure that they have access to concrete equipment such as counters or base 10 equipment to support them in this task.

ASSESSMENT CHECKPOINT Can children partition numbers in different ways to divide them? Do children notice the relationship between the divisor, multiples and remainders?

ANSWERS Answers for the **Practice** part of the lesson appear in the separate **Practice and Reflect answer guide**.

Reflect

WAYS OF WORKING Independent thinking

IN FOCUS Children should now be able to differentiate between calculations with remainders and calculations without remainders. Encourage them to reason and to explain the answer, using concrete equipment and part-whole models to support them. Children should be able to reason that the greatest remainder they can have is 1 less than the divisor.

ASSESSMENT CHECKPOINT Can children predict whether a calculation will have a remainder or not?

ANSWERS Answers for the **Reflect** part of the lesson appear in the separate **Practice and Reflect answer guide**.

After the lesson 

- Do children know that some situations will leave a remainder?
- Can they explain why the same number can be partitioned in different ways but with the same result?
- Can they explain why the greatest possible remainder when dividing by 4 is 3, and so on?

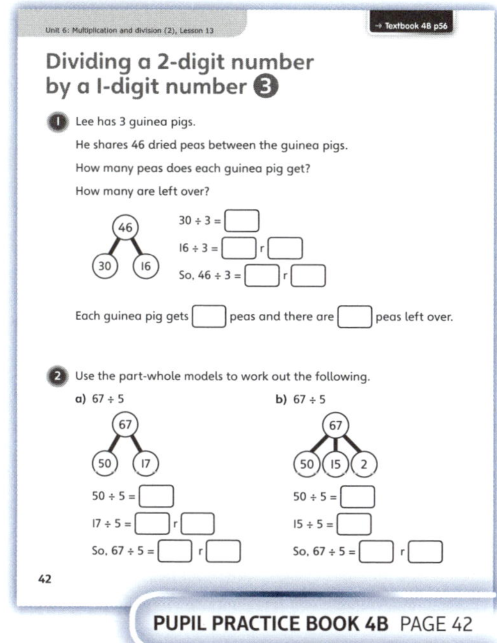

PUPIL PRACTICE BOOK 4B PAGE 42

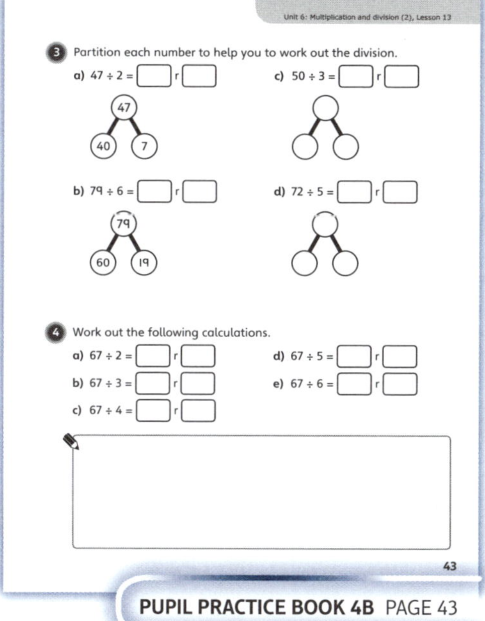

PUPIL PRACTICE BOOK 4B PAGE 43

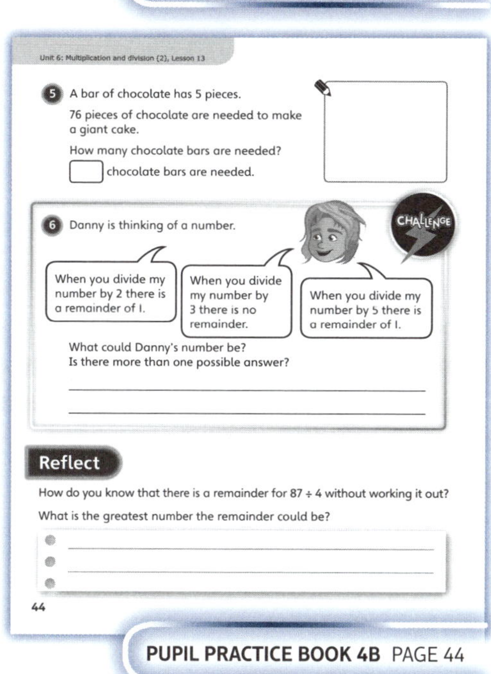

PUPIL PRACTICE BOOK 4B PAGE 44

Dividing a 3-digit number by a 1-digit number

Learning focus

In this lesson, children will use partitioning to divide a 3-digit number by a 1-digit number.

Small steps

→ Previous step: Dividing a 2-digit number by a 1-digit number (3)
→ **This step: Dividing a 3-digit number by a 1-digit number**
→ Next step: Problem solving – division

NATIONAL CURRICULUM LINKS

Year 4 Number – Multiplication and Division

Use place value, known and derived facts to multiply and divide mentally, including: multiplying by 0 and 1; dividing by 1; multiplying together three numbers.

ASSESSING MASTERY

Children can explain and demonstrate the mental calculation for dividing 3-digit numbers, using their understanding of partitioning from the previous lesson.

COMMON MISCONCEPTIONS

Children may think that when dividing 'big numbers' they will need to use a written method, such as $128 \div 4 = ?$ They may forget that division has distributive properties too. To address these misconceptions, ask:
• *Can you partition the number? Can you divide each part by 4?*

Some children may know that $12 \div 4 = 3$; however they do not think of 120 as a multiple of 10 and are unable to divide 120 by 4. Some children treat division and multiplication as discrete and separate operations and forget that they are inverse operations. Remind children that multiplication and division both have distributive properties and that they can use multiplication to check whether their calculations are reasonable.

STRENGTHENING UNDERSTANDING

Ensure that children have access to concrete equipment such as counters or cubes and show them how to carry out both grouping and sharing procedures for division. Children should realise that the answer remains the same regardless of the way the number is partitioned. If necessary, use smaller numbers to begin with to help to secure understanding.

GOING DEEPER

To encourage deeper thinking, ask children to work back and solve missing number problems such as $\_ \div 4 = 52$ and $155 \div 5 = \_$. Children may then tackle problems such as $\_ \div \_ = 23$. Expand their thinking further by asking if they can find several answers to this problem. Can they link multiplying and dividing?

KEY LANGUAGE

In lesson: how many, in total, partitioning, divide, division

Other language to be used by the teacher: dividing, multiplying, equal, sharing, grouping, array, whole, inverse operations

STRUCTURES AND REPRESENTATIONS

part-whole model

RESOURCES

Mandatory: place value counters, cubes, print-outs of an 11×12 square grid

 In the eTextbook of this lesson, you will find interactive links to a selection of teaching tools.

Before you teach 

• Do children know how to use a part-whole model?
• Do they know how to partition a number?
• Do they know all their multiplication facts?

Discover

WAYS OF WORKING Pair work

ASK

- Question ❶ a): *How many squares are there in total on the plan of the field? How many sheep can graze in each square? How can you partition the field? Can you think of a different way?*
- Question ❶ b): *How many cows can graze in each square? What number are you dividing and by what other number? How can you partition the number to make the calculations easier?*

IN FOCUS Children practise dividing 3-digit numbers. They need to multiply the number of rows and columns to find the total number of squares in the field, and then have to think how to partition 132 squares so that they can divide each group by 3. Encourage children to use the part-whole model to partition 132 into 60 + 60 + 12. They could use place value counters to support them further. In question ❶ b), children need to divide 132 by 4. Some children may think of dividing by 4 as halving and then halving again. Other children may partition 132 into 80, 40 and 12.

PRACTICAL TIPS Ensure that children have print-outs of an 11 × 12 square grid. Can they split it up in three different ways so that each part is a multiple of 3?

ANSWERS

Question ❶ a): 11 × 12 = 132 and 132 ÷ 3 = 44
44 sheep can graze in the field.

Question ❶ b): 132 ÷ 4 = 33
33 cows can graze in the field.

Share

WAYS OF WORKING Whole class teacher led

ASK

- Question ❶ a): *What is Flo saying? Is this true? In what ways can you partition 132? Could you have predicted that there was not going to be a remainder? How?*
- Question ❶ b): *Can you explain why this is a division? Can you use another way to partition 132? Would the result differ?*

IN FOCUS As a class, discuss how you would approach dividing 132 by 3. Model this for children with cubes, counters or other objects. Ask them how they partitioned the number. Did they all partition it in the same way? Did they all get the same answer? In question ❶ b), children need to partition into numbers that can easily be divided by 4. The numbers chosen to partition 132 are all multiples of 4. Model for children how the answer will be the same even if they partition in a different way, for example, if they partition 132 as 100, 24 and 8.

PUPIL TEXTBOOK 4B PAGE 60

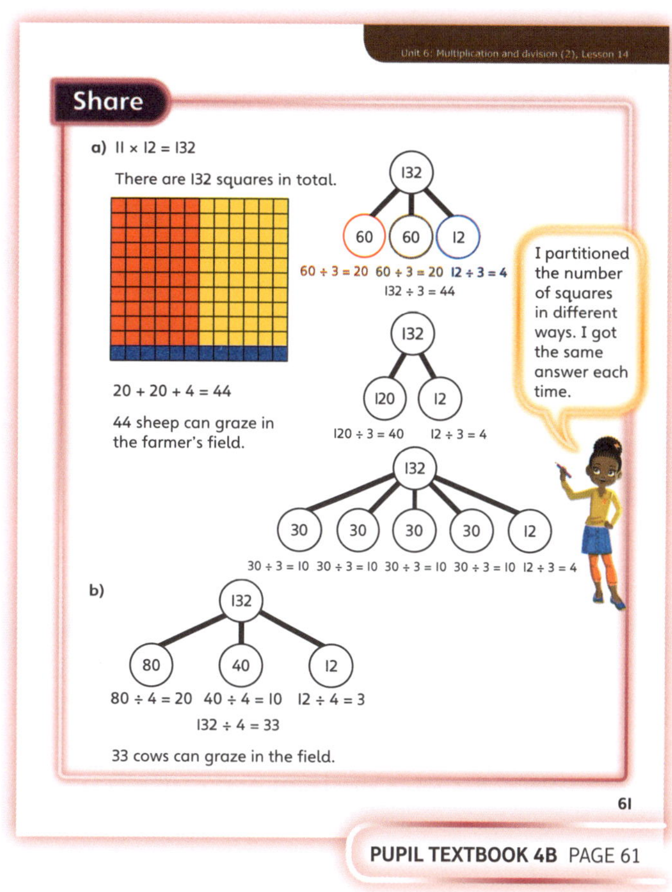

PUPIL TEXTBOOK 4B PAGE 61

Think together

WAYS OF WORKING Whole class teacher led (I do, We do, You do)

ASK

- Question **1**: *Why has 146 been partitioned as 100, 40 and 6? Can you think of a different way to do it? Would the answer to the division change?*
- Question **2**: *How does the part-whole model help you to divide by 5? How else might you partition 185 to help you to share it into 5 equal groups?*
- Question **3**: *Which of the part-whole models is most helpful when dividing 168 by 6? Can you partition in a different way? What do all the numbers have in common? Will the answer change depending on how you partition the number?*

IN FOCUS Question **1** is an opportunity for children to practise dividing a 3-digit number by a 1-digit number, where the structure they need to follow is provided. Question **2** moves on children's thinking, showing them that sometimes it makes sense to partition a 3-digit number into two parts instead of three. Children must choose the calculations they will use and write the correct number statement. In question **3**, children choose their favourite way to partition the numbers from the three models provided. This should lead them to generalise that all the numbers are multiples of the divisor and the answer will not change.

STRENGTHEN To support understanding of dividing 3-digit numbers, children can use concrete equipment to partition the numbers and complete the divisions. Ask children to carry out the partitioning first.

DEEPEN To expand thinking in this section, build on question **3** by asking: *Can you give me an example of partitioning that will not be very helpful for this question? Why is it not good to partition in this way?* Prompting children to think about how a number cannot be usefully partitioned encourages deeper thought about multiples of 3-digit numbers.

ASSESSMENT CHECKPOINT Children can partition numbers in different ways and are able to divide 3-digit numbers by a 1-digit number. They know how to write their answer and understand what each number in the answer statement means, for example, which number represents the number of sheep in the field.

ANSWERS

Question **1**: $146 \div 2 = 73$

Question **2**: $185 \div 5 = 37$

Question **3** a): Children choose one model to find the answer, such as:
$168 = 120 + 48$
$120 \div 6 = 20$ and $48 \div 6 = 8$
$20 + 8 = 28$ and so $168 \div 6 = 28$

Question **3** b): 246 could be partitioned as follows: $240 + 6$, $120 + 120 + 6$ or $180 + 60 + 6$

Question **3** c): Children use one model to find the answer, such as:
$180 \div 3 = 60$, $60 \div 3 = 20$ and $6 \div 3 = 2$
$60 + 20 + 2 = 82$ and so $246 \div 3 = 82$

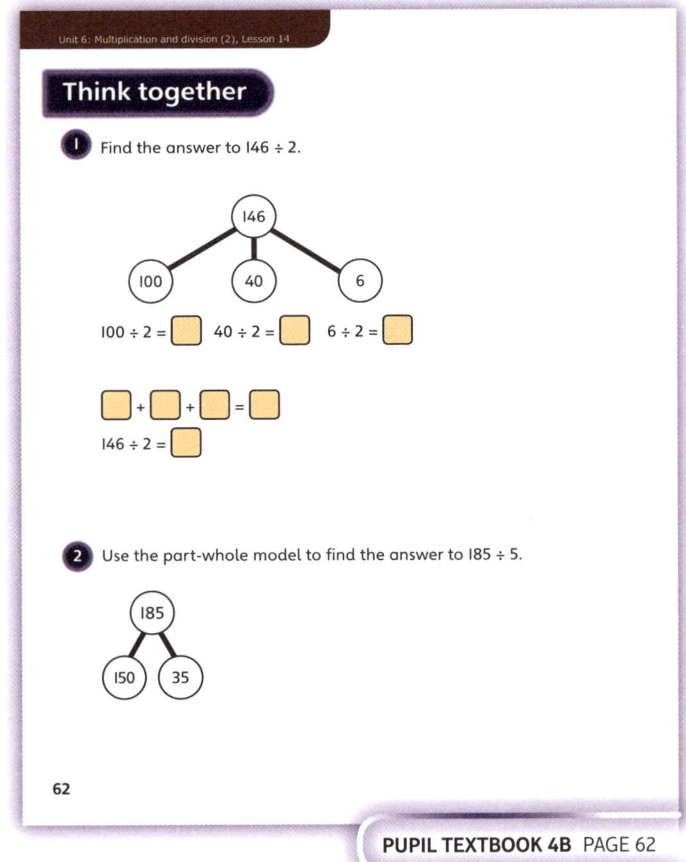

PUPIL TEXTBOOK 4B PAGE 62

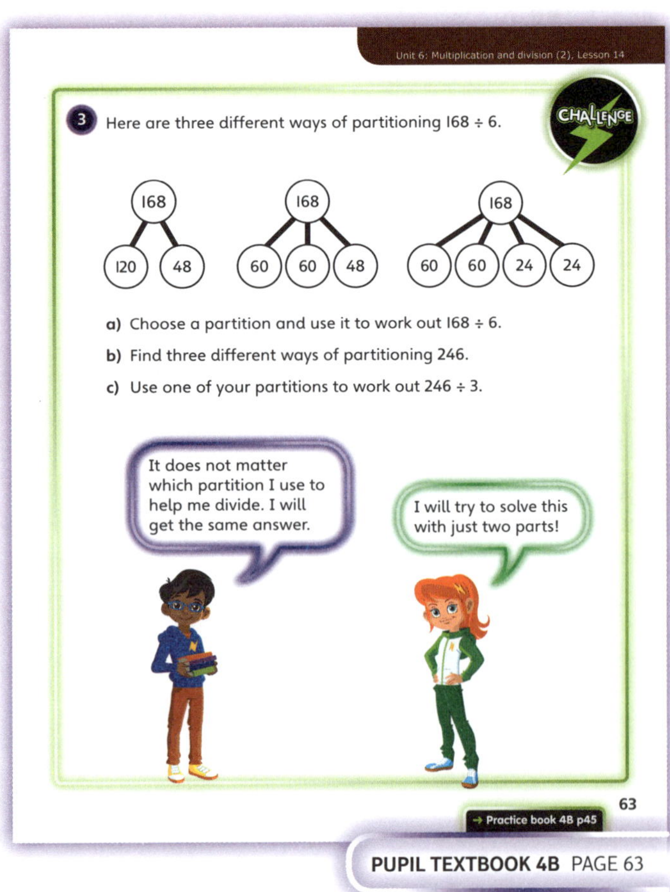

PUPIL TEXTBOOK 4B PAGE 63

Practice

WAYS OF WORKING Independent thinking

IN FOCUS Question ❶ provides suitable partitions so that children can focus on securing the method of dividing each part and then adding. In question ❷, scaffolding is gradually reduced and children begin to think about how they are going to partition the numbers. Question ❸ provides abstract practice, so children need to decide how to partition each number before dividing it. Question ❺ challenges children to find three different ways to partition 584. If necessary, encourage children to focus on what partitions will help them most, given that the divisor is 4.

STRENGTHEN To support understanding of dividing a 3-digit number by a 1-digit number, children can use concrete equipment and physically partition the numbers in different ways. In question ❹ b), children may need support with how many counters they need to take and the number of groups they need to make, or the number in each group. Children need to see the clear link between the numbers in the question and the objects they are using to aid their understanding.

DEEPEN To deepen thinking, ask children to create their own problems similar to that in question ❺, where they have to work out different ways to partition a number. For example, ask children to find examples where _ ÷ _ = 132. How many examples can they find? Encourage children to check their answers by completing the relevant multiplication statements.

THINK DIFFERENTLY Question ❹ requires children to think differently about the concept of division, starting with the parts and using these numbers to work out what division they represent. The question will help children to think deeply about which partitions make sense for different divisors.

ASSESSMENT CHECKPOINT Children know how to partition a number in different ways so that they can make the division easier.

ANSWERS Answers for the **Practice** part of the lesson appear in the separate **Practice and Reflect answer guide**.

Reflect

WAYS OF WORKING Independent thinking

IN FOCUS This question requires children to explain how a 3-digit number can be divided by a 1-digit number, and encourages them to reflect on what they have learnt in this lesson. Children may use a part-whole model to explain their answer and partition 172 in different ways. They should be able to explain that, regardless of the way the number is partitioned, the answer will be the same.

ASSESSMENT CHECKPOINT Children can partition a 3-digit number efficiently and know that a number can be partitioned in different ways.

ANSWERS Answers for the **Reflect** part of the lesson appear in the separate **Practice and Reflect answer guide**.

After the lesson 

- Can children divide a 3-digit number by a 1-digit number?
- Can they find different ways to partition a 3-digit number?
- Do they understand that, although numbers can be partitioned in different ways, the answer will be the same?
- Can children use multiplication to check their answers?

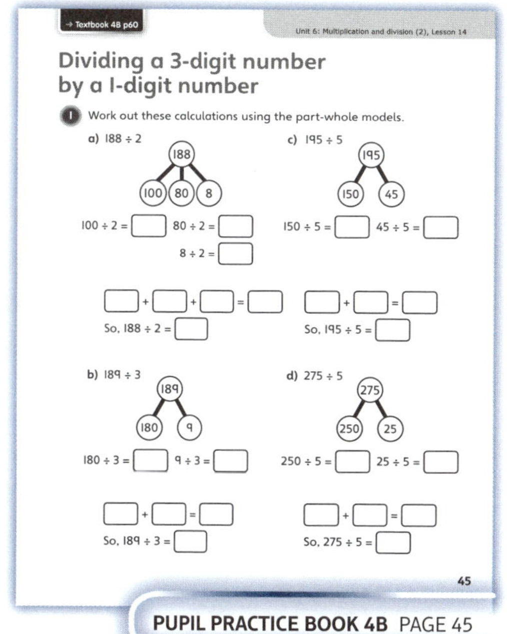

PUPIL PRACTICE BOOK 4B PAGE 45

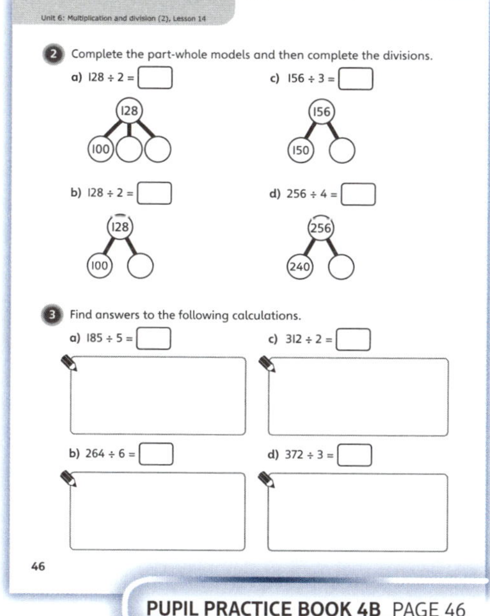

PUPIL PRACTICE BOOK 4B PAGE 46

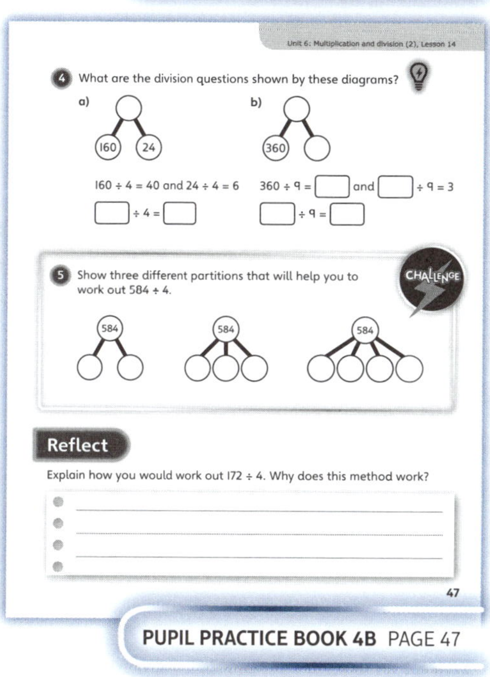

PUPIL PRACTICE BOOK 4B PAGE 47

Problem solving – division

Learning focus

In this lesson, children will apply the multiplying and dividing methods they have learnt in this unit to solve problems.

Small steps

→ Previous step: Dividing a 3-digit number by a 1-digit number
→ **This step: Problem solving – division**
→ Next step: What is area?

NATIONAL CURRICULUM LINKS

Year 4 Number – Multiplication and Division

Solve problems involving multiplying and adding, including using the distributive law to multiply two digit numbers by one digit, integer scaling problems and harder correspondence problems such as *n* objects are connected to *m* objects.

ASSESSING MASTERY

Children can solve a range of multiplication and division problems. They are able to use key information in the question to draw diagrams and to decide whether they need to multiply or divide.

COMMON MISCONCEPTIONS

Children may not be sure whether to multiply or divide and they may guess the operations that they need to use. Encourage them to read the question step by step and to use a part-whole model to represent the question. Ask:

• *What are the keywords in the question? Can you describe the question? What will the part-whole model look like? Do you know how many parts it should have? What is the value of one part? What does each part represent? What does each number in the question mean?*

STRENGTHENING UNDERSTANDING

If children are unsure what to do, read the question aloud, changing your intonation at the keywords. If necessary, ask children to represent the question with counters or cubes; this should support their understanding of the more abstract questions. Some children may not be able to recall multiplication facts quickly. If they do not know the answer, encourage them to use counting up, grouping or sharing strategies to help them to work it out.

GOING DEEPER

Expand children's thinking by asking them to find different ways to work out the answers to the questions in this lesson. Ask them to partition numbers in different ways and to explore which way is the easiest, and which is the hardest. Encourage children to explain their reasoning to a partner.

KEY LANGUAGE

In lesson: how many, how much, left over, divides equally, array, in total, groups, share, remainder, division

Other language to be used by the teacher: times-tables, recall, sharing, grouping, repeated addition, multiply, word problem

STRUCTURES AND REPRESENTATIONS

part-whole model, arrays

RESOURCES

Mandatory: counters, cubes

 In the eTextbook of this lesson, you will find interactive links to a selection of teaching tools.

Before you teach 

• Do children know all the multiplication facts?
• Do they know that multiplication is commutative and distributive?
• Do they know how to partition numbers?

Discover

 Pair work

- Question ❶ a): *How many dancers are on stage? How will the dancers be grouped? How many dancers are there in each group? What division facts can you see?*
- Question ❶ b): *What is the question asking? How could the children be grouped? How can you work out how many groups to have? Can you think of a different way to solve it?*

 Question ❶ a) requires children to explain why 72 cannot be divided equally by 5. Ensure children have access to 72 counters that they can use to represent the question. Question ❶ b) asks children to use the multiplication facts they know to find out which numbers are factors of 72.

 To further strengthen understanding of question ❶ b), ask children to think about different ways to group the counters and therefore partition the number. Ask: *In how many different ways can the counters be grouped into equal amounts?*

 Ensure there are enough counters or cubes for children to be able to make an array of 72 counters.

Question ❶ a): Mo knows that 72 cannot be shared equally into 5 groups because 72 is not in the 5 times-table. There would be 2 children left over.

Question ❶ b): Children can stand in groups of 2, 3, 4, 6, 8, 9 and 12 without any being left over.

Share

 Whole class teacher led

- Question ❶ a): *How can you partition 72? Is there a different way to do it? What do you think Sparks means? Do all multiples of 5 end in 0 or 5? How do you know?*
- Question ❶ b): *What does 'group size' mean? What is the question asking? What does the array look like for 8 × 9 = 72? What is the 8, what is the 9, and what is the 72? How can you use the array to find all the answers?*

 In question ❶ a), children use the part-whole model to partition 72 into the parts 50, 20 and 2. Discuss with children how the answer should be written. Question ❶ b) requires children to use the multiplication facts they know to work out which numbers are a multiple of 72. Explain that an array can show the different ways in which 72 counters can be arranged. Ensure that at each stage you make it clear what each number corresponds to (for example, the total number of counters, the number of columns, the number of rows), to reinforce understanding. Children should make it clear what the group size is and how many groups are formed each time.

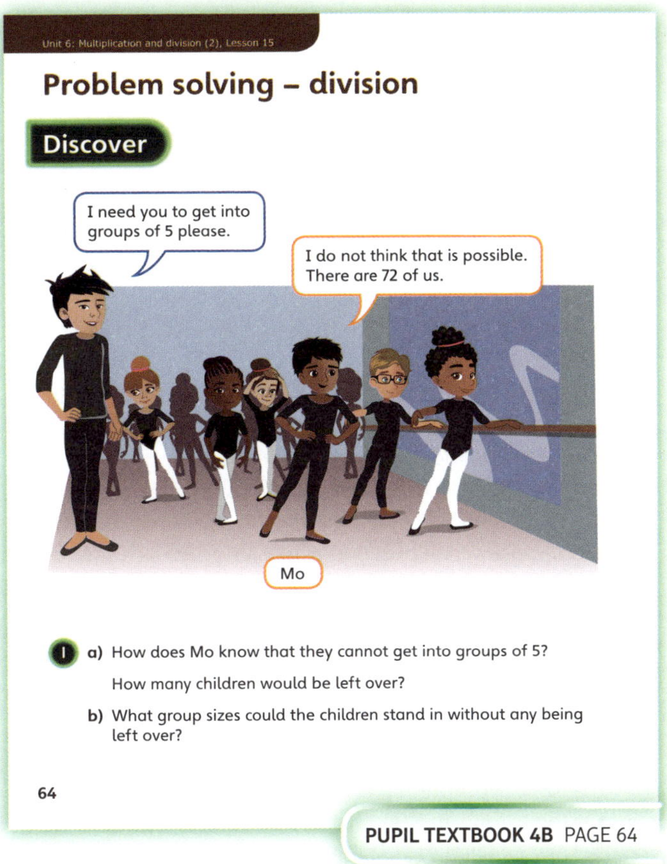

PUPIL TEXTBOOK 4B PAGE 64

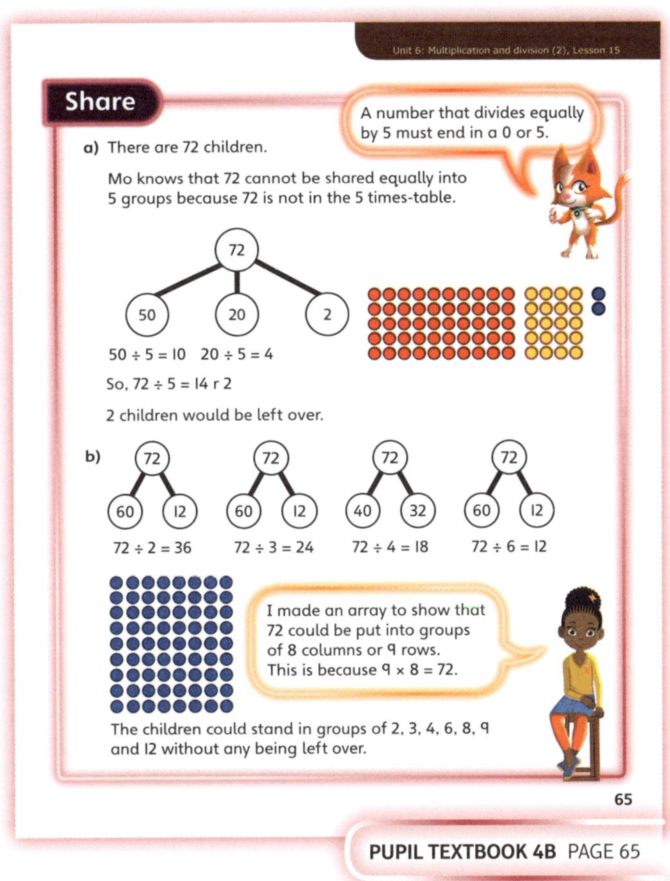

PUPIL TEXTBOOK 4B PAGE 65

Think together

 Whole class teacher led (I do, We do, You do)

- Question ❶ b): *How can you work out how many children are left? What calculation can you do to work out how to share this number into groups of 3?*
- Question ❷ b): *If there was no remainder, what could the question have been? Can you think of a different answer? And another? If there was a remainder of 1, what could the question have been? What about a remainder of 2? What is the largest remainder you could have had?*
- Question ❸: *How long is the piece of paper? How long is one small piece of coloured paper? How many coloured pieces of paper can cover one whole length?*

 This section takes children through a series of word problems. The key aim is for children to decide whether they need to multiply or divide, or do both. In question ❸, children need to look for a connection between the dimensions of the piece of paper used in the art project and the small coloured pieces of paper. Some children may decide to complete a row first; others may think of completing a column first. Discuss both methods and the results they will achieve. Some children may need to use counters to support them with this.

 Discuss Astrid's idea in question ❸ and help children to make the diagram on their desks. Ask: *How can the diagram help? What will you do first? What calculations can you use? In how many ways can you solve the question?*

 In question ❷ a), encourage children to find different ways to partition the number. Challenge them in question ❷ b) to make their own word problem with the answer of 8 remainder 2. What could the problem be if children had to multiply first, and then divide to solve the problem? Listen for children who can generalise that the number they are dividing must be 2 more than a multiple of the divisor.

 Children can solve division word problems. They can use manipulatives and draw diagrams to decide whether they need to multiply or divide.

Question ❶ a): 5 × 4 = 20; there are 20 children in groups of 4.

Question ❶ b): 59 − 20 = 39, 39 ÷ 3 = 13; there are 13 groups of 3 children.

Question ❷ a): 40 ÷ 4 = 10, 32 ÷ 4 = 8, 10 + 8 = 18.
So 73 ÷ 4 = 18 r 1

Question ❷ b): Accept all correct answers, such as 42 ÷ 5 = 8 r 2

Question ❸: The number of pieces in each row: 60 ÷ 4 = 15 coloured pieces.
The number of rows: 60 ÷ 3 = 20 rows.
The number of pieces altogether = 20 × 15 = 300 pieces.

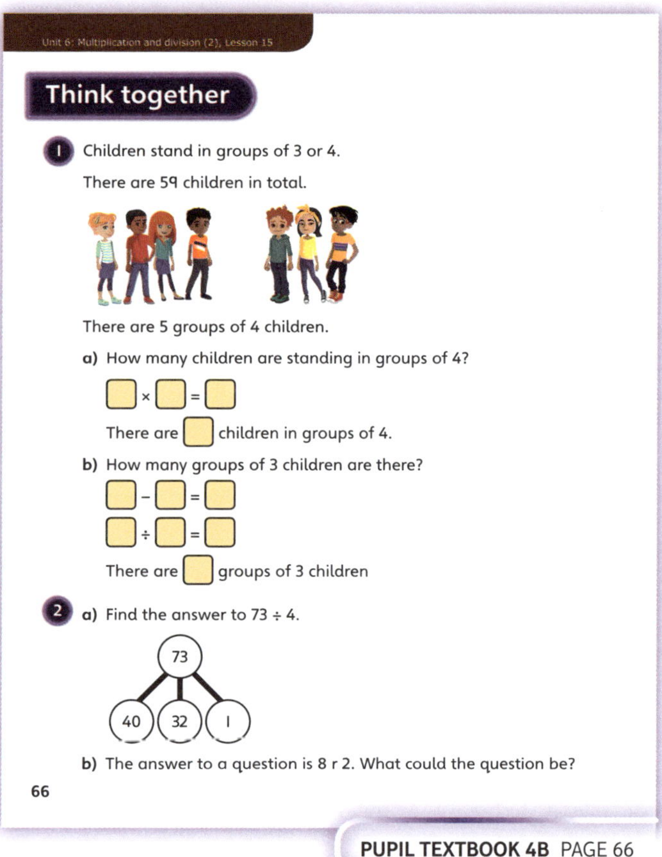

PUPIL TEXTBOOK 4B PAGE 66

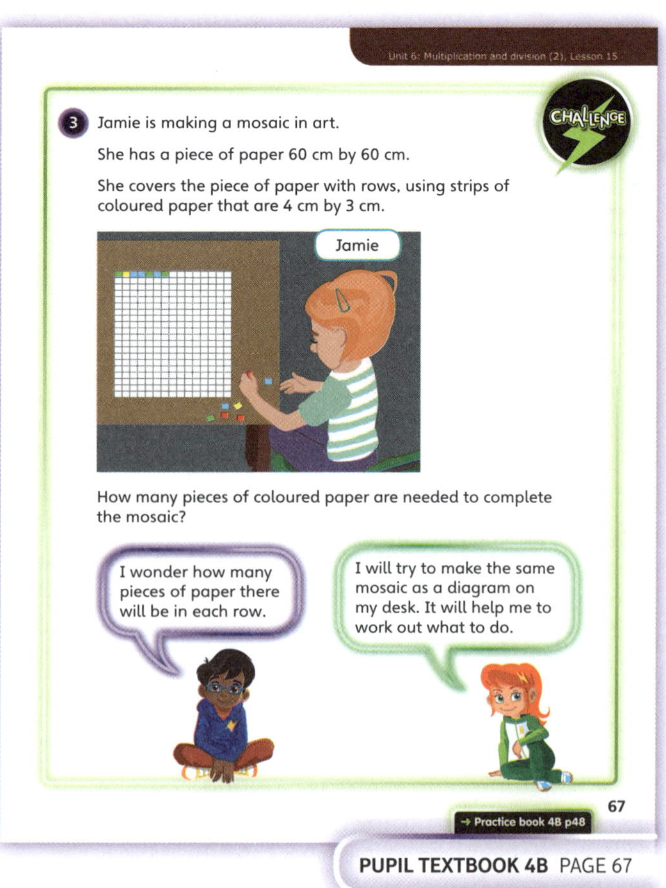

PUPIL TEXTBOOK 4B PAGE 67

Practice

WAYS OF WORKING Independent thinking

IN FOCUS In question **3**, the total number has not been partitioned and children will need to do this first. Do they realise that because this calculation has a remainder of 4, and because the 4 represents 4 children, these 4 children will also need a bench to sit on?

Question **4** asks children to compare division facts to work out whether the statement is true or false. Some children may be able to work this out without solving the divisions.

STRENGTHEN Encourage children to work through the questions slowly and step by step, reading them aloud if this is helpful. Prompt them to think about what information they have and how this information can help them to work out the answers. For example, have they been given the number of groups, or the number of objects in a group?

DEEPEN Deepen thinking in question **6** by discussing what Dexter is saying. Can children predict whether all the children can be in pairs without actually doing the division? Extend thinking further by asking: *What is the division if children are grouped in pairs? What numbers can be divided by 2 without a reminder?*

THINK DIFFERENTLY Question **5** requires children to work backwards from the answer to find three different division calculations that give the same answer. To support children here, ask them to think about what the numbers could be if there was no remainder. From this, build on their knowledge of multiplication facts by asking: *What if the remainder is 1, or 2? What if the remainder is 3?*

ASSESSMENT CHECKPOINT Can children solve a multi-step word problem? Can they decide which numbers to divide? Can they predict whether the answer will have a remainder or not?

ANSWERS Answers for the **Practice** part of the lesson appear in the separate **Practice and Reflect answer guide**.

Reflect

WAYS OF WORKING Independent thinking

IN FOCUS Children demonstrate that they know the properties of multiples of 3 or 5 and that they understand when a number divided by 3 or 5 has a remainder and when it does not. Children use their knowledge of times-table facts to work out what the remainder (if any) will be and explain their reasoning clearly.

ASSESSMENT CHECKPOINT Can children predict if a division will have a remainder or not?

ANSWERS Answers for the **Reflect** part of the lesson appear in the separate **Practice and Reflect answer guide**.

After the lesson ⏸

- Can children solve a range of division word problems?
- Can children identify the key information in a problem?
- Can they use this information to determine the operations they will use to solve the problem?

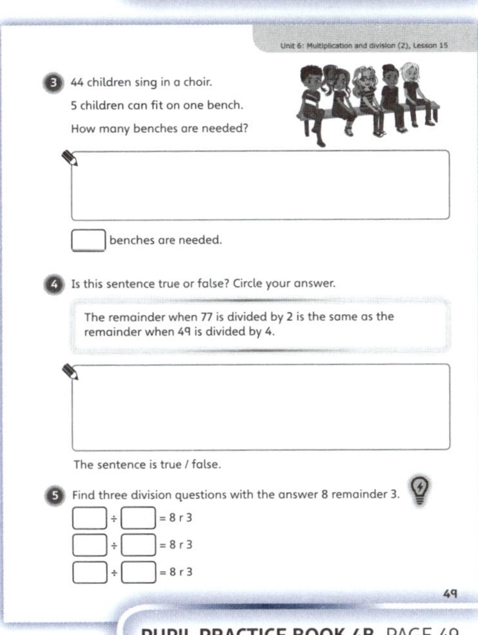

PUPIL PRACTICE BOOK 4B PAGE 48

PUPIL PRACTICE BOOK 4B PAGE 49

PUPIL PRACTICE BOOK 4B PAGE 50

End of unit check

Don't forget the *Power Maths* unit assessment grid on p26.

WAYS OF WORKING Group work adult led

IN FOCUS These questions cover the multiplication and division methods from this unit. They are designed to draw out misconceptions or misunderstandings.

Questions ❸, ❹ and ❺ look at which strategy children adopt. In question ❺, for example, do children just work out all the answers or do they reason correctly why D must be the answer (40 and 38 are not multiples of 3)?

For question ❻, ask children to read and understand what the question is asking before deciding whether to multiply or divide to work out the answer.

ANSWERS AND COMMENTARY Children who have mastered the concepts in this unit will not only know their multiplication tables but will also understand what the facts mean and use the multiplication facts to work out others. They see division as both sharing and grouping and also understand that division is the inverse of multiplication. They are able to use the distributive law to partition numbers and create equivalent calculations. They are confident in using arrays to show equivalences and derive new facts.

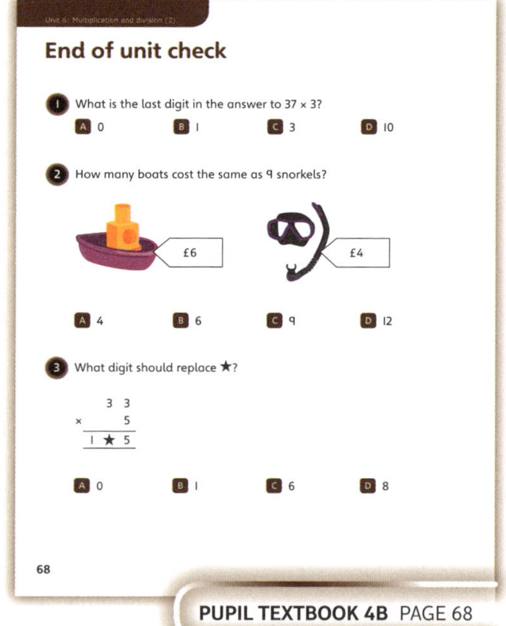

PUPIL TEXTBOOK 4B PAGE 68

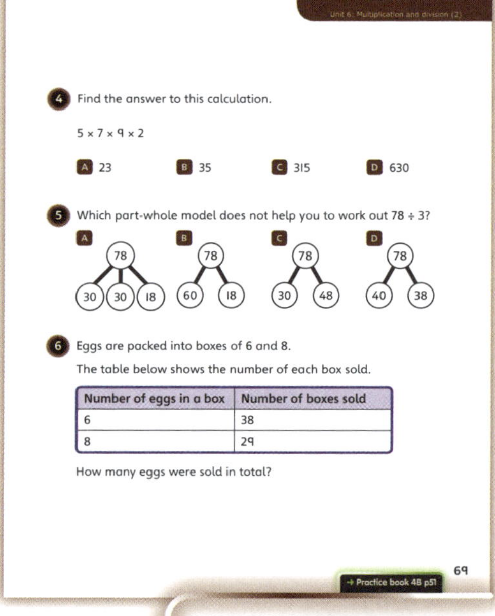

PUPIL TEXTBOOK 4B PAGE 69

Q	A	WRONG ANSWERS AND MISCONCEPTIONS	STRENGTHENING UNDERSTANDING
1	B	A indicates the child has added the final digits.	For question 1, children can use a written method. Alternatively, they can partition 37 into 30 and 7 and multiply each part by 3, before adding the results.
2	B	A, C or D shows that the child is not fluent with their multiplication facts, or they may have misinterpreted the question.	For question 2, children can use a bar model, which indicates whether it is a multiplication or a division.
3	C	This question checks fluency of times-tables and using the written method of multiplication. Children may choose D if they add 5 and 3 instead of multiplying.	In question 5, children can use a part-whole model, and then divide each part by 3.
4	D	A, B and C show children don't understand that they need to multiply.	
5	D	A, B and C all show parts that are multiples of 3.	For question 6, children can use a bar model to represent the number of eggs. They can also use two different-coloured counters to visualise the question better.
6	460	Children have to work out which numbers to multiply and which ones to add.	

My journal

WAYS OF WORKING Independent thinking

ANSWERS AND COMMENTARY

1) $45 \times 7 = 315$; $132 \times 6 = 792$; $78 \div 6 = 13$; $94 \div 5 = 18$ r 4

2) Children should be able to reason effectively about the similarities and differences between short and long multiplication. They should use language such as columns, hundreds, tens and ones, and note that using long multiplication requires addition as well as multiplication.

This activity draws together the written and mental strategies children have learnt for multiplying and dividing; they need to divide and multiply 2- and 3-digit numbers, and explain the difference between long and short formal multiplication.

- Look out for children who still have misconceptions, such as multiplying the digit, rather than the number: for example, when multiplying 20 by 3, they might just multiply 2 by 3.
- Children should remember to carry 1 ten when multiplying 6 by 3 and then add 1 to the answer of 2 tens × 3.
- When dividing, look out for children who use the part-whole model correctly. Pay attention to the way they partition the number.

Power check

WAYS OF WORKING Independent thinking

ASK

- *Do you know all the multiplication facts off by heart?*
- *Do you know how to position the numbers when using the written methods?*
- *Do you know how to check that your answer is correct?*

Power puzzle

WAYS OF WORKING Pair work

IN FOCUS Use this Power puzzle to test fluency with using times-tables and with dividing 2-digit numbers with and without remainders. Support children in understanding what multiplication facts mean: for example, 12 × 4 can be visualised as 12 groups of 4 or a 12 × 4 array.

ANSWERS AND COMMENTARY

1) 1, 1, 1, 4, 1, 0, 1

2) 0, 2, 2, 0, 2, 1, 2

3) 1, 0, 3, 1, 3, 2, 3

4) Look out for any patterns noticed by children.

5) Lee started with 27. Some children may be able to look for much larger numbers in the 3 times-table that Lee could also have started with.

Look for children who need to write out their working out and encourage them to visualise the grouping mentally. Is extra times-table practice required? Ensure children choose a 2-digit number when asked to create their own table.

After the unit ⏸

- Do children know how to apply times-tables to division problems?
- Can they predict whether a division has a remainder or not?

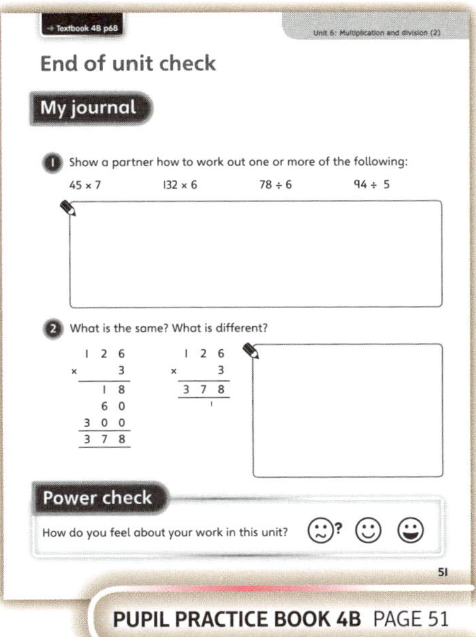

PUPIL PRACTICE BOOK 4B PAGE 51

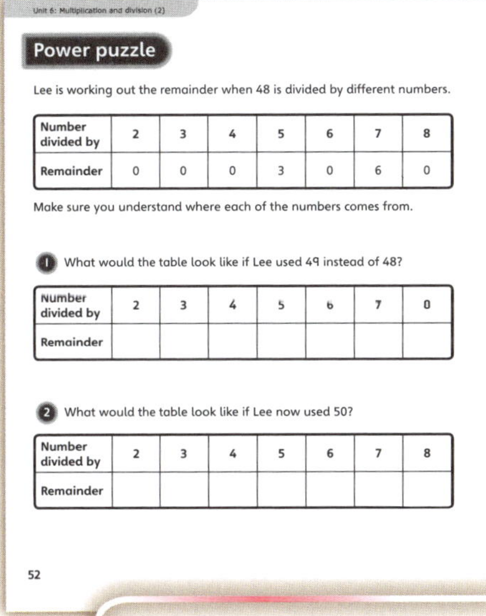

PUPIL PRACTICE BOOK 4B PAGE 52

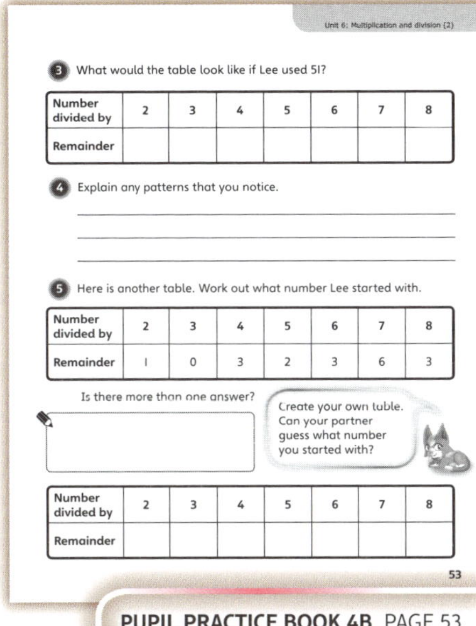

PUPIL PRACTICE BOOK 4B PAGE 53

Strengthen and **Deepen** activities for this unit can be found in the *Power Maths* online subscription.

Unit 7
Measure – area

WHY THIS UNIT IS IMPORTANT

This unit introduces the concept of area to children, giving them a tangible way to measure and compare a shape's size. Until now, children will have been able to say whether a shape is longer or shorter, wider or narrower, and will have been able to measure a shape's length and width. This unit provides them with the tools to measure the space that a shape takes up.

Area is introduced by children using non-standard units and seeing how many of these will fit inside a shape. Children will be made aware of the need for standard units of measurement and will progress to measuring area by counting the number of centimetre squares that fit within a shape. The unit helps children to apply their knowledge in problem-solving and investigative contexts, such as exploring different shapes that all have the same area. Although the relationship between length, width and area is not expounded in this unit, children may recognise informal links between the three concepts from their activities when counting squares. These links will provide a foundation for further development of the concept of area in later years.

WHERE THIS UNIT FITS

→ Unit 6: Multiplication and division (2)

→ **Unit 7: Measure – area**

→ Unit 8: Fractions (1)

This unit builds on children's understanding of the properties of squares, rectangles and rectilinear shapes. It extends children's basic comprehension of shapes being 'bigger' or 'smaller' than one another and gives them a tangible way of measuring this. Children already know how to measure the distance around a shape and now are taught how to measure the space inside it.

Before they start this unit, it is expected that children:
- understand what is meant by a 2D shape and are able to identify the space inside it
- understand simple properties of squares and rectangles.

ASSESSING MASTERY

Children who have mastered this unit will understand that the area of a 2D shape is the space inside it. They will recognise the importance of using standard units and will be able to confidently measure area by counting squares. They will be able to apply their knowledge to find solutions involving squares, rectangles and rectilinear shapes, including exploring shapes with the same area and comparing shapes based on their areas.

COMMON MISCONCEPTIONS	STRENGTHENING UNDERSTANDING	GOING DEEPER
Children may not recognise the 'conservation of area' (when the squares that form a rectangle are rearranged into different shapes, the area remains the same).	Build links between concrete and pictorial representations by providing squares for children to arrange into the rectilinear shapes shown in their books.	Encourage children to consider half squares and whether they can draw a triangle and work out its area.
Children may assume a shape is automatically larger if it is longer or taller, rather than finding the answer by counting squares.	Give children opportunities to cover areas in different ways. Make links between the area of the shape and the size of the unit of measurement.	Children could be challenged to find the area of rectangular objects using squared paper by drawing around the object, then counting the number of squares within it.

Unit 7: Measure – area

WAYS OF WORKING

Use these pages to introduce the concept of the area of 2D shapes to the whole class, checking their understanding of larger or smaller in relation to shapes.

STRUCTURES AND REPRESENTATIONS

Squared paper overlaid with 2D shapes or counters.

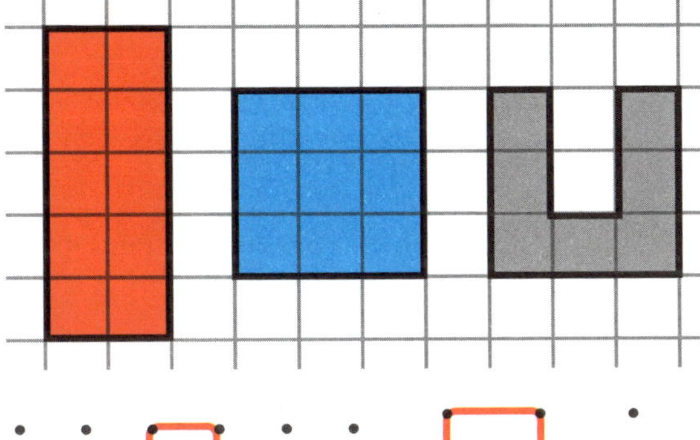

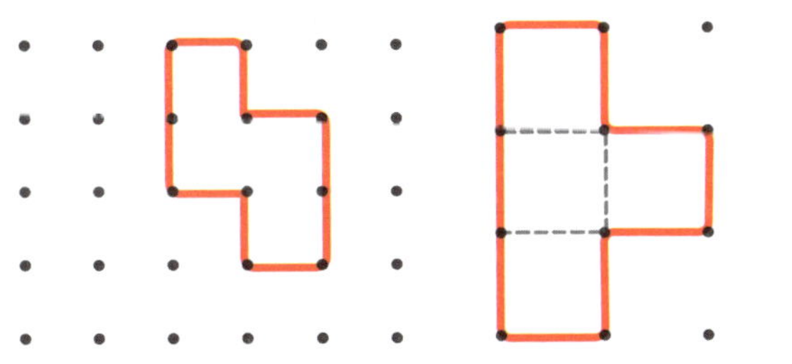

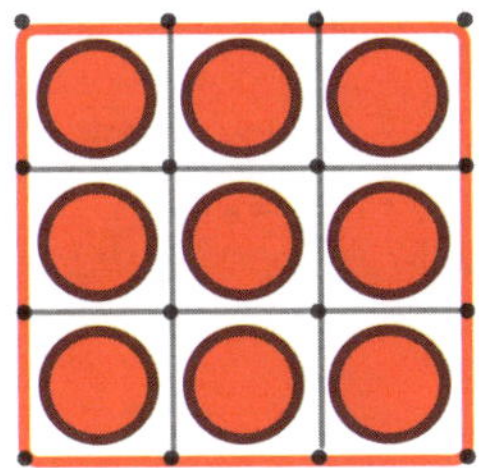
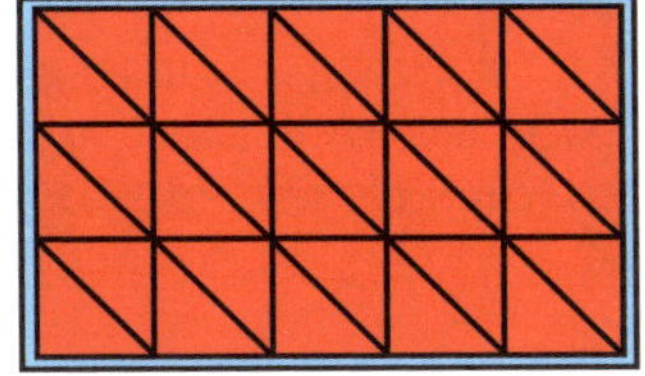

KEY LANGUAGE

There is some key language that children will need to know as part of the learning in this unit:

- area, space, inside, units, rows
- length, width, measure
- shape, triangle, square, rectangle, trapezium, rectilinear shape, 2D shapes
- larger, more area, smaller, less area, least area, greatest area
- right angle
- counting, subtraction
- reflection, rotation
- compare, order, size

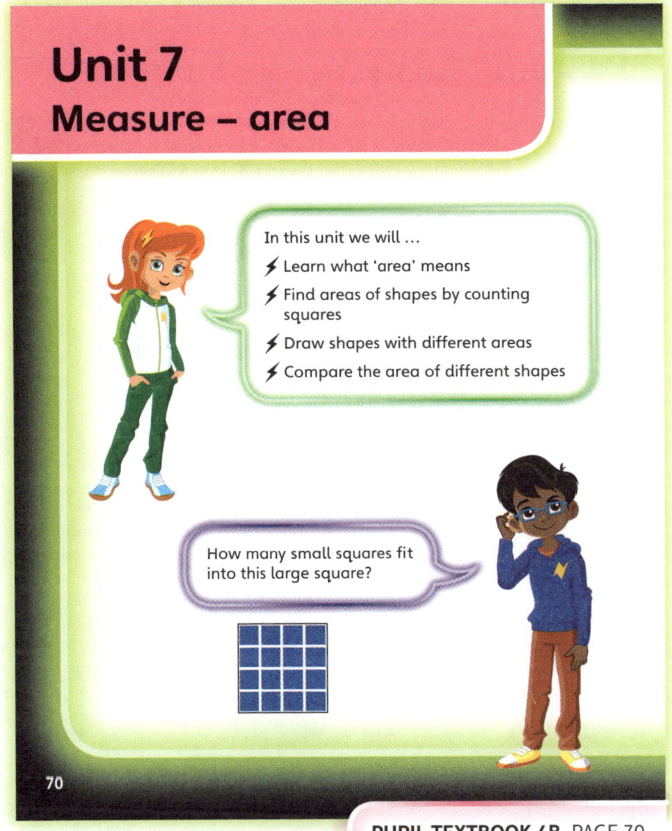

PUPIL TEXTBOOK 4B PAGE 70

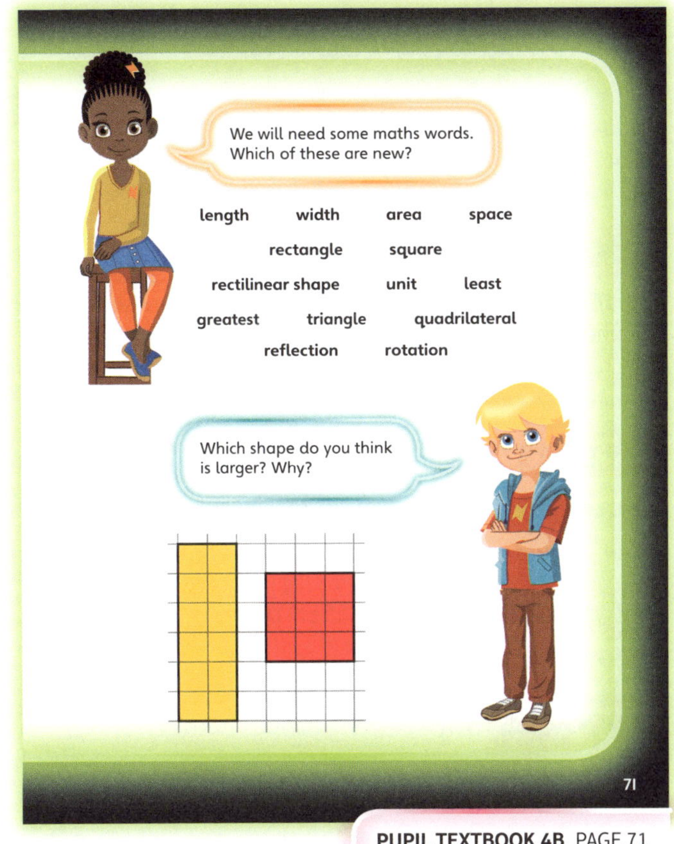

PUPIL TEXTBOOK 4B PAGE 71

What is area?

Learning focus

In this lesson, children will be introduced to the concept of the area of a 2D shape. They will measure this by counting non-standard units that fit within squares and rectangles.

Small steps

→ Previous step: Problem solving – division
→ **This step: What is area?**
→ Next step: Counting squares (1)

NATIONAL CURRICULUM LINKS

Year 4 Measurement
- Find the area of rectilinear shapes by counting squares.
- Estimate, compare and calculate different measures, including money in pounds and pence.

ASSESSING MASTERY

Children can confidently explain that the area of a 2D shape is the space it takes up. Children can express area by measuring this 'space' using non-standard units.

COMMON MISCONCEPTIONS

Children may link a shape's height (or width) to its size, so will assume that a taller shape is necessarily larger than a shorter shape. Ask:
- *Which shape is larger? Why do you think this?*

Children may use different types of non-standard units of measure and compare them or may use one type inconsistently (for example, leaving gaps in between them). Ask:
- *Should your units of measure cover the whole shape? Why? What should you remember when choosing the units of measure to use? Does it matter how you arrange the shapes?*

STRENGTHENING UNDERSTANDING

To strengthen understanding, give children simple tasks that involve them exploring area more generally. Ask: *How many children can sit on the mat? How many playing cards will fit on a window ledge? Will a newspaper page fit on a desk-top?* Let children cover shapes cut out from paper or 2D maths shapes with plastic counters. Ask: *How many counters cover this shape? Which of these two shapes needs more counters to cover it?*

GOING DEEPER

Ask children to set up measuring scenarios where they have made a mistake on purpose. For example, showing a book covered in different-sized coins or where the units overlap and/or have gaps between them. Can children spot the mistakes? Ask children to answer the question: *Is the area the same whatever units you use to measure it? Why not?* Provide a variety of non-standard units and an object to measure.

KEY LANGUAGE

In lesson: area, triangle, square, rectangle, trapezium

Other language to be used by the teacher: 2D shape, space inside, units of measurement

STRUCTURES AND REPRESENTATIONS

2D shapes

RESOURCES

Mandatory: small counters (16 mm preferably), a variety of flat, non-standard units to measure with (flat coloured squares or triangles, playing cards, coins)

Optional: a variety of squares and rectangles to measure (book covers, newspaper pages, paper or card)

 In the eTextbook of this lesson, you will find interactive links to a selection of teaching tools.

Before you teach 

- Can you think of any misconceptions that children may have when measuring area?
- How can you use your school environment to introduce and reinforce the concept of area?

Discover

WAYS OF WORKING Pair work

ASK

- Question **1** a): *Which shape looks larger? Why do you think this? Is there a way you could measure the shapes to see which is larger?*
- Question **1** b): *How should you place the counters? Should you use counters to measure both shapes? Why?*

IN FOCUS Question **1** a) encourages children to begin talking about what makes a shape 'larger'. At this point, their experience of measure will involve length and width, so these measurements will be at the forefront of their answers. Challenge them by asking: *What makes a shape larger than another one? What if it is tall but thin, or wide but not very tall?* Tease out the concept of the space that a shape takes up and that its area is the measure of this space.

PRACTICAL TIPS Encourage children to begin thinking about the space that shapes take up through practical scenarios. For example, ask: *If these were plates, how would you know which was larger? If these were lawns you were mowing, how would you know which was larger? If these were car parks, how would you know which was smaller?*

ANSWERS

Question **1** a): The window is larger because it has a larger area.

Question **1** b): A different type of object (unit of measurement) could be used to measure the areas before comparing.

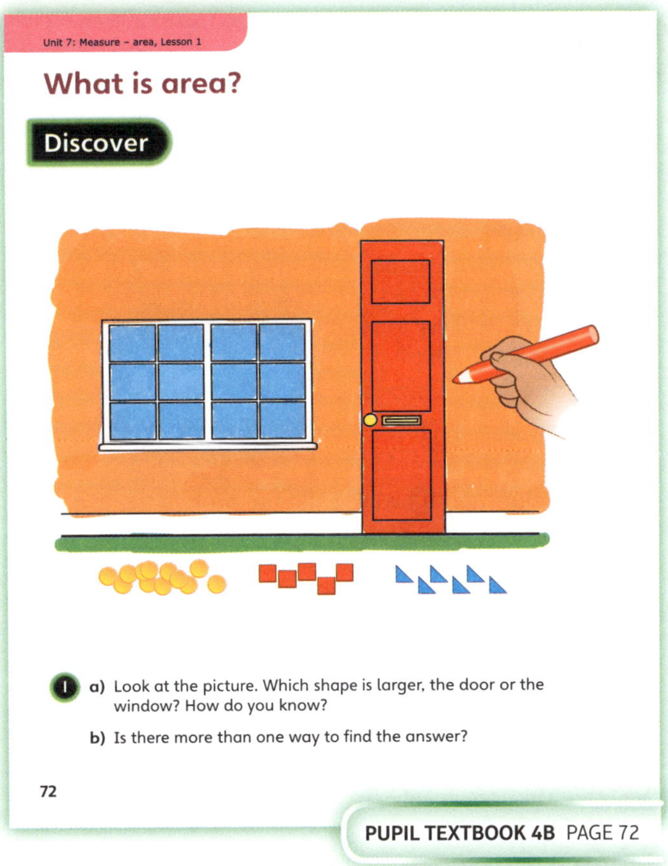

PUPIL TEXTBOOK 4B PAGE 72

Share

WAYS OF WORKING Whole class teacher led

ASK

- Question **1** a): *What does the word 'area' mean? When have you heard it used before? (penalty or local or nature areas)*
- Question **1** a): *What do you notice about the way the counters have been arranged on the rectangles? Why do you think they have been placed like this?*
- Question **1** a): *What do you think would be a wrong way to arrange the counters on the rectangles?*
- Question **1** a): *What makes a shape larger or smaller than another shape?*

IN FOCUS In question **1**, children will see that they can use a variety of non-standard measures to find the area and compare the sizes of two shapes. In this case, counters and plastic triangles. Avoid using more 3D objects such as cubes, as these will be used to measure capacity in Year 5.

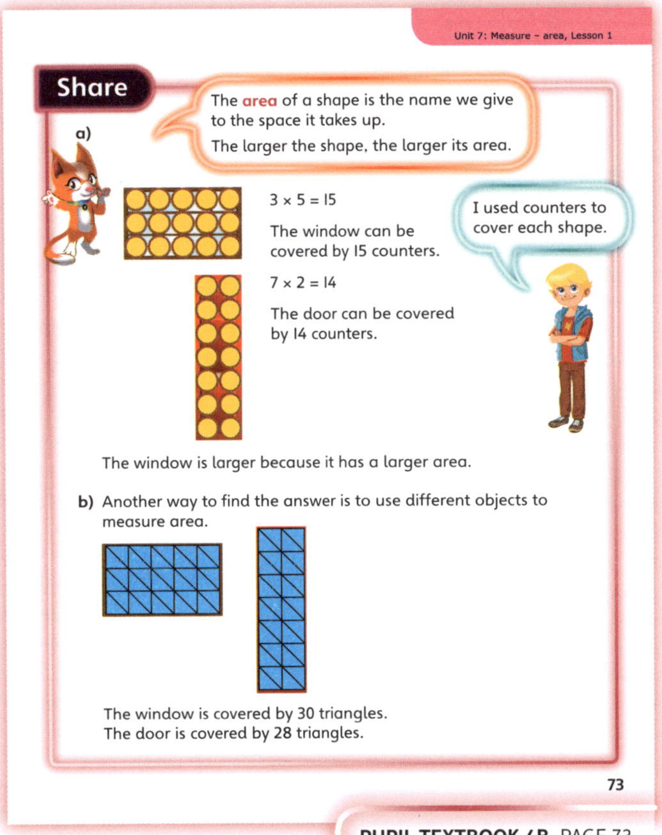

PUPIL TEXTBOOK 4B PAGE 73

Think together

WAYS OF WORKING Whole class teacher led (I do, We do, You do)

ASK

- Question **1**: *How do you need to arrange the counters on the shapes? Is there a right way and a wrong way to do this? Which shape has the larger area? How do you know? Do you think you would have found the same answer if you had used squares instead?*
- Question **2**: *Is there a quick way to find the area of this shape? Can you draw a second rectangle with counters on it to show that it is larger or smaller than this one?*

IN FOCUS Question **1** leads children through the various concepts of area. They find out how many counters fit inside the two shapes. They are then asked to relate their findings to statements about the shape with the larger space inside and the shape with the larger area. Ensure children understand that these concepts are the same.

STRENGTHEN For questions **1** and **2**, give children opportunities to discuss how to arrange the units of measurement to cover the space inside the shape as best they can. This may involve rearranging the counters to find the best arrangement.

DEEPEN In question **3** ask children to consider the difference between the two rectangles. Discuss the possible effect on their results of using different-sized units of measurement. Ask: *How can you change things to make a fairer way to compare the areas?*

ASSESSMENT CHECKPOINT Children should understand that 'area' is a measure of the size of a shape. They should understand which lengths are relevant to measuring an area, and should be able to confidently suggest ways to measure an area using non-standard units of measurement.

ANSWERS

Question **1**: Answers depend on the size of the counters used. (Rectangle = approximately 24 16 mm counters; trapezium = approximately 7–8 16 mm counters.) The shape with the larger space inside (the larger area) is the rectangle.

Question **2**: 30 triangles fit inside the rectangle. The area of the rectangle is 30 triangles.

Question **3**: Rectangle B has the larger area. The units of measure are different sizes so cannot be compared. Use either plastic squares or coins for both shapes, then the areas can be compared.

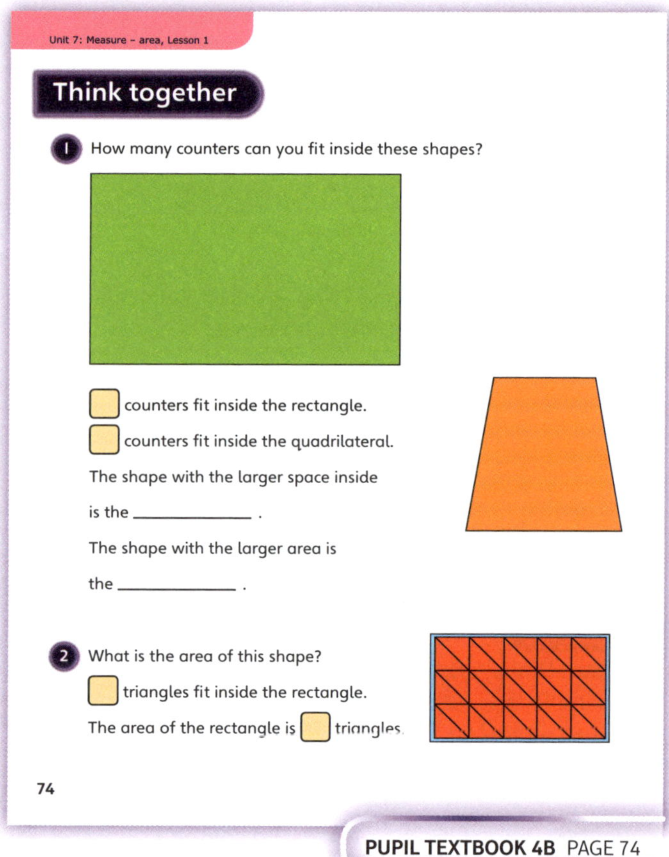

PUPIL TEXTBOOK 4B PAGE 74

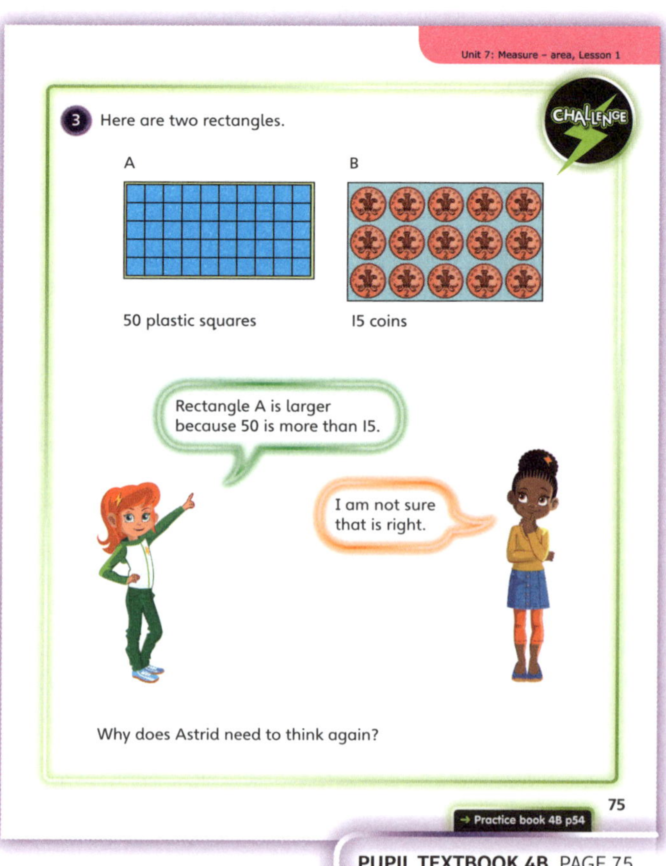

PUPIL TEXTBOOK 4B PAGE 75

Practice

WAYS OF WORKING Independent thinking

IN FOCUS Questions ❸ and ❹ are designed to assess children's understanding of the definition of area as the space inside a 2D shape.

STRENGTHEN Provide children with some actual objects when answering question ❸, asking them to use simple non-standard units to find each shape's area.

DEEPEN In question ❻, children consider the relationship between a shape's perimeter and its area. Encourage children to investigate different shapes to see that more objects do not necessarily fit inside a shape than around it.

THINK DIFFERENTLY Question ❺ encourages children to consider appropriate choices when deciding which non-standard units of measurement to use to measure area. They should begin to recognise that shapes that tessellate are better than those that do not, as they cover the space more effectively. This concept is a useful precursor to Lesson 2 where they begin to use squares to measure area.

ASSESSMENT CHECKPOINT Children should be able to arrange and count small objects to find a shape's area. They should start to understand that some non-standard units of measurement are better than others (ones that fill a shape more efficiently).

ANSWERS Answers for the **Practice** part of the lesson appear in the separate **Practice and Reflect answer guide**.

PUPIL PRACTICE BOOK 4B PAGE 54

PUPIL PRACTICE BOOK 4B PAGE 55

Reflect

WAYS OF WORKING Independent thinking

IN FOCUS This question provides an opportunity to check children's methodology. How do they go about looking for an appropriate shape? How do they arrange the counters? Encourage them to explain how they found the area.

ASSESSMENT CHECKPOINT Are children able to define and measure the area of a shape correctly?

ANSWERS Answers for the **Reflect** part of the lesson appear in the separate **Practice and Reflect answer guide**.

After the lesson 

- Could children have been given more opportunities to explain their reasoning in this lesson?
- Do you feel that children will understand why squares are the standard unit of measurement when they are introduced in the next lesson?

PUPIL PRACTICE BOOK 4B PAGE 56

Counting squares ❶

Learning focus

In this lesson, children will begin to use squares as a standard unit of measuring the area of squares and rectangles.

Small steps

→ Previous step: What is area?
→ **This step: Counting squares (1)**
→ Next step: Counting squares (2)

NATIONAL CURRICULUM LINKS

Year 4 Measurement

Find the area of rectilinear shapes by counting squares.

ASSESSING MASTERY

Children can recognise why squares are used as standard units of measurement of area and can confidently apply their knowledge of area by counting squares in shapes made of up to nine squares. Children can write a shape's area as X squares (units) and understand what this means.

COMMON MISCONCEPTIONS

Children may confuse the distance around a shape (perimeter) with the space inside it (area). Ask:
• *Can you show me the area of this shape? Now show me the perimeter.*

Children may use standard units inconsistently (for example, leaving gaps between squares) or not understand why squares are a good measure of area (they tessellate leaving no gaps). Ask:
• *Why do you think squares are a good measure of area? How should they be arranged?*

STRENGTHENING UNDERSTANDING

To strengthen understanding, give children four flat squares and ask them to rearrange them into different positions. Ask: *What shapes can you make using these squares? What is the area of your shapes?* Repeat with other numbers of squares. Explain that the squares need to tessellate (fit together with no gaps).

GOING DEEPER

Challenge children to demonstrate why squares are an efficient unit for measuring area. Ask: *Can you show why squares are a better measure of area than plastic counters?* Encourage children to consider why squares are better than rectangles as a standard unit of measurement. Ask: *Both squares and rectangles tessellate, so why do you think squares are used as standard units of measurement?*

KEY LANGUAGE

In lesson: units, area, size, shape, measure, squares, rectangle

Other language to be used by the teacher: tessellate, unit of measurement, standard unit, efficient

STRUCTURES AND REPRESENTATIONS

rectilinear shapes

RESOURCES

Mandatory: squared paper

Optional: geoboards

 In the eTextbook of this lesson, you will find interactive links to a selection of teaching tools.

Before you teach ❚❚

• Do children know what the area of a shape is?
• Can you think of squares that occur in your classroom environment that you could use to help children visualise area (such as tiles on a wall)?

Discover

 Pair work

ASK

- Question **1** a): *Do you remember what the word 'area' refers to? How might you find the area of these shapes? Do you think squares are a good way to cover a shape? Why or why not?*

IN FOCUS Question **1** a) requires children to visualise a way to cover the space inside the shapes. The fact that both shapes are on geoboards gives them a pictorial indication that the shapes can be split into uniform squares. Observe those children who can see this from the picture and those who will need to cover the area with counters or other non-standard units as they did in lesson 1.

PRACTICAL TIPS Use enlarged square dotted paper to replicate the shapes in the picture. Ask children to trace the outline of both shapes. Ask: *Can you see any shapes within these shapes? How many small squares can you see?* Encourage children to draw coloured lines connecting the dots to form squares. Some children may need to number the squares to count them.

ANSWERS

Question **1** a): Shape A has an area of 9 squares (units). Shape B has an area of 2 squares (units).

Question **1** b): Children should draw a shape with an area of between 2 and 9 squares.

Share

WAYS OF WORKING Whole class teacher led

ASK

- Question **1** a): *Last lesson you measured area by placing units of measurement inside a larger shape. How is this method the same? How is it different?*
- Question **1** a): *How can you make sure you count all of the squares but do not count any twice or miss any out?*
- Question **1** a): *How would you complete this sentence? 'The shape with the larger area is the one that …'*

IN FOCUS For question **1** a) discuss with children how they might count the squares that fill each shape. Share ideas and provide opportunities for them to try these methods. Children could cover each square with a square of coloured card or colour in each square as it is counted. Using card squares is a particularly effective way of making the link between the pictorial and the concrete.

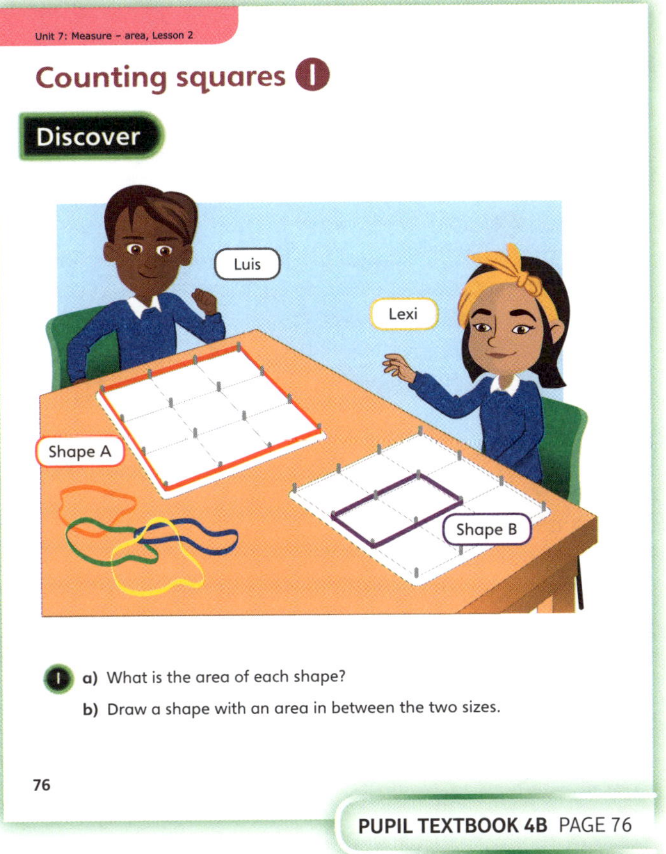

PUPIL TEXTBOOK 4B PAGE 76

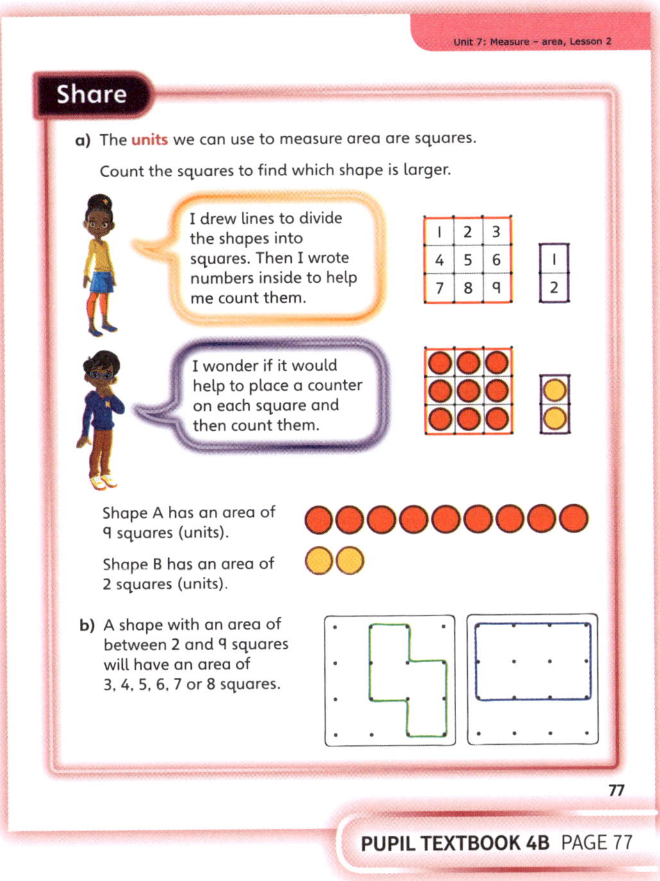

PUPIL TEXTBOOK 4B PAGE 77

Think together

WAYS OF WORKING Whole class teacher led (I do, We do, You do)

ASK

- Question **1**: *How is the area of these shapes measured? Which shape has the largest area? How do you know? What if you put small flat squares inside each shape? Would you get the same or a different answer?*
- Question **2**: *What is different and what is the same about this question? How can you work out the area of the final shape?*

IN FOCUS Questions **1** and **2** progress from shapes where the inner squares are shown to a shape where children need to visualise the squares that might fit within it. Ask: *Which shape has the smallest area? How could you alter Shape A so that it has an area that is the same as Shape C?*

STRENGTHEN Provide children with paper squares the same size as the squares shown in the book. Invite them to completely cover each shape with squares. Ask: *How many squares cover the shape? What is its area?* This practice will be particularly useful when finding the area of Shape C in questions **1** and **2**.

DEEPEN In question **3**, ask children to explain the difference in the way the rectangles have been measured. Ask: *What is the same? What is different? Are the rectangles the same size? Why is counting the squares not a helpful way to compare the areas at the moment?* Ensure children recognise that the squares need to be aligned, so there are no gaps between them and they fill the space completely.

ASSESSMENT CHECKPOINT Children should be able to confidently count the number of squares that fill a simple rectilinear shape, giving this value as its area. Children should know how to arrange squares to measure a shape's area efficiently. Question **3** will raise the issue of what happens when an exact number of squares does not fit. You could discuss how Ash could have tried to see if an extra half square would fit in each column, or used a different method: measuring the sides and multiplying.

ANSWERS

Question **1**: The area of Shape A is 5 squares.
The area of Shape B is 4 squares.
The area of Shape C is 6 squares.

Question **2**:

Shape	Area
A	4 squares
B	5 squares
C	7 squares

Question **3**: Various answers are possible. Explanations might mention that the paper squares have not been lined up correctly, there are gaps between them and they do not cover all of the space inside the shape.

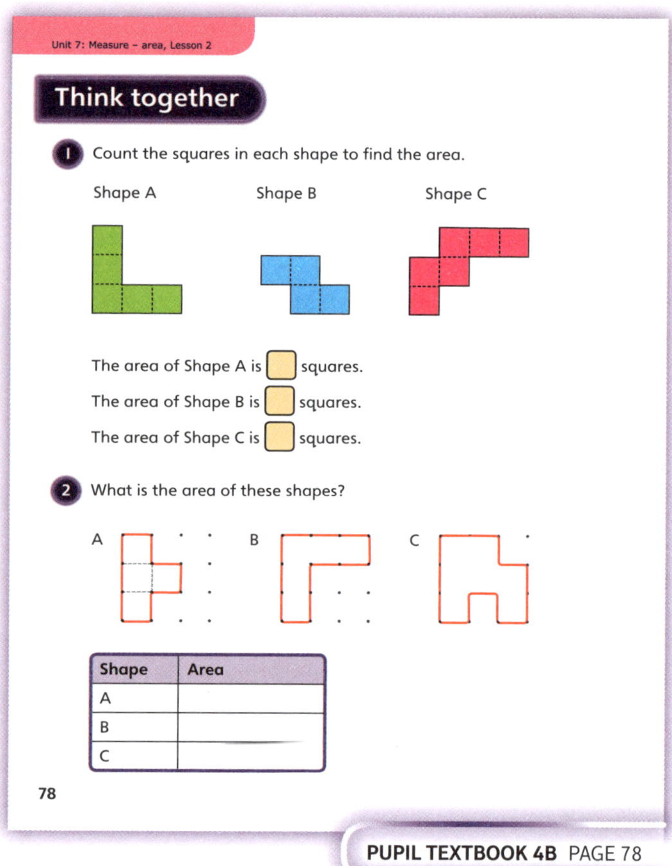

PUPIL TEXTBOOK 4B PAGE 78

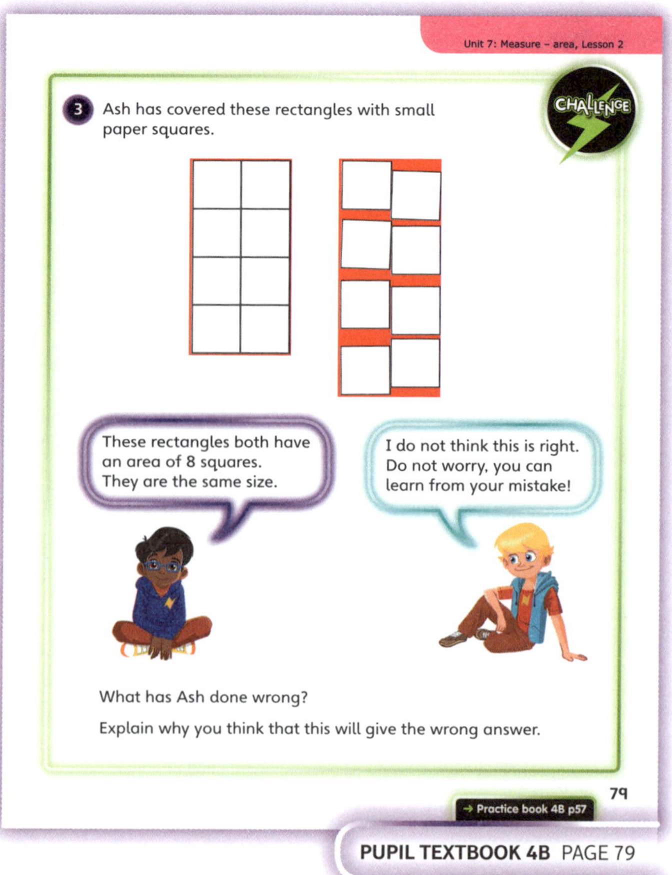

PUPIL TEXTBOOK 4B PAGE 79

Practice

WAYS OF WORKING Independent thinking

IN FOCUS Question **4** addresses one of the main errors children make when measuring area. It reinforces the fact that all units should be arranged so that there are no gaps.

STRENGTHEN Allow children to apply their growing knowledge of area by providing them with square tiles and asking them to make some of the shapes that feature in the questions. Ask: *How many tiles did you use? What is the shape's area? Can you move the tiles to make the same shape, but in a different position? Has the area changed? Can you move some of the tiles to make a completely new shape? Has the area changed?*

DEEPEN In question **6**, children are presented with a sequence of squares. Children can explore the pattern in the areas on several levels: by counting the number of small squares within each larger square and then recognising visually how the pattern continues; by drawing or predicting the next squares in the sequence; and by recognising the numerical pattern (the area increases by the next odd number each time: +3, +5, +7 and so on).

THINK DIFFERENTLY Question **5** provides children with a code to crack. They are asked to work out the area of each shape, then find the relevant letter using the key. When they write the code letter, a phrase will be revealed. Encourage them to find the area of that object (table top) using plastic squares.

ASSESSMENT CHECKPOINT Children should be able to find areas of simple shapes by counting squares. Some children may need manipulatives, others will be able to count squares in images and some may visualise the shapes described. They should recognise that a shape is the same size as another if it has the same area.

ANSWERS Answers for the **Practice** part of the lesson appear in the separate **Practice and Reflect answer guide**.

Reflect

WAYS OF WORKING Independent thinking

IN FOCUS This question has been designed to assess how children's methodology has progressed over the course of the lesson. Children should now describe how to find area using standard units, by counting squares.

ASSESSMENT CHECKPOINT Look for children who are able to confidently and clearly describe how they can find the area of simple shapes.

ANSWERS Answers for the **Reflect** part of the lesson appear in the separate **Practice and Reflect answer guide**.

After the lesson ⏸

- Are children confident in finding areas by counting squares?
- Do children recognise why squares are an efficient unit of measurement to use when finding area?

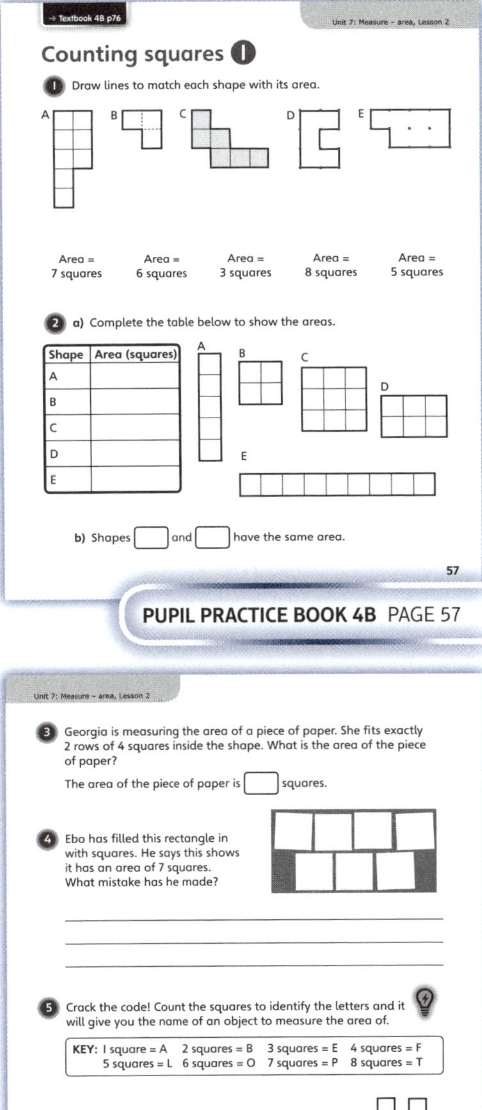

PUPIL PRACTICE BOOK 4B PAGE 57

PUPIL PRACTICE BOOK 4B PAGE 58

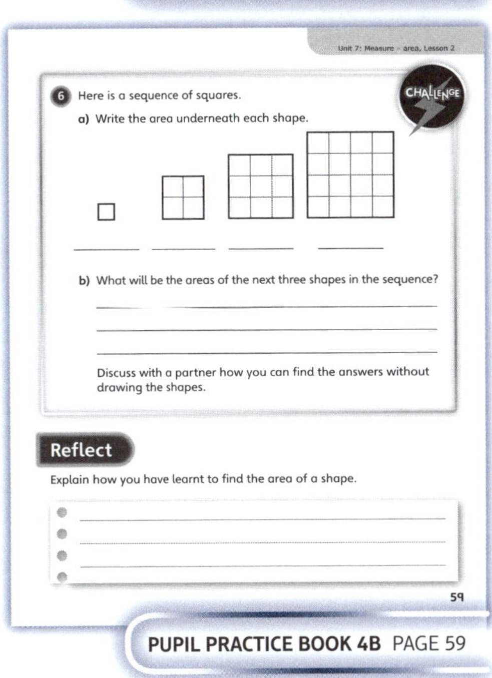

PUPIL PRACTICE BOOK 4B PAGE 59

Counting squares ❷

Learning focus

In this lesson, children will find areas of more complex rectilinear shapes (including those drawn on squared grids) by counting squares.

Small steps

→ Previous step: Counting squares (1)
→ **This step: Counting squares (2)**
→ Next step: Making shapes

NATIONAL CURRICULUM LINKS

Year 4 Measurement

Find the area of rectilinear shapes by counting squares.

ASSESSING MASTERY

Children can confidently find the area of rectilinear shapes drawn on a square grid by counting squares, write the area as __ squares (units) and understand what this means. Children can apply their knowledge of area to find areas in different contexts.

COMMON MISCONCEPTIONS

Children often have difficulties in identifying and relating the area of rectilinear shapes to the rectangles that make up the shape. Similarly, with a shape made from a rectangle with an inner rectangle removed to show a hole, they may ignore the hole and simply work out the area of the complete rectangle, or they may count the overlap of the rectangle twice or forget to subtract the 'hole'. Ask:

• *Is this part of the shape or not? Why is this section not part of the area of this shape?*

STRENGTHENING UNDERSTANDING

Provide children with concrete opportunities to recognise and measure the area of rectilinear shapes around school. Use masking tape around floor tiles in a corridor or use chalk around square paving slabs to show simple rectilinear shapes. Ask: *Which part of this shape is its area? What is the area of this shape?* Explore different ways of showing that each square has only been counted once (such as placing a sticky note inside it, drawing a chalk dot inside it, standing inside it). The activity could be framed as an area-based scavenger hunt.

GOING DEEPER

Set children the challenge of designing their own ideal bedroom or creating a classroom plan on squared paper. Limit the objects within their plans to rectilinear shapes. Ask them to devise questions based on their plans (for example: *What is the area of …? Which is larger, the … or the …? How much free space is in the room?*). By writing down the expected answer, children are applying their knowledge of area in this context.

KEY LANGUAGE

In lesson: area, space, squares, rows, rectilinear shape, right angle, rectangle, counting, subtraction

Other language to be used by the teacher: 2D shape, units of measurement (or 'units'), plan

STRUCTURES AND REPRESENTATIONS

rectilinear shapes

RESOURCES

Mandatory: squared paper

 In the eTextbook of this lesson, you will find interactive links to a selection of teaching tools.

Before you teach 

• How will you define the term 'rectilinear' and describe rectilinear shapes (particularly as squares and rectangles are both rectilinear shapes)?
• In your school or classroom environment, where are there squares arranged in grids (such as tiles) that you could use to illustrate real-life rectilinear shapes?

Discover

 Pair work

ASK

• Question ❶: *Can you say both questions in a different way using the word area? Which parts of the shapes do you need to measure? Why do you think the bedroom plans have been drawn on squared paper?*

IN FOCUS In both parts of question ❶, children may notice that there are different strategies they can use to count squares to find the area. Some children may use their knowledge of arrays (from Year 1) by counting the number of rows and columns. In question ❶ b), children may note that the space in each bedroom is the same as the total area minus the size of the bed. This concept is revisited in **Think together** question ❸.

PRACTICAL TIPS Provide children with flat plastic squares. Ask them to arrange these to make the same shape as each bed. Link these concrete examples of rectangles with the images in the book. Children could also make shapes to represent other items found in the bedroom, such as a chest of drawers or a table.

ANSWERS

Question ❶ a): 12 squares is a larger area than 10 squares, so Aki's bed is larger.

Question ❶ b): Kate has 26 squares of empty space. Aki has 23 squares of empty space.

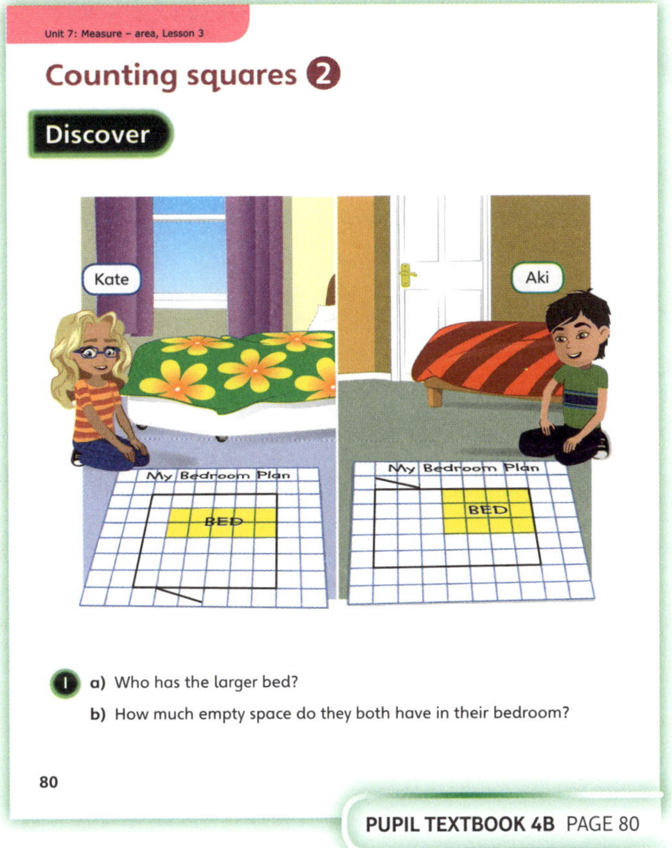

PUPIL TEXTBOOK 4B PAGE 80

Share

 Whole class teacher led

ASK

• Question ❶ a): *What makes a shape 'larger' than another shape? Kate's bed looks longer than Aki's. Does that make it larger? Why not?*
• Question ❶: *What can you do to find the area of the different shapes? How many squares are there in one row? How many rows are there? How does this information help?*

IN FOCUS Encourage children to model how they would find each rectangle's area accurately, without double-counting or missing any squares. When considering question ❶ b), challenge children to consider the relative sizes of the shapes – does the fact Aki's bed is larger mean that the space in his room will always be less? Does it depend on something else? Some children may find the answer in different ways. For example, they can find the total number of squares in the room, and subtract the size of the bed.

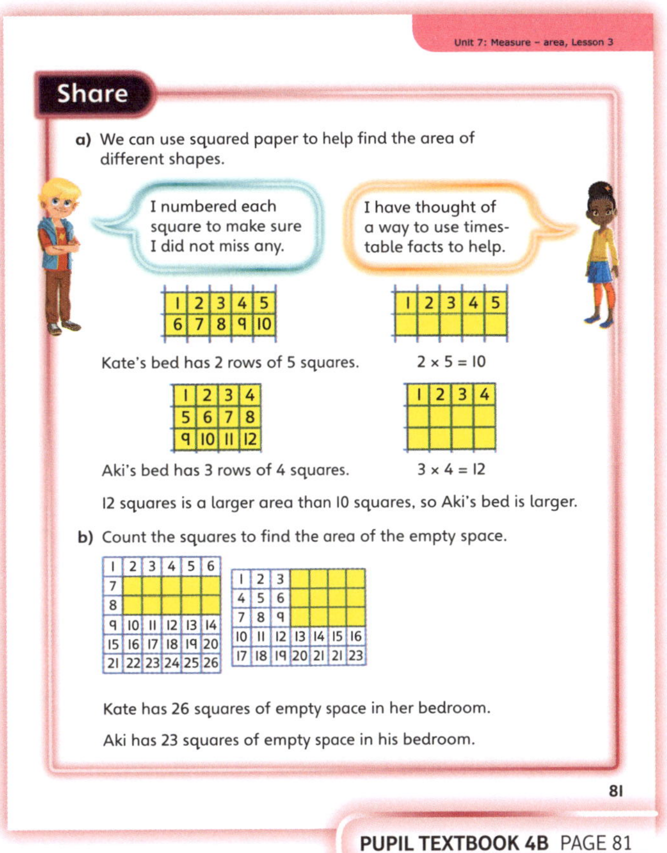

PUPIL TEXTBOOK 4B PAGE 81

Think together

 Whole class teacher led (I do, We do, You do)

ASK

- Question **1**: *Which shape looks like it has the largest area? Is there more than one way to find the area of these shapes?*
- Question **2**: *Compare these shapes with squares and rectangles. What is the same? What is different? Is finding the area of these shapes the same as finding the area of a square or a rectangle? How?*

IN FOCUS Question **2** introduces children to the concept that area is not limited to squares and rectangles. Encourage them to describe each shape in their own words. Explain that if a rectilinear shape looks like two rectangles put together, they could work out the area of the whole shape by adding the areas of the rectangles together. Give them opportunities to draw some rectilinear shapes on the board.

STRENGTHEN Question **2** features a shaded shape presented on a squared grid so that no grid lines are visible inside. To support children with this, place a large rectangle of coloured card on to an enlarged squared grid. Ask children to draw the grid lines across the rectangle, so that they can count the squares. When children can do this confidently, present further examples where they are expected to visualise the squares, using the grid lines around the shapes to help.

DEEPEN Question **3** is a shape with a part missing from the centre. Ask: *Which part of the shape do you need to use to find its area? Should you include the square in the centre?* Ensure children understand that the missing square is **not** part of the shape's area. Give them examples to help visualise this. For example, when painting a wall, the window is not counted as part of the area that has been painted. Refer to Flo's comment to discuss how subtraction might be used to find the area. Ask children to draw similar shapes on squared paper with a square or rectangular part missing and find their areas.

ASSESSMENT CHECKPOINT At this point, children should be able to confidently find the area of these larger rectilinear shapes made up of squares. They should do so either by counting the visible squares inside the shape or by using the squared paper around the shape to visualise the number of squares it covers. In question **2**, children may notice that the areas can be found as the sum of two rectangles, in the form $a \times b + c \times d$.

ANSWERS

Question **1**: Rectangle A = 20 squares
Rectangle B = 18 squares
Rectangle C = 18 squares

Question **2**: The area of Shape A is 30 squares.
The area of Shape B is 30 squares.

Question **3**: The area of this shape is 38 squares.
Children's methodology will vary, but they should be able to confidently explain how they found the answer. For example, add all squares then take away the 4 squares that are cut out.

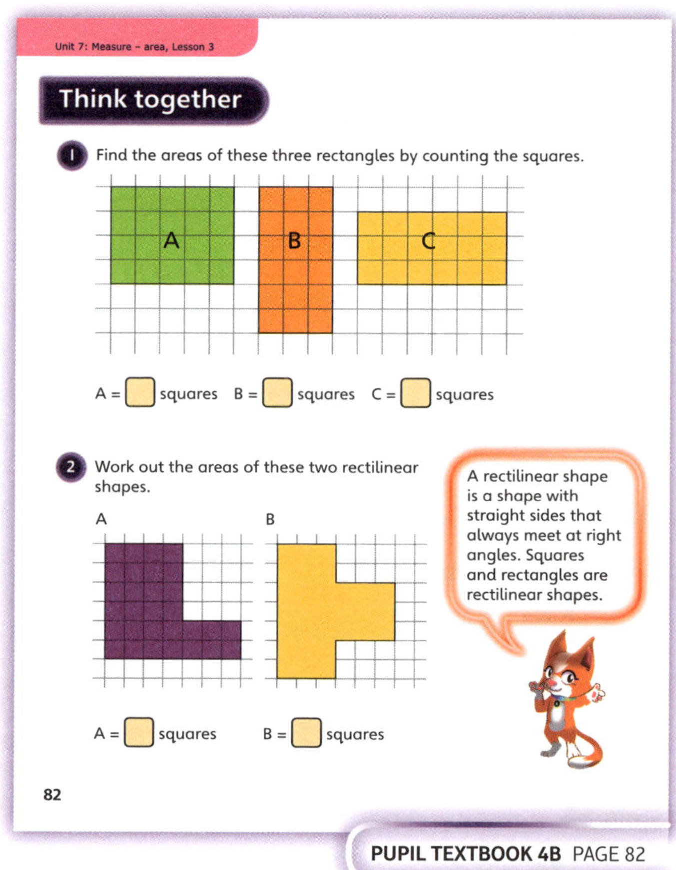

PUPIL TEXTBOOK 4B PAGE 82

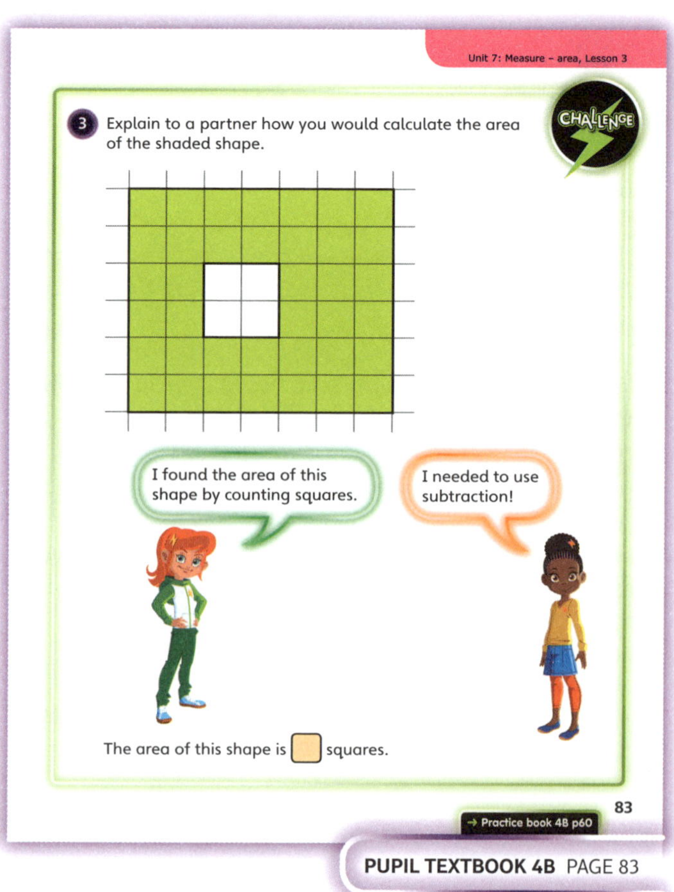

PUPIL TEXTBOOK 4B PAGE 83

Practice

WAYS OF WORKING Independent thinking

IN FOCUS Question ❷ shows two rectangles adjacent to each other. Children are guided to find the area of each individual rectangle to find the total area. This is an important model to help children recognise that rectilinear shapes can be split into smaller squares and rectangles to find their area. If time allows, ask children to describe how some of the rectilinear shapes in question ❶ could be split into smaller squares and rectangles too.

STRENGTHEN Encourage children to write on the grids in their **Practice Books** and devise a way of ensuring that they count the number of squares correctly. They might choose to do this by marking each square as it is counted (tick, colour or dot) or by writing numbers in each square. As well as ensuring that their counting is accurate, this will also help children to visualise the area of the shape.

DEEPEN Question ❺ is a challenge requiring several steps. Lead children though the process. Ask: *How can you find the number of squares around the farmhouse? (Count: 15) This total area is 15 squares, and there are 5 fields, so how many squares will each field be? (3) Which different shapes could you draw with an area of 3 squares? Can you think of a way to divide the land so that all the fields are the same shape?*

ASSESSMENT CHECKPOINT Children should be able to find the area of any rectilinear shape drawn on a grid, including examples where rectilinear shapes have been split into separate rectangles or squares, those that have 'holes' and those that are shaded so that the inner grid lines cannot be seen.

ANSWERS Answers for the **Practice** part of the lesson appear in the separate **Practice and Reflect answer guide**.

Reflect

WAYS OF WORKING Independent thinking

IN FOCUS In this question, children are given an opportunity to explain how they would use squared paper to find the area of a cardboard shape. Some may suggest drawing around the shape and then counting the squares on the squared paper. Others may suggest placing the cardboard on to the paper and using the grid lines to help work out how many squares it is covering. Either way, children should demonstrate their understanding that the area of the shape is the number of squares that fit inside it or that it covers.

ASSESSMENT CHECKPOINT Identify those children who have a correct methodology, demonstrating a confident understanding of how to use squared paper to find the area of a shape.

ANSWERS Answers for the **Reflect** part of the lesson appear in the separate **Practice and Reflect answer guide**.

After the lesson ⏸

- Do children completely understand the concept of a shape's area (as opposed to other measurements: length, width, perimeter)?
- Were children able to find the area of simple composite shapes made up of squares and rectangles?

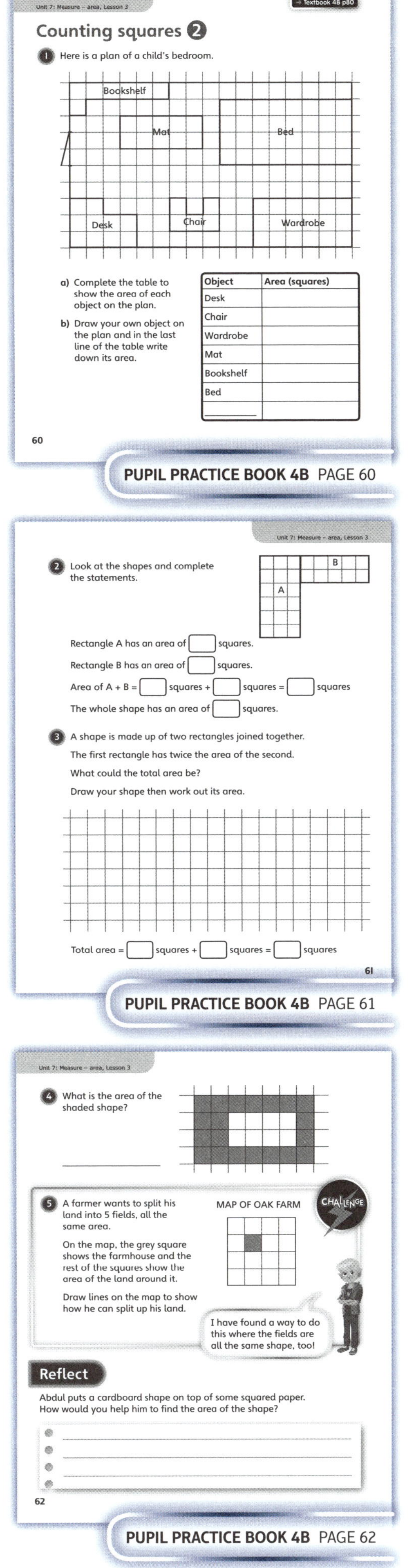

115

Making shapes

Learning focus

In this lesson, children will be given opportunities to apply their understanding of the concept of area by making shapes with given areas.

Small steps

→ Previous step: Counting squares (2)
→ **This step: Making shapes**
→ Next step: Comparing area

NATIONAL CURRICULUM LINKS

Year 4 Measurement

Find the area of rectilinear shapes by counting squares.

ASSESSING MASTERY

Children can apply their knowledge of area by making and drawing rectilinear shapes with given areas (expressed in terms of squares). Children are able to find all possibilities of shapes with given areas, recognising where rectilinear shapes are the same and have simply been rotated or reflected.

COMMON MISCONCEPTIONS

Children may not recognise the 'conservation of area' (for example, that when a rectangle is rearranged into a different shape it still has the same area as the original rectangle). Children may also see a long, thin combination of squares as having a different area to, for example, the same number of squares arranged in a block. Ask:

- *What is the area of this shape? Can you rearrange the squares into a different shape? What is its area now? Is that the same or a different area to the first shape?*

Children may think that they have arranged squares to make a new rectilinear shape, when it is a reflection or rotation of one already found. Ask:

- *Can you turn this shape so that it looks the same as this one?*

STRENGTHENING UNDERSTANDING

Take children to an area of the school where there are squares arranged in a grid (such as, tiles by the sink, concrete slabs around the playground or carpet tiles). Let them mark the outline of different shapes with masking tape or chalk, asking them to make a shape with a given area. Concentrate on whether the shapes have the given area, rather than whether any are repeated or whether all possibilities have been found. Take digital photos of the shapes that children make and use them as a display.

GOING DEEPER

Give each team of 2–4 children a set of 5–8 large squares and challenge them to make as many different shapes as they can in 1 minute, using all the squares each time. If they are making shapes that are reflections or rotations, ask them to connect the squares so that they can be rotated or flipped over to see if they can spot the duplicates. Encourage them to start with a rectangle and move one square, then two squares and so on.

KEY LANGUAGE

In lesson: shape, area, rectilinear shape, rectangle, square(s), reflection, rotation

Other language to be used by the teacher: duplicate, rotate, reflect, flip

STRUCTURES AND REPRESENTATIONS

2D shapes

RESOURCES

Mandatory: 1 cm paper squares or counters, squared paper, square dotted paper

Optional: chalk, masking tape

 In the eTextbook of this lesson, you will find interactive links to a selection of teaching tools.

Before you teach ⏸

- How will you ensure children recognise that, although a shape looks different to another, they may have the same area?
- What opportunities will you give children to use reasoning during the lesson?

Discover

 Pair work

ASK

- Question ❶ a): *How has the patio in the picture been made? What is the area of the patio that is shown?*
- Question ❶ b): *Do you think there is just one way or more than one way to arrange the squares?*

IN FOCUS In part a), children are given a practical problem to solve; then, in part b), they are asked to consider their methodology and use reasoning to explain it. Some children may assume that the patio must be rectangular, and only find two solutions. Once all five have been identified, consolidate understanding that they all have the same area. There may be children who create shapes with overlapping squares. Ask: *Do these have the same area too?* Some children may count reflections as separate shapes. Discuss that they would be the same if you put a mirror next to them. Allow sufficient time on part b) for children to discuss methodologies. For example, they might think about how many ways they can make a group of three slabs first; then how many ways they can add an extra slab.

PRACTICAL TIPS Ask children whether they have any experience of arranging squares to make a shape – perhaps they have seen their parents making a patio or tiling a wall. They may have played a board game where they had to place squares next to each other to make different shapes.

ANSWERS

Question ❶ a): There are five possible shapes that the patio can be. These are:

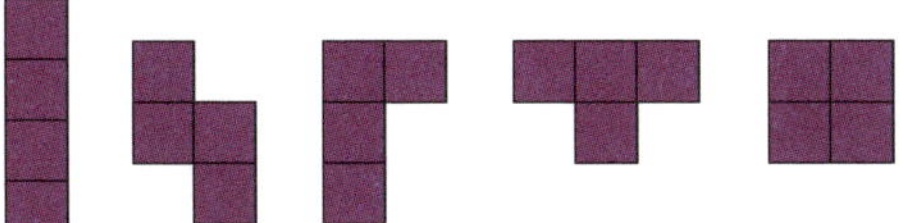

Question ❶ b): Various answers are possible. Children should be able to explain a systematic method of ensuring that they find all the possible shapes.

Share

WAYS OF WORKING Whole class teacher led

ASK

- Question ❶: *How would you describe each of these shapes? Is it true that there can be lots of different shapes that all have the same area? Why?*
- Questions ❶ b): *What mistake has Ash made? How can you check this?*

IN FOCUS For question ❶ b), encourage children to consider systematic ways of working, so they can be sure they have found all possibilities. These will vary, but it is sensible to begin from a clear starting point (a column or row of four in this case) and then change one element at a time. The diagrams should help them.

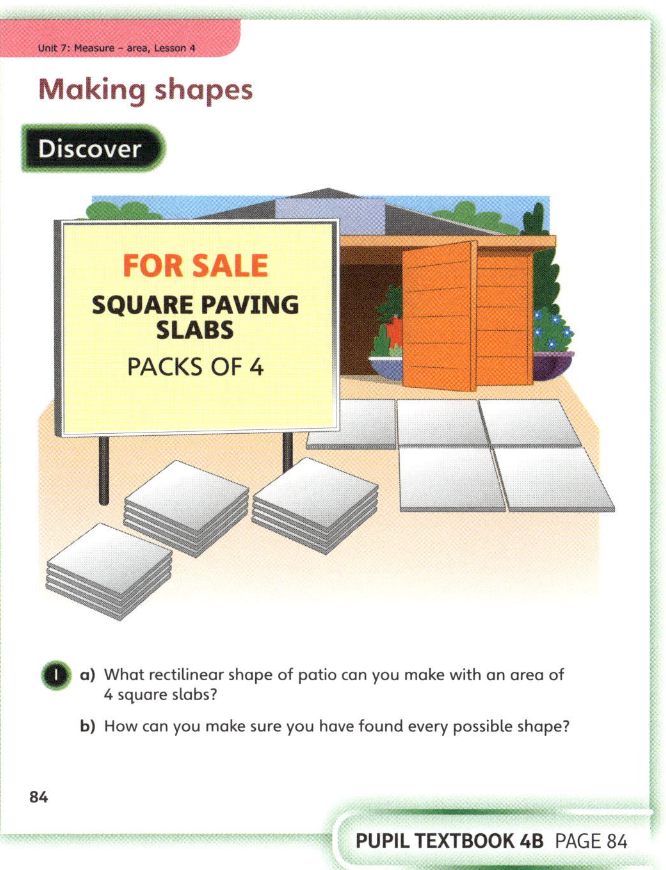

PUPIL TEXTBOOK 4B PAGE 84

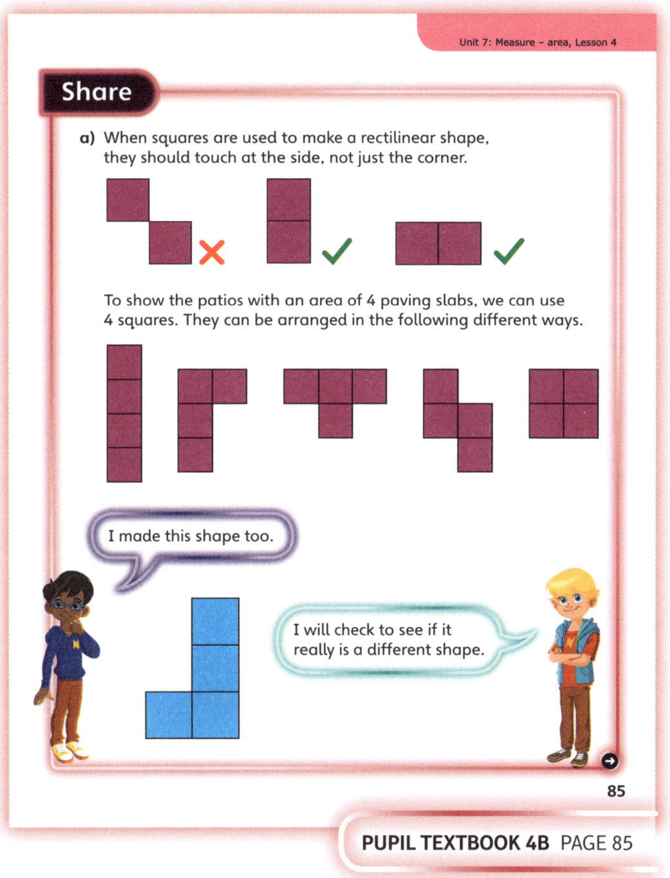

PUPIL TEXTBOOK 4B PAGE 85

Think together

WAYS OF WORKING Whole class teacher led (I do, We do, You do)

ASK

- Question **1**: *How could you use your answer to **Discover** part a) to help you answer this question? How would you describe the shapes you have made?*
- Question **2**: *How could you use your answer to **Discover** part 1 b) to help you answer this question? Do you think there will be more, fewer or the same number of possible shapes with eight squares?*

IN FOCUS In questions **1** and **2** children explore making rectilinear shapes with a given area. Observe those children who arrange eight square objects and then draw around them, and those children who can draw their ideas directly on to squared paper. Ask: *How do you know that your shape has an area that is the same as two packs of concrete slabs? What is the longest patio you can make? Can you make a rectangular patio? Can you make a patio that looks like a letter of the alphabet?*

STRENGTHEN For questions **1** and **2** provide children with squares and encourage them to arrange them to make different shapes. Let them use a new set of squares for each shape they find, so they can compare the shapes they have already found and decide whether their new shape differs. Ask: *What is different about your shapes? What is the same?*

DEEPEN In question **3**, encourage children to consider the reasons why each shape is (or is not) a possible patio. Rectilinear shapes are made by joining squares at the edges, not their corners. A square cannot be formed by eight smaller squares. Discuss how patio A could be changed to make it a possible patio (moving squares so edges are joined). Ask: *How could Holly make patio B correct?* (Add a square.)

ASSESSMENT CHECKPOINT Can children make shapes with a given area, understanding that shapes may look different but have the same area? Assess whether children work systematically, and whether their suggestions meet the given criterion about area. Can they avoid suggesting shapes that are simply reflections or rotations of each other?

ANSWERS

Question **1**: Both shapes should be appropriate rectilinear shapes with an area of 8 squares.

Question **2**: Both shapes should be formed by moving as few squares as possible. They should again be appropriate rectilinear shapes with an area of 8 squares.

Question **3** Shape A cannot be the patio because the slabs are touching at the corners, not edges.

Shape B cannot be the patio because it has an area of 7 squares, not 8.

Shape C could be the patio.

Shape D cannot be the patio because a square cannot be made out of 8 smaller squares.

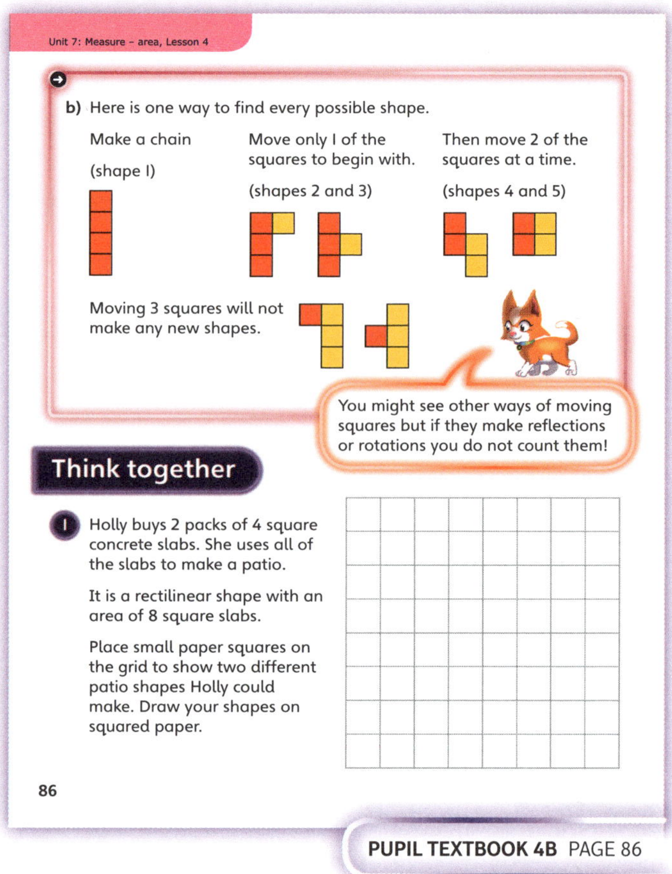

PUPIL TEXTBOOK 4B PAGE 86

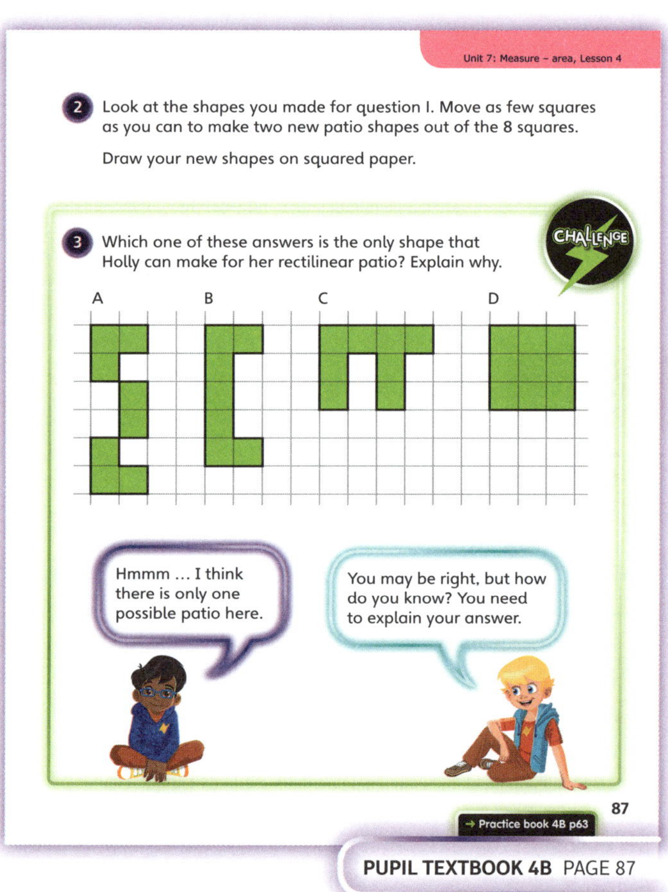

PUPIL TEXTBOOK 4B PAGE 87

Practice

WAYS OF WORKING Independent thinking

IN FOCUS Question **4** addresses the misconception that a rotated or reflected shape is a different shape. Children need to identify that only two different shapes are shown in the group of five. Discuss possible strategies they could use to deal with the misconception (turning the page to view shapes from different positions or cutting a shape out of paper and seeing if it fits into the area of another shape).

STRENGTHEN Question **3** is about visualising whether certain shapes can be made with a given area. Provide children with four paper or cardboard squares, so they can physically explore which of the shapes can be made. In question **5** children could just draw their initials.

DEEPEN Question **5** asks children to investigate the areas of their names. Encourage children to draw the letters using squares (no diagonal lines). Most names will contain 'holes'. Ask: *Is the hole part of the area of the shape? Should you count it or not? Do any letters have the same area? Do they look the same size?*

ASSESSMENT CHECKPOINT Can children make different shapes with a given area? Can they recognise when two shapes are the same (rotations or reflections of each other) or different? Can children work systematically to find all solutions?

ANSWERS Answers for the **Practice** part of the lesson appear in the separate **Practice and Reflect answer guide**.

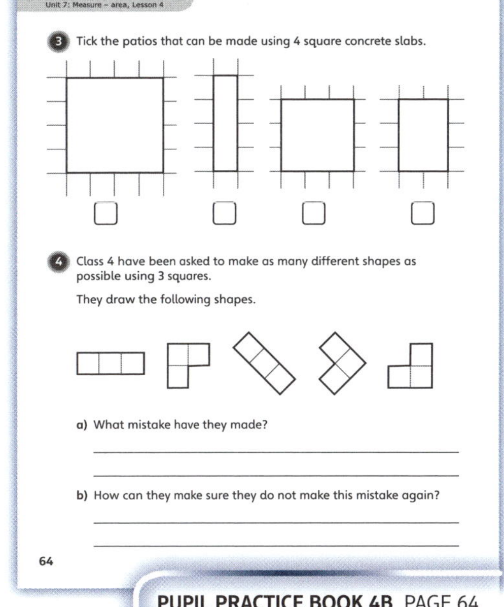

PUPIL PRACTICE BOOK 4B PAGE 63

PUPIL PRACTICE BOOK 4B PAGE 64

Reflect

WAYS OF WORKING Independent thinking

IN FOCUS Children are asked to devise three rules to help when making different rectilinear shapes out of the same number of squares. Their answers may differ, but should demonstrate an understanding of how to make shapes from a given area.

ASSESSMENT CHECKPOINT Look for children who give appropriate 'rules' that clearly describe the methodology of making shapes with a given area.

ANSWERS Answers for the **Reflect** part of the lesson appear in the separate **Practice and Reflect answer guide**.

After the lesson ⏸

- Are children able to confidently make shapes with a given area?
- What other cross-curricular opportunities can be provided to encourage children to work systematically?

PUPIL PRACTICE BOOK 4B PAGE 65

Comparing area

Learning focus

In this lesson, children will learn how to compare shapes according to their areas.

Small steps

➜ Previous step: Making shapes
➜ **This step: Comparing area**
➜ Next step: Tenths and hundredths (1)

NATIONAL CURRICULUM LINKS

Year 4 Measurement

Estimate, compare and calculate different measures, including money in pounds and pence.

ASSESSING MASTERY

Children can confidently compare the area of rectilinear shapes by a) measuring their area accurately by counting squares and then b) comparing both values to see which is the larger. Children can apply this skill to order several shapes according to their areas.

COMMON MISCONCEPTIONS

Children may compare a shape's area visually, by the way it looks rather than by counting squares. As area is dependent on two dimensions (length and width), children may find it difficult to compare shapes when both dimensions differ. Ask:
• *What do you need to do to work out which shape is larger?*

STRENGTHENING UNDERSTANDING

Play a variation on the game in the **Discover** section. Display a grid on the board and divide the group into a number of teams, each team with different coloured squares. They take it in turns to put a coloured square onto the board. They can try to build a shape, but can also block each other by placing their coloured square where other teams might want to go. When all the board is covered ask:
• *Which shape has the largest area? How could you order the shapes in terms of area?*

Draw the shapes away from the board and show them in order.

GOING DEEPER

Challenge children to devise their own area-based games that include both measurement of the area by counting squares and comparison of area. Children could then play these games to practise these skills.

KEY LANGUAGE

In lesson: area, compare, order, size

Other language to be used by the teacher: multiply, total, subtract, count, rectangle, rectilinear shape, grid, measure, accurately

STRUCTURES AND REPRESENTATIONS

2D shapes

RESOURCES

Optional: a large grid, coloured squares (or pens to colour in the squares), manipulatives (counters, small squares)

 In the eTextbook of this lesson, you will find interactive links to a selection of teaching tools.

Before you teach 

• What misconceptions have children shown when comparing numbers? Could they make the same mistakes when comparing areas?
• How can you explain the concept of winning a game by having the largest area to your class? Can you think of any games where this is the case?

Discover

 Pair work

- Question **1** a): *Can you predict who is winning the game just by looking? Why do you think this? Which shape is taller? Does this mean it will have the larger area?*
- Question **1** b): *How can you compare the areas?*

  Both parts of this question encourage children to compare the areas of two different shapes. They will already have plenty of experience of looking at two shapes and choosing the one that looks larger. Now they are required to show the one that is larger, finding each area by counting squares and then comparing the values.

  Play a similar game to the one shown in **Discover**. Display a small grid on the board and divide the group into two teams. They take it in turns to colour any square on the board. If they manage to 'trap' squares of their opponent's colour between two squares of their own colour, they can change the trapped squares into their colour (this is the same concept as the game 'Othello'). After children have had several turns, pause the game and ask: *Which shape has the largest area?* Only include rectilinear shapes.

Question **1** a): 17 > 16, so Olivia is winning the game.

Question **1** b): The area of the board that is not covered is larger.

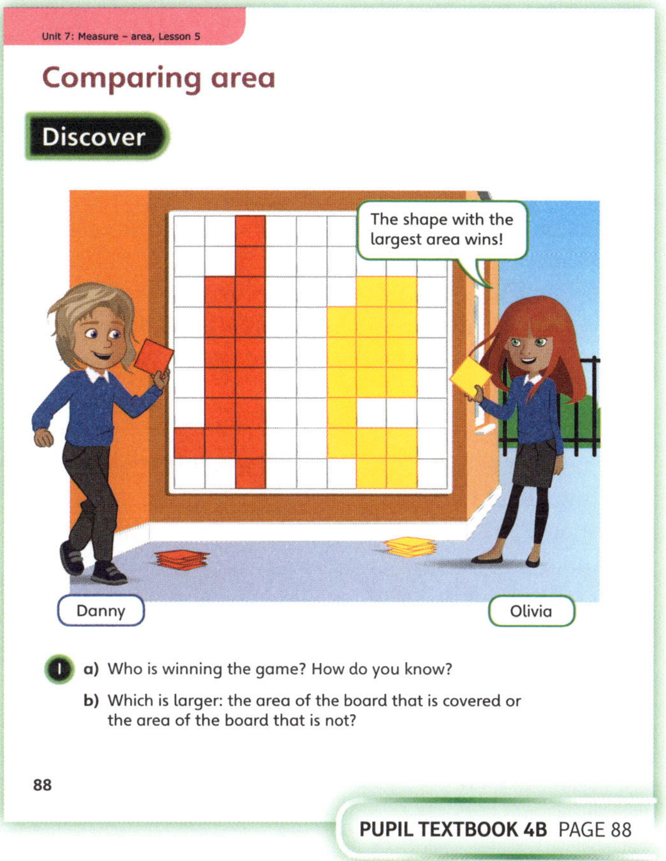

PUPIL TEXTBOOK 4B PAGE 88

Share

  Whole class teacher led

- Question **1** a): *Can you tell who is winning just by looking? Why or why not? How can you check which shape has the larger area? If Danny has beaten Olivia at the game, what can you tell me about the area of Danny's shape?*
- Question **1** b): *One way to find the number of white squares is to count all the white squares. Is there another way to find the same answer?*

  Encourage children to work out question **1** b) using both methods to check their answer. Firstly, counting all the white squares, then working out that there are 9 × 9 = 81 squares on the board and subtracting the total number of coloured squares. Discuss which method they found easier.

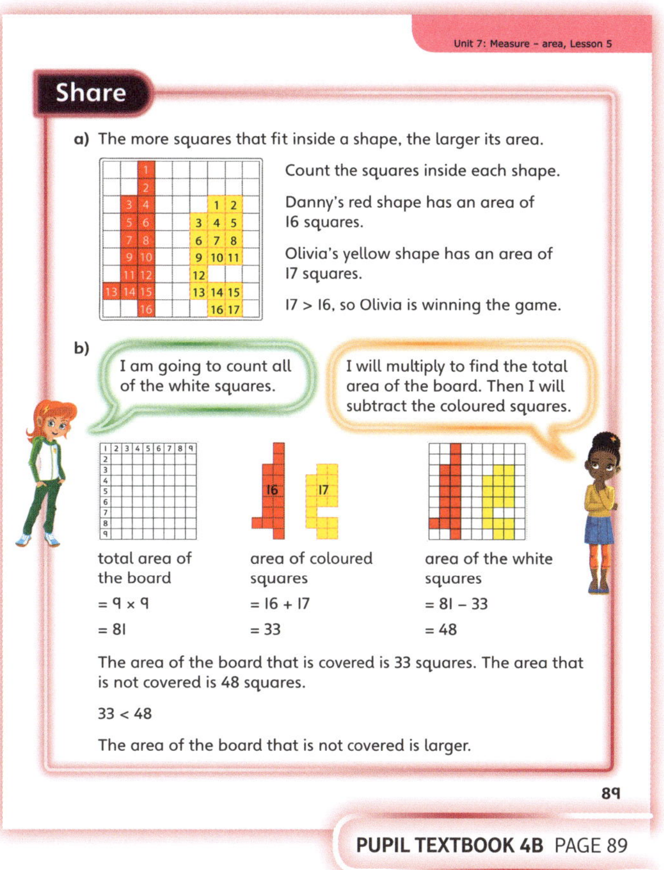

PUPIL TEXTBOOK 4B PAGE 89

Think together

WAYS OF WORKING Whole class teacher led (I do, We do, You do)

ASK

- Question **1**: *What is the only way to compare areas accurately?*
- Question **2**: *Are the squares inside the middle of the letter a part of its area? Why or why not? What steps do you need to go through to put these letters into the correct order of size?*

IN FOCUS Questions **1** and **2** take children through the steps needed to compare the area of rectilinear shapes. First they should find the areas by counting squares. Then they should compare the values to see which is larger. In question **2**, children are expected to use their comparison to write the areas in order.

STRENGTHEN For questions **1** and **2**, give children manipulatives to place on the shapes (one for each square). They should count these and place each set of manipulatives in a line. The longer line shows a concrete representation of the greater area. Move on from this to examples where children simply locate the values on a number line. Ask: *Which number is larger? How do you know?*

DEEPEN In question **3**, discuss the mistake Astrid has made. Ask: *Is this a common mistake to make?* Remind children that we all make visual judgements about measurements all the time. Ask: *What is the only way to make sure you compare the areas accurately?* Ensure children can also explain how to find the area of both shapes given that there are no visible squares inside them.

ASSESSMENT CHECKPOINT Children should understand that they cannot rely on visual cues to compare shapes, and that the only accurate way is to find their areas and then compare the values.

ANSWERS

Question **1**: Area of A = 18 squares
Area of B = 19 squares
19 > 18
Shape B has the larger area.

Question **2** a): Letter a = 23 squares
Letter s = 21 squares
Letter t = 17 squares

Question **2** b): t, s, a

Question **3**: Flo's rectangle has a greater area than the purple shape. Children should mention the fact that the only way to compare areas is to count the number of squares and then compare these numbers. A shape can be taller and/or wider, but still have a smaller area.

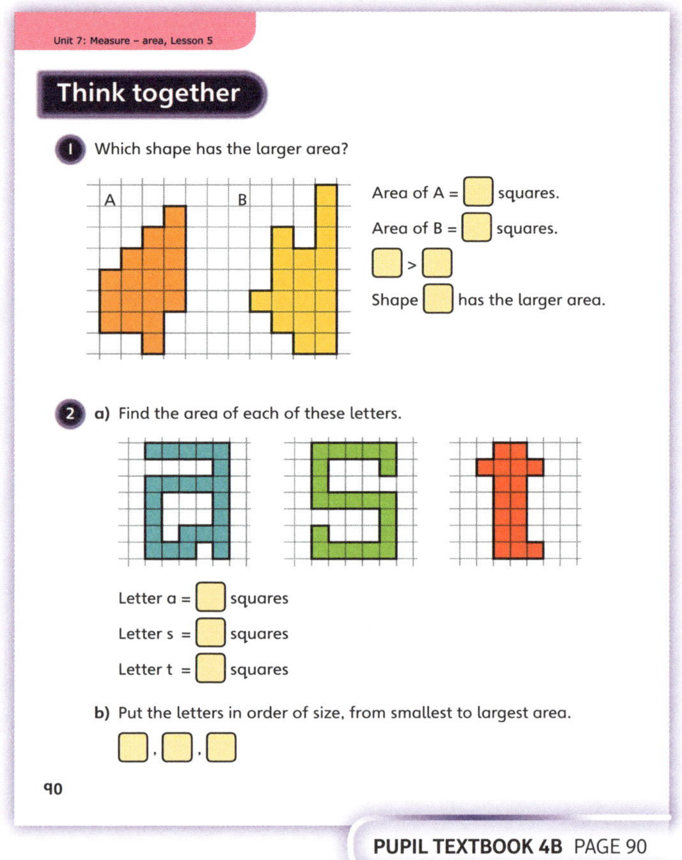

PUPIL TEXTBOOK 4B PAGE 90

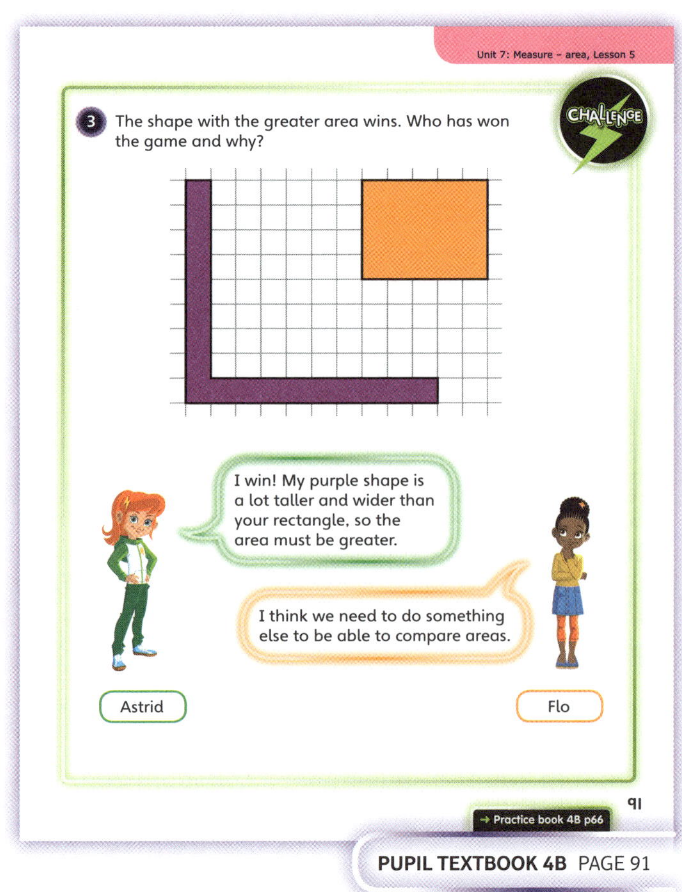

PUPIL TEXTBOOK 4B PAGE 91

Practice

WAYS OF WORKING Independent thinking

IN FOCUS Question **3** shows a picture sequence so it may be interesting to ask children what they notice. Direct their attention to what is happening to the left-hand shapes as the sequence progresses, then to the right-hand shapes. Ask: *How many squares are added to this shape each time? Where have they been drawn?* When the sequence starts, the left-hand shape in each pair has the greater area, but this changes at part c) (same area) and reverses at part d). Ask: *If you continued the sequence, would the shape on the left ever be larger again?*

STRENGTHEN To help children count accurately in the larger shapes, they could write the total of each line as they go, then add the lines to find the overall area of the shape.

DEEPEN Question **4** involves reasoning about the relationship between the length and width of a shape and its area. When two shapes are the same shape (such as two rectangles, two triangles), the taller, wider shape always has the greater area, but this is not true for all pairs of shapes. For example, a 4 cm wide × 3 cm tall triangle has an area of 6 cm^2 but a 3 cm wide × 2·5 cm tall rectangle has an area of 7·5 cm^2. Ask: *Can you draw a tall wide shape that has a smaller area than a shorter, narrower one?*

ASSESSMENT CHECKPOINT Children should be working confidently to find areas by counting squares and then comparing and ordering those areas. Children should be able to use reasoning to explain their answers and make generalisations.

ANSWERS Answers for the **Practice** part of the lesson appear in the separate **Practice and Reflect answer guide**.

Reflect

WAYS OF WORKING Independent thinking

IN FOCUS This question asks children to explain how they could compare the area of two shapes. Their answers should explain that they need to calculate both areas first (by counting squares) and then compare these areas, with the larger number denoting the larger area.

ASSESSMENT CHECKPOINT Look for children who describe a clear method, including the two steps required (calculating and then comparing).

ANSWERS Answers for the **Reflect** part of the lesson appear in the separate **Practice and Reflect answer guide**.

After the lesson

- How confident are children when comparing the area of rectilinear shapes?
- Are children now less reliant on comparison by sight and do they understand why it is important to measure before comparing the size of shapes?

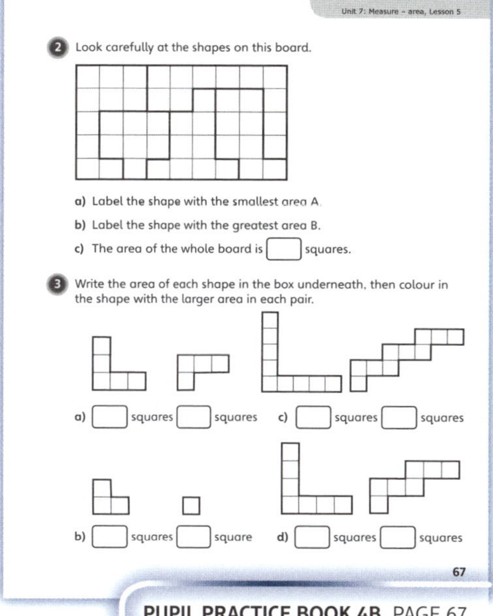

PUPIL PRACTICE BOOK 4B PAGE 66

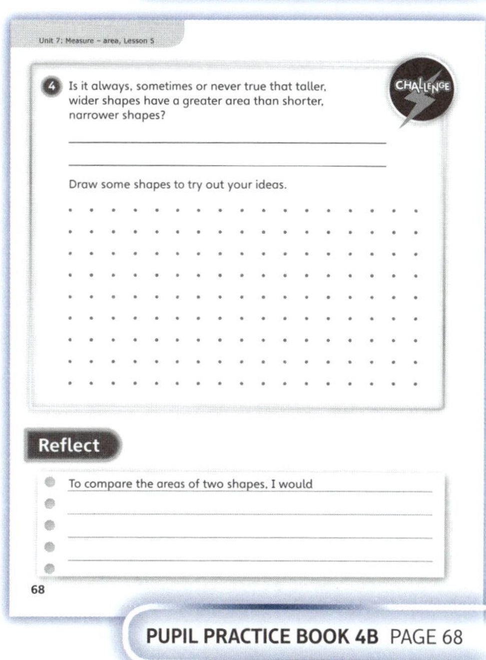

PUPIL PRACTICE BOOK 4B PAGE 67

PUPIL PRACTICE BOOK 4B PAGE 68

End of unit check

Don't forget the *Power Maths* unit assessment grid on p26.

WAYS OF WORKING Group work adult led

IN FOCUS Question **1** assesses children's ability to define the area of a shape.

Question **2** assesses children's ability to find the area of a rectangle drawn on squared paper by counting squares within it.

Question **3** assesses children's ability to find the area of solid squares and rectangles drawn on squared paper by using the squares around them to work out the number of squares within them.

Question **4** assesses children's ability to compare the area of rectilinear shapes.

Question **5** assesses children's ability to make different shapes with a given area.

Question **6** is a SATs-style question, assessing children's ability to identify a shaded area as the difference of two products.

ANSWERS AND COMMENTARY Children who have mastered the concepts in this unit will be able to define the term 'area' and calculate the area of rectilinear shapes by counting the squares within them. They will be able to apply their knowledge to find solutions involving solid shapes drawn on grids, comparing areas and making different shapes with a given area.

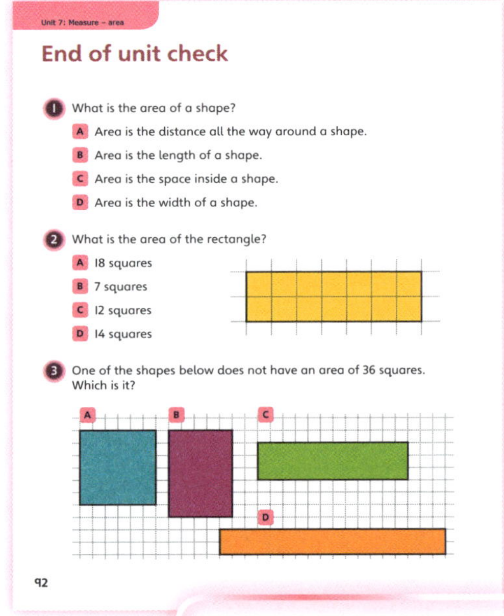

PUPIL TEXTBOOK 4B PAGE 92

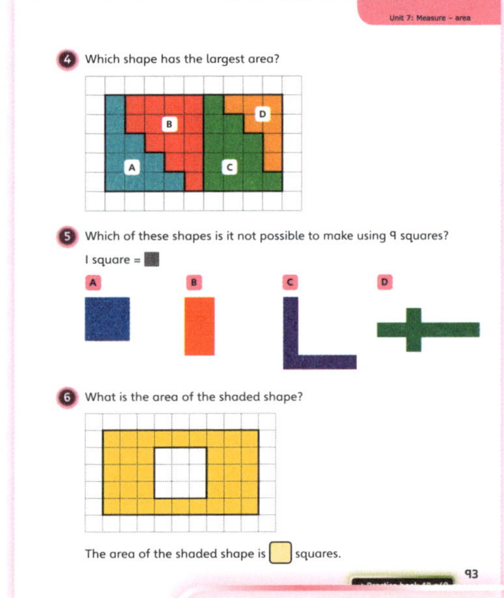

PUPIL TEXTBOOK 4B PAGE 93

Q	A	WRONG ANSWERS AND MISCONCEPTIONS	STRENGTHENING UNDERSTANDING
1	C	B or D suggest children link area with one dimensional measurements (whether a shape is longer or wider).	Some of the misconceptions children may have about area will be a result of not understanding what the area of a shape is. Strengthen their understanding of the general concept of area by providing opportunities to use concrete resources (such as flat plastic squares, squares cut out of paper, acetate squared grids) to measure the number of squares that fit inside a shape. Allow them to draw rectilinear shapes on squared paper, then cover the shapes with coloured paper squares.
2	D	A is the perimeter; B suggests children counted the columns; C suggests inaccurate counting.	
3	B	A, C and D suggest children are unsure how to use the squares around a solid shape to help derive its area.	
4	B	A, C and D suggest children have either miscounted squares or compared areas incorrectly.	
5	B	A, C or D suggest children have not visualised the possible shapes correctly.	
6	36 squares	45 squares is the area of the shaded rectangle without removing the inner white area.	

My journal

WAYS OF WORKING Independent thinking

ANSWERS AND COMMENTARY Each shape should be drawn on the squared paper provided and should have an area of 12 squares.

Possible answers are:
- rectangles with dimensions 1×12, 4×3 or 3×4.
- various other rectilinear shapes (not including squares as it is not possible to make a square using 12 squares).

To help children explain how to choose their measurements, ask:
- *Draw a rectangle. Does it have an area of 12 squares? If not, how could you adjust it so that it does? If so, how could you adjust it so that it still has an area of 12 squares?*
- *Is it possible to draw a square with an area of 12 squares? How do you know?*

Power check

WAYS OF WORKING Independent thinking

ASK

- *What did you know about area before you began this unit?*
- *What do you know now?*
- *Do you think you would be able to find the area of any rectilinear shape drawn on squared paper on your own?*

Power play

WAYS OF WORKING Pair work or small groups

IN FOCUS Use this Power play to see if children can explore combinations of the given rectilinear shapes to make two rectangles with different areas. Children should be able to use their knowledge of the properties of rectangles to derive both shapes before calculating their areas.

ANSWERS AND COMMENTARY

a) The two rectangles formed should have the dimensions 3×7 and 5×4 and look as follows: 

b) The areas of the chocolate bars are 21 squares and 20 squares.

c) Children should mention the fact that the bar containing 21 squares has a larger area (whether they would choose this bar depends on whether or not they like chocolate!).

Completing the Power play shows that they understand the properties of rectangles and can find their area by counting the squares. If they are unable to complete the question, check whether they have simply been unable to make the rectangles or, having made them, are unsure how to calculate or compare the areas.

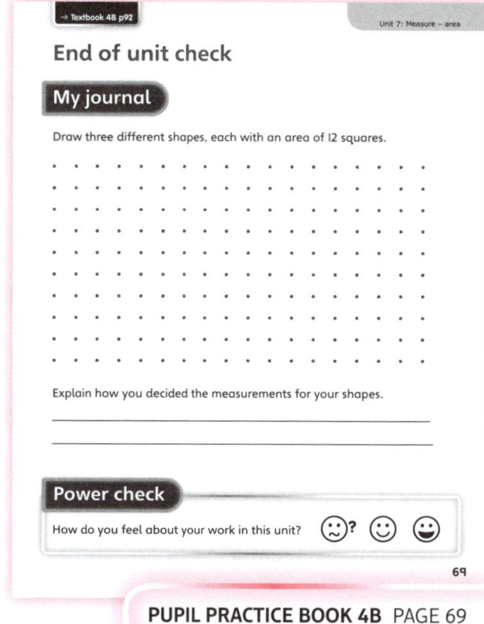

PUPIL PRACTICE BOOK 4B PAGE 69

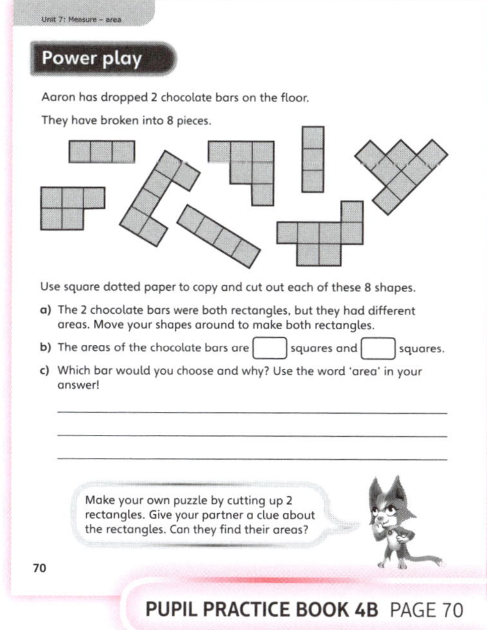

PUPIL PRACTICE BOOK 4B PAGE 70

After the unit ⏸

- What will you do differently next time you teach this unit?
- How did children respond to the materials and approaches used? Were they adequately (or excessively) challenged by them?

Strengthen and **Deepen** activities for this unit can be found in the *Power Maths* online subscription.

Unit 8
Fractions ❶

WHY THIS UNIT IS IMPORTANT

This unit introduces the concept of hundredths for the first time. It builds on children's knowledge of tenths from previous learning and provides opportunity for children to apply this learning to hundredths. Children will need to use their previous understanding of fractions and apply this to fractions that are greater than 1.

At first, children will explore the relationship between tenths and hundredths using models to support understanding. They will then move on to equivalent fractions using fraction strips and a fraction wall to aid understanding. Following this, children will further develop their understanding of how to simplify fractions. Finally, children will look at fractions that are greater than 1 and explore what happens when a fractional number is more than a whole.

WHERE THIS UNIT FITS

➜ Unit 7: Measure – area

➜ **Unit 8: Fractions (1)**

➜ Unit 9: Fractions (2)

This unit builds on work done in Year 3 on fractions. It introduces children to hundredths and then develops their understanding of equivalent fractions, before introducing them to fractions greater than 1 in the form of mixed numbers and improper fractions. The next unit builds on these concepts to calculate with fractions.

Before they start this unit, it is expected that children:

• know what whole and part mean
• know how to find simple equivalent fractions
• can use a number line.

ASSESSING MASTERY

Children who have mastered this unit will recognise the link between tenths and hundredths (for example, that 1 tenth is equal to 10 hundredths). They will be able to use strip diagrams to show equivalent fractions and will be able to simplify fractions to their simplest form. Children will also be able to count beyond 1 using a number line and understand that fractions greater than 1 can be written in different ways.

COMMON MISCONCEPTIONS	STRENGTHENING UNDERSTANDING	GOING DEEPER
Children may not be able to see the relationship between tenths and hundredths.	Provide children with a hundredths grid and counters. Allow children to explore hundredths and see what $\frac{1}{100}$ looks like.	Ask children to give their answers in hundredths, tenths or a combination of both (where applicable).
Children may not be able to see that fractions are not equivalent when the numerators are the same, but the denominators are different.	Provide children with fraction strips so that they can directly compare fractions with the same and different numerators and denominators.	Ask children to provide more than one equivalent fraction. Challenge them to draw their equivalent fractions in ways other than using strip diagrams, for example on number lines.

Unit 8: Fractions ❶

WAYS OF WORKING

Use these pages to introduce the unit focus to children. Use the characters to discuss concepts and phrases that children have not heard before. Ensure children understand what is meant by equivalent fractions.

STRUCTURES AND REPRESENTATIONS

Hundredths grid: This will allow children to see the connection between tenths and hundredths.

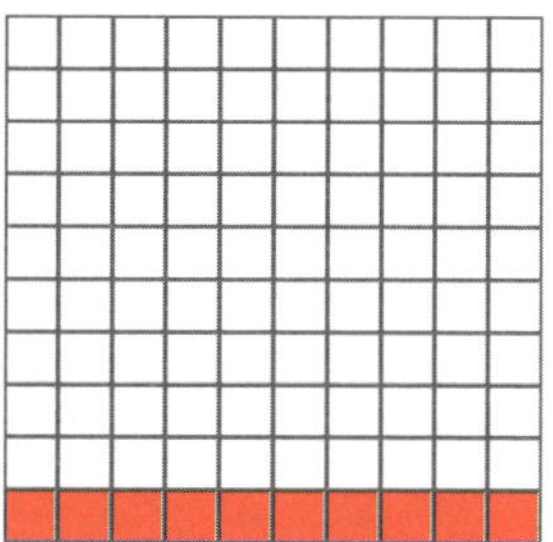

Part-whole model: This will allow children to visualise how hundredths are made up of tenths and hundredths, and how mixed numbers are made of wholes and fractions.

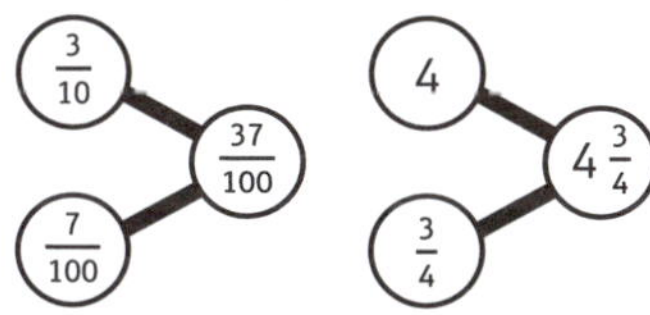

Fraction strips and fraction wall: These will allow children to visualise fractions and identify equivalent fractions.

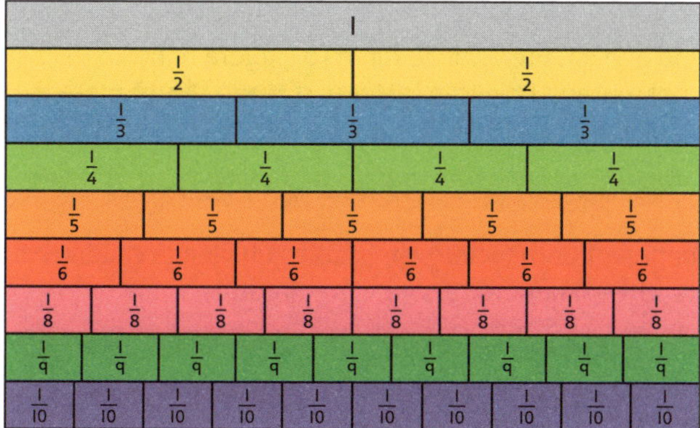

Number line: This will allow children to count in fractions, on number lines showing mixed numbers or improper fractions.

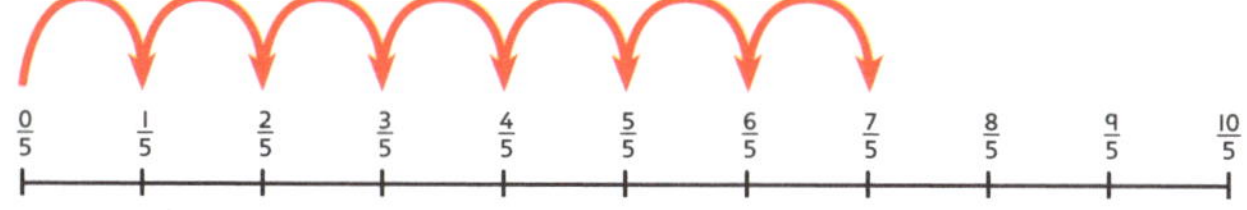

KEY LANGUAGE

There is some key language that children will need to know as part of the learning in this unit:

➜ tenth, hundredth

➜ equivalent fraction

➜ improper fraction, mixed number

➜ simplify, simplest fraction

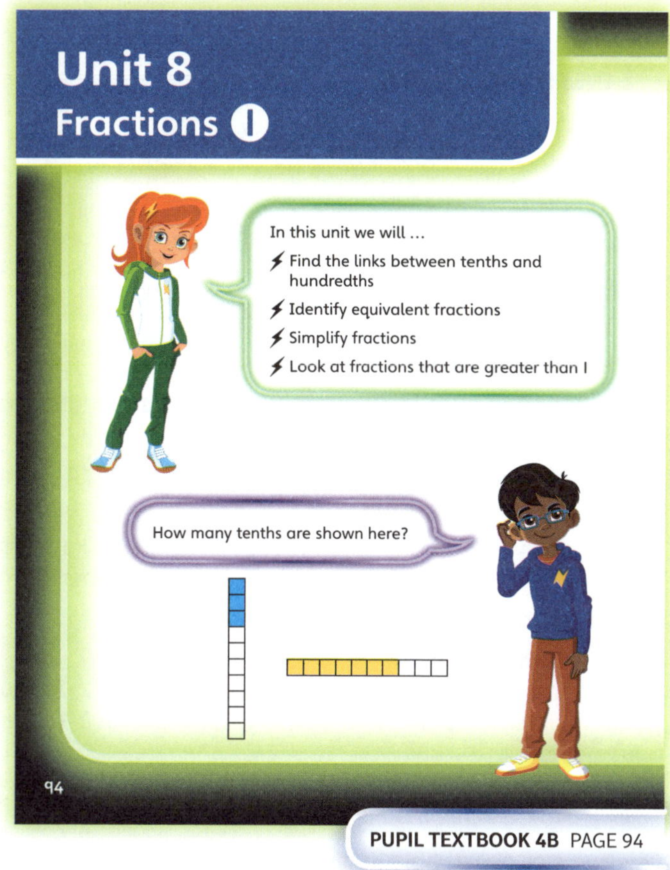

PUPIL TEXTBOOK 4B PAGE 94

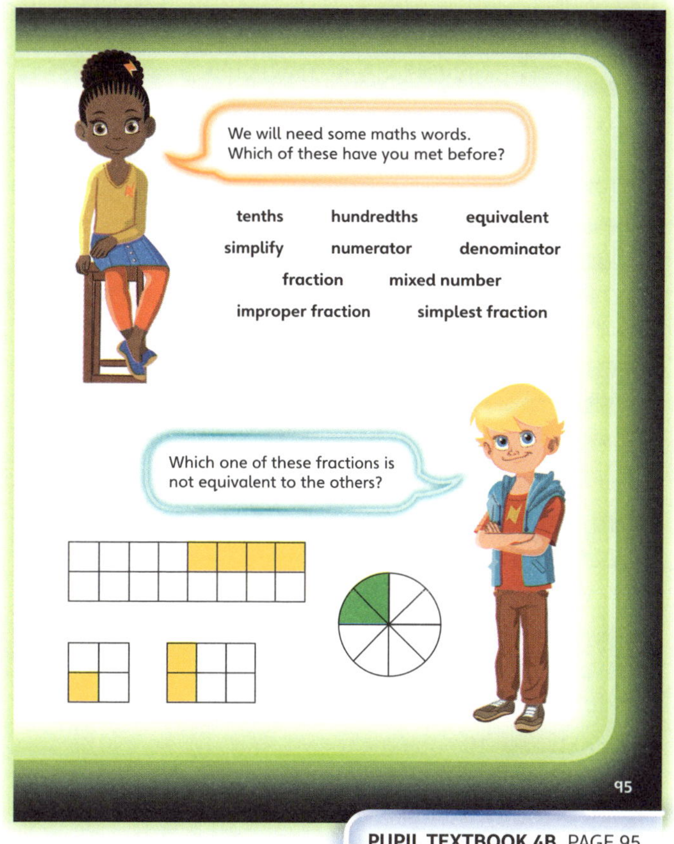

PUPIL TEXTBOOK 4B PAGE 95

Tenths and hundredths 

Learning focus

In this lesson, children will learn how to find hundredths as fractions. They will also recognise that 1 tenth is the same as 10 hundredths.

Small steps

→ Previous step: Comparing area
→ **This step: Tenths and hundredths (1)**
→ Next step: Tenths and hundredths (2)

NATIONAL CURRICULUM LINKS

Year 4 Number – Fractions (Including Decimals)

Count up and down in hundredths; recognise that hundredths arise when dividing an object by one hundred and dividing tenths by ten.

ASSESSING MASTERY

Children can visualise what a hundredth looks like and understand the connection between tenths and hundredths, and that $\frac{1}{10}$ is equivalent to $\frac{10}{100}$. They can show this relationship through pictorial methods, such as hundredths grids.

COMMON MISCONCEPTIONS

Children may not be able to visualise what a hundredth is and what a tenth is. Ask:
- *How many squares are there in total?* (100) *What would the value of one square be?* (1 out of 100) *Can you write this as a fraction?*
- *If 10 squares were covered, what would this be as a fraction?* ($\frac{10}{100}$) *Is there another way to write this?* ($\frac{1}{10}$) *This shows that one of the ten rows or columns has been covered.*

Children may not be able to see the relationship between tenths and hundredths (for example, that $\frac{25}{100}$ can be seen as $\frac{2}{10}$ and $\frac{5}{100}$). Ask:
- *Show $\frac{25}{100}$ on the grid. How many full rows are there? What fraction is this? How many are left over? What fraction is this?*

STRENGTHENING UNDERSTANDING

To strengthen understanding, provide children with a hundredths grid and counters. Allow them to explore hundredths and see what $\frac{1}{100}$ looks like. Ask children how many counters will fit on a full row and how this shows $\frac{1}{10}$. Extend this to more than one row being filled.

GOING DEEPER

Ask children to give their answers in hundredths, tenths or a combination of both (where applicable). Challenge children to create their own sentences to describe the fraction (for example, 2 tenths + 4 hundredths = 24 hundredths).

KEY LANGUAGE

In lesson: fraction, tenth, **hundredth**

Other language to be used by the teacher: numerator, denominator, part, whole, equivalent, simplify

STRUCTURES AND REPRESENTATIONS

hundredths grid

RESOURCES

Mandatory: counters, large hundredths grid

Optional: base 10 equipment

 In the eTextbook of this lesson, you will find interactive links to a selection of teaching tools.

Before you teach 

- Are children able to write fractions?
- Do children know what the numerator and denominator are?
- What resources can you use so that children can visualise hundredths?

Discover

WAYS OF WORKING Pair work

ASK

- Question ❶ a): *How many squares are on the board in total? What would the value of one square be? (1 out of 100.) Can you write this as a fraction? How many squares has Danny covered?*
- Question ❶ b): *How many more counters does Danny have than Lexi? Who has covered a larger fraction of the board? How do you know?*

IN FOCUS Question ❶ a) should allow children to see that the game is being played on a 10×10 grid and that each square is worth one hundredth. Lexi's counters clearly show that $\frac{10}{100}$ have been covered and this is a good starting point for the lesson as it allows children to see hundredths as well as tenths.

PRACTICAL TIPS Use hundredths grids and counters for children to recreate the board. This will also provide a structure for further questions to deepen and support understanding. Use the boards to show the equivalence of $\frac{10}{100}$ and $\frac{1}{10}$ by looking at rows and columns.

ANSWERS

Question ❶ a): Each square is worth $\frac{1}{100}$ or one hundredth.

Danny has covered $\frac{17}{100}$ of the board.

Question ❶ b): Lexi has covered $\frac{10}{100}$ or $\frac{1}{10}$ of the board.

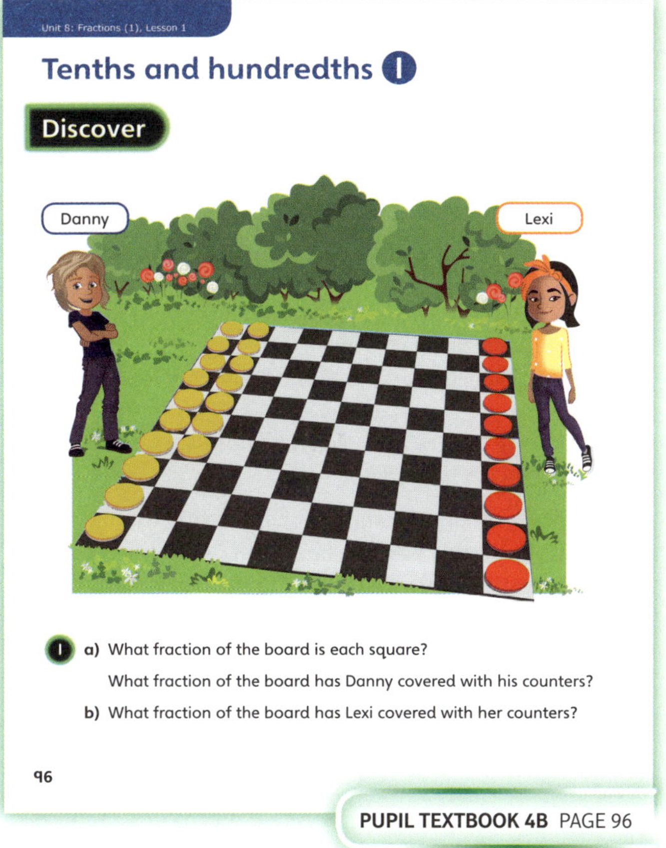

PUPIL TEXTBOOK 4B PAGE 96

Share

WAYS OF WORKING Whole class teacher led

ASK

- Question ❶: *What can you say about each row in the 10×10 board game?*
- Question ❶ a): *What has Sparks noticed about the numerator and the denominator? Can you simplify Danny's fraction? Why not? Can you write Danny's fraction as tenths and hundredths?*

IN FOCUS Question ❶ b) shows children that $\frac{10}{100}$ is the same as $\frac{1}{10}$. Encourage children to look carefully at the numerator and denominator and say what they notice. This links back to their learning on equivalent fractions from Year 3, Unit 10. Explain that this process is called simplifying fractions.

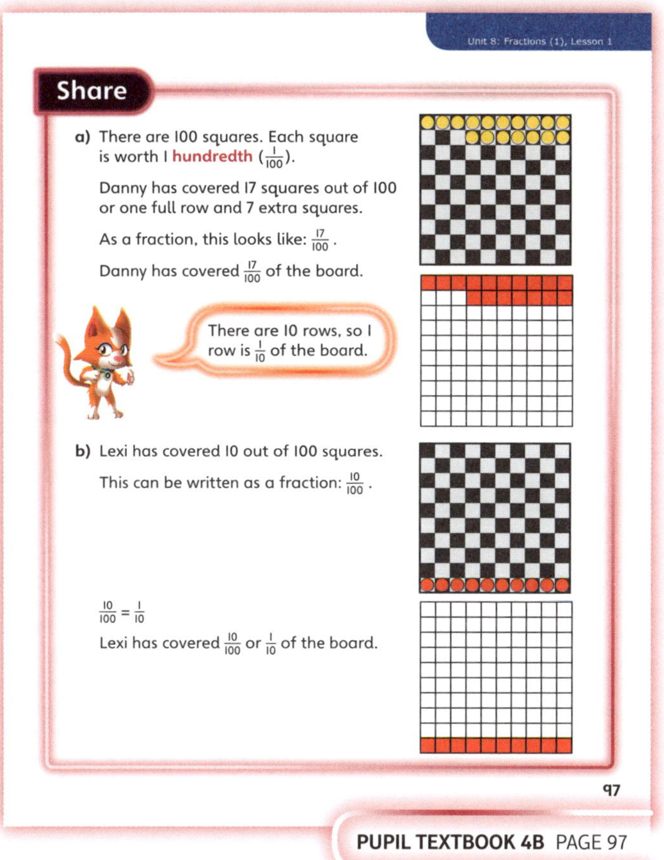

PUPIL TEXTBOOK 4B PAGE 97

Think together

WAYS OF WORKING Whole class teacher led (I do, We do, You do)

ASK

- Question **1** a): *How many tenths have both players covered? What is the same about each player's counters? What is different?*
- Question **1** b): *How many full rows has Ambika filled? How many counters are in the final row?*
- Question **3** b): *Can you write the number of full rows as tenths?*
- Question **3** c): *How many full rows has Richard filled? Can you write this as hundredths?*

IN FOCUS Questions **1** and **2** give children the opportunity to apply what has been taught in the **Share** section. Questions **1** b) and **2** b) show fractions that can be simplified. In question **2** c), children can apply their knowledge of tenths and hundredths from a hundredths grid to a line of 10 squares with 9 shaded; encourage children to give their answer in tenths and hundredths.

STRENGTHEN Children can use a hundredths grid and counters to represent hundredths and tenths. Creating the boards in the classroom will help cement that 25 counters on a board of 100 squares is 25 hundredths.

DEEPEN Encourage children to give their answers in hundredths, tenths (if possible) and a combination of tenths and hundredths. Ask children to explain why an answer can be given in tenths and hundredths. Encourage them to explore which fractions can only be given in hundredths and which fractions can only be given in tenths.

ASSESSMENT CHECKPOINT Use questions **1** and **2** to assess whether children can express hundredths as fractions. Questions **1** b) and **2** b) can also be used to see which children can simplify and use knowledge of multiples. Check whether children give their answers in hundredths, tenths or a combination of hundredths and tenths.

ANSWERS

Question **1** a): Max has covered 23 squares.
Max has covered $\frac{23}{100}$ of the board.

Question **1** b): Ambika has covered 25 squares.
Ambika has covered $\frac{25}{100}$ of the board.

Question **1** c): Together they have covered $\frac{48}{100}$ of the board.

Question **2** a): $\frac{33}{100}$

Question **2** b): $\frac{70}{100}$ or $\frac{7}{10}$

Question **2** c): $\frac{90}{100}$ or $\frac{9}{10}$

Question **3** a): Children's counters on a grid should show $\frac{34}{100}$ and $\frac{60}{100}$.

Question **3** b): Amelia scored $\frac{3}{10}$ and $\frac{4}{100}$.

Question **3** c): Richard scored $\frac{60}{100}$.

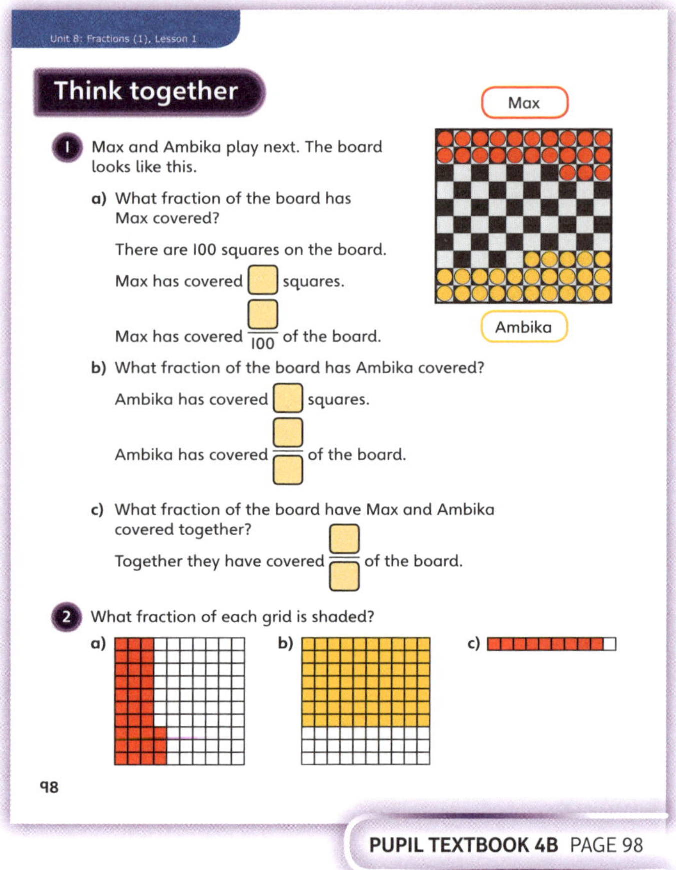

PUPIL TEXTBOOK 4B PAGE 98

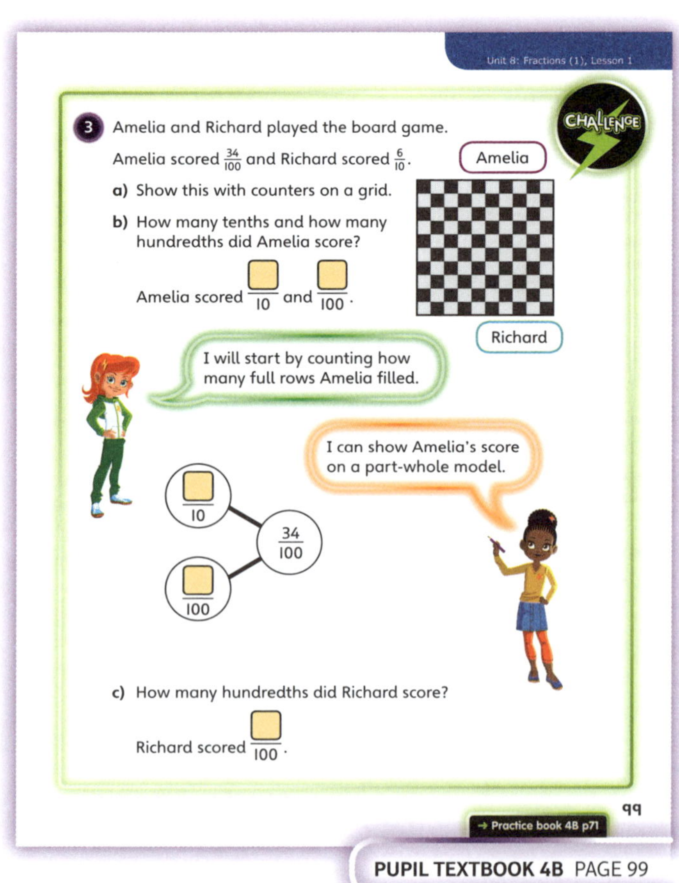

PUPIL TEXTBOOK 4B PAGE 99

Practice

WAYS OF WORKING Independent thinking

IN FOCUS Question ❶ progresses from hundredths grids showing tenths to hundredths grids showing tenths and hundredths, and finally to a tenths grid with 7 squares shaded, to prompt children to identify hundredths in a tenths image. Question ❷ develops children's skills by asking them to represent a fraction on a grid, instead of writing down the fraction shown on a grid. This question also asks children to write the fraction that is not shaded.

STRENGTHEN Provide children with base 10 equipment in 1s, 10s and 100s. They can use this to represent fractions. This will also help to reinforce the relationship between tenths and hundredths.

DEEPEN Allow children the opportunity to give their answers in tenths, hundredths and a combination of the two. Ask children to explain the similarities and differences between tenths and hundredths.

THINK DIFFERENTLY Question ❸ provides children with an opportunity to show their understanding of the relationship between tenths and hundredths by explaining why all three children are correct. They need to give clear reasons in their answer.

ASSESSMENT CHECKPOINT Use question ❶ to assess whether children can express hundredths as fractions and have understood the connection between tenths and hundredths. Use question ❷ to assess whether they can represent a fraction on a grid. Check whether children can provide their answers in tenths (where applicable), hundredths and a combination of tenths and hundredths.

ANSWERS Answers for the **Practice** part of the lesson appear in the separate **Practice and Reflect answer guide**.

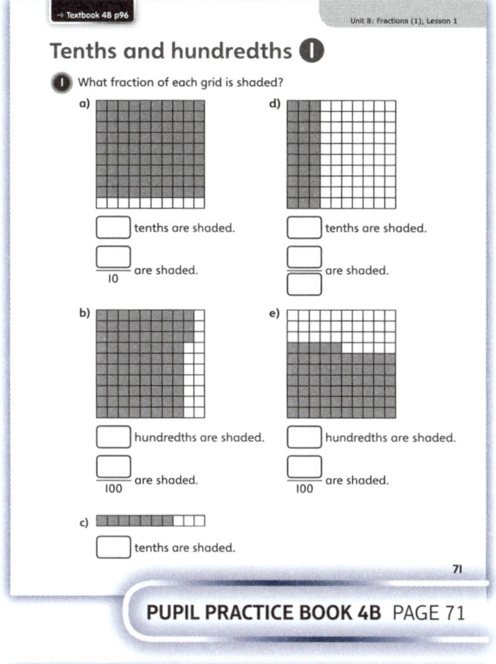

PUPIL PRACTICE BOOK 4B PAGE 71

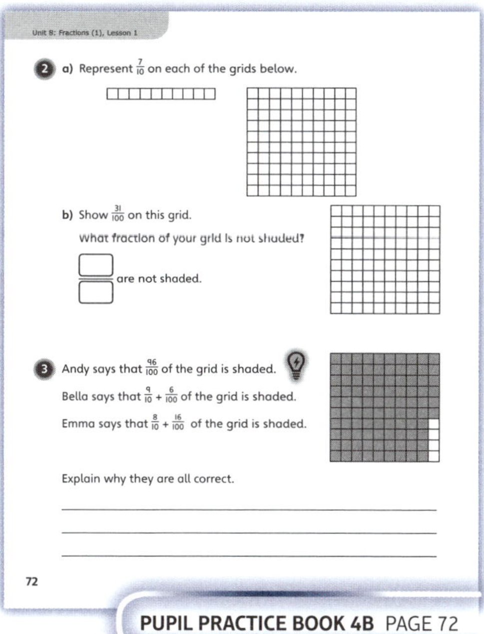

PUPIL PRACTICE BOOK 4B PAGE 72

Reflect

WAYS OF WORKING Pair work

IN FOCUS This question provides children with the opportunity to articulate what a tenth is and what a hundredth is. It also allows children to identify the similarities and differences between tenths and hundredths. This activity will identify those who still have misconceptions about how to identify tenths and hundredths and so would benefit from additional help.

ASSESSMENT CHECKPOINT Assess whether children can correctly define a tenth and a hundredth, and show the relationship between them.

ANSWERS Answers for the **Reflect** part of the lesson appear in the separate **Practice and Reflect answer guide**.

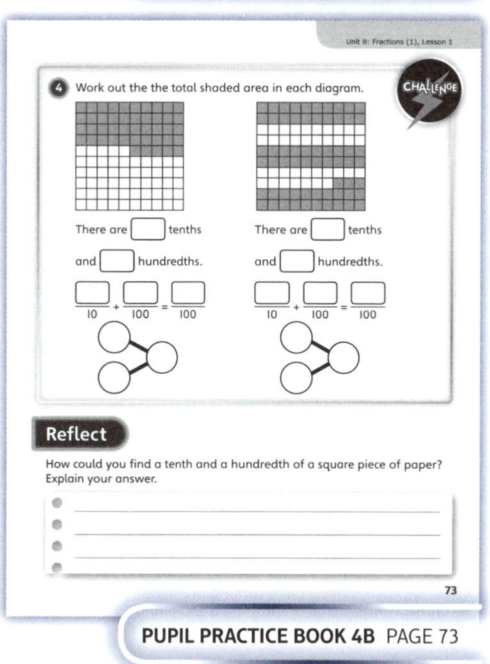

PUPIL PRACTICE BOOK 4B PAGE 73

After the lesson 

- Are children confident finding hundredths and tenths?
- Can children identify the relationship between tenths and hundredths?
- Did children need to use a hundredths grid and counters to help support understanding?

Tenths and hundredths ②

Learning focus

In this lesson, children will continue to see the link between tenths and hundredths using a number line.

Small steps

→ Previous step: Tenths and hundredths (1)
→ **This step: Tenths and hundredths (2)**
→ Next step: Equivalent fractions (1)

NATIONAL CURRICULUM LINKS

Year 4 Number – Fractions (Including Decimals)

Count up and down in hundredths; recognise that hundredths arise when dividing an object by one hundred and dividing tenths by ten.

ASSESSING MASTERY

Children can use a number line to identify fractions. They understand that a tenth is 10 times greater than a hundredth.

COMMON MISCONCEPTIONS

Children may not be able to see that the greater the denominator, the smaller the fraction. Ask:

- *Show $\frac{1}{10}$ on the grid. Show $\frac{1}{100}$ on the grid. Which is smaller?*

Children may not be able to see that a tenth is 10 times smaller than a hundredth. Ask:

- *If 10 pieces of cake out of a hundred were eaten what would that be as a fraction? ($\frac{10}{100}$) Is this similar to any other fractions you have seen before? ($\frac{1}{10}$) How many hundredths are in a tenth?*

STRENGTHENING UNDERSTANDING

Provide children with a fraction number line so they can count the fractions themselves.

GOING DEEPER

Ask children to give their fraction answers in different ways: on a fraction number line (blank or marked), on a hundredths square, using shapes and written as a fraction.

KEY LANGUAGE

In lesson: tenths, hundredths

Other language to be used by the teacher: fraction, numerator, denominator, equivalent, equal, greater

STRUCTURES AND REPRESENTATIONS

hundredths grid, fraction number line

RESOURCES

Optional: number lines, hundredths grid

 In the eTextbook of this lesson, you will find interactive links to a selection of teaching tools.

Before you teach 

- Do children know how to write a fraction?
- Can children use a number line?
- Can children visualise fractions?

Discover

 Pair work

ASK

- Question **1** a): *What does the entire wedding cake represent? What does one piece of the wedding cake represent as a fraction?*
- Question **1** b): *How many hundredths have been eaten?*

IN FOCUS Children should be able to see that once the cake has been cut into 100 pieces, each piece of cake is worth one hundredth. Question **1** b) allows children to see the link between tenths and hundredths and that 13 hundredths can be written as $\frac{13}{100}$ or $\frac{1}{10}$ and $\frac{3}{100}$.

PRACTICAL TIPS Provide children with a hundredths grid and counters so that they can physically create the question being asked. This will allow them to fully visualise the tenths and hundredths.

ANSWERS

Question **1** a): The cake can be cut into 10 equal rows and 10 equal columns.

Each piece is worth one hundredth (or $\frac{1}{100}$).

Question **1** b): $\frac{13}{100}$ has been eaten (or $\frac{1}{10}$ and $\frac{3}{100}$).

There is $\frac{87}{100}$ of the cake left (or $\frac{8}{10}$ and $\frac{7}{100}$).

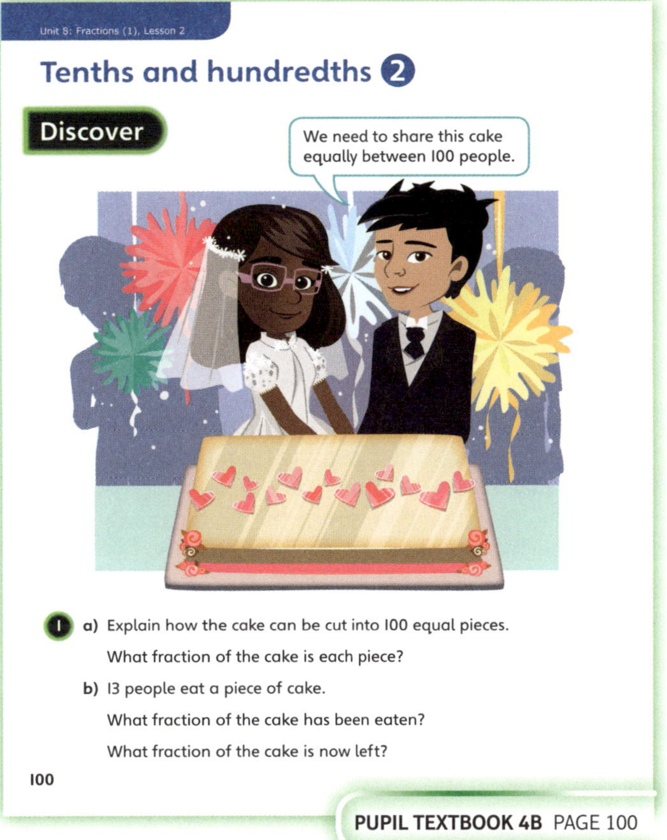

PUPIL TEXTBOOK 4B PAGE 100

Share

 Whole class teacher led

ASK

- Question **1** a): *What fraction of the wedding cake would a full row or column represent?*
- Question **1** b): *How can you write the 13 pieces of eaten cake as a fraction? Can you write it in different ways? How many tenths are there? How many hundredths are there?*

IN FOCUS Question **1** b) helps children to see $\frac{13}{100}$ and make the connection between tenths and hundredths.

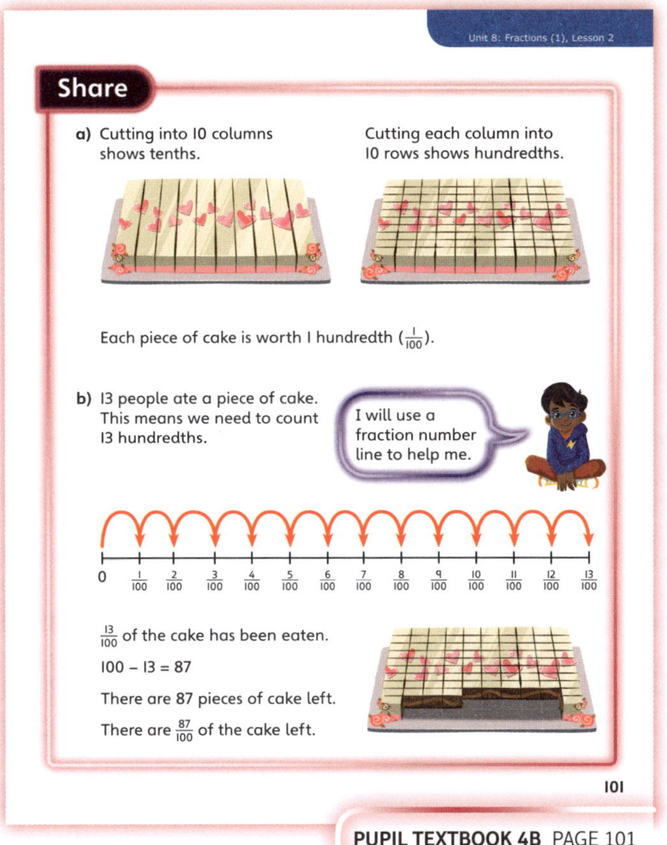

PUPIL TEXTBOOK 4B PAGE 101

Think together

Unit 8: Fractions (1), Lesson 2

Think together

WAYS OF WORKING Whole class teacher led (I do, We do, You do)

ASK

- Question **1**: *What steps is the number line going up in? How else can you write the fraction? How many tenths are in the fraction? What is the connection between tenths and hundredths?*
- Question **2**: *What fractions can you see on the number line? How can this help you?*

IN FOCUS Questions **1** and **2** provide children with an opportunity to apply their learning from the **Share** section to see if they can use a number line to work out the value of fractions. Question **3** extends this understanding of fractions on number lines to a real-life context.

STRENGTHEN Provide children with number lines so that they can count the fractions themselves.

DEEPEN Challenge children to represent and write fractions in multiple ways.

ASSESSMENT CHECKPOINT Use questions **1** and **2** to assess whether children can use a number line to work out missing fractions. Also, use question **1** to check whether they can count on and back on the number line to add and subtract small fractions.

ANSWERS

Question **1** a): $\frac{65}{100}$ was eaten by 11 pm (or $\frac{6}{10}$ and $\frac{5}{100}$).

Question **1** b): $\frac{69}{100}$ was eaten by midnight (or $\frac{6}{10}$ and $\frac{9}{100}$).

Question **1** c): $100 - 69 = 31$

There is $\frac{31}{100}$ of the cake left at midnight (or $\frac{3}{10}$ and $\frac{1}{100}$).

Question **2**: $\frac{4}{10}, \frac{5}{10}, \frac{6}{10}, \frac{7}{10}, \frac{8}{10}, \frac{9}{10}$.

$\frac{40}{100}, \frac{50}{100}, \frac{60}{100}, \frac{70}{100}, \frac{80}{100}, \frac{90}{100}$

Question **3** a):

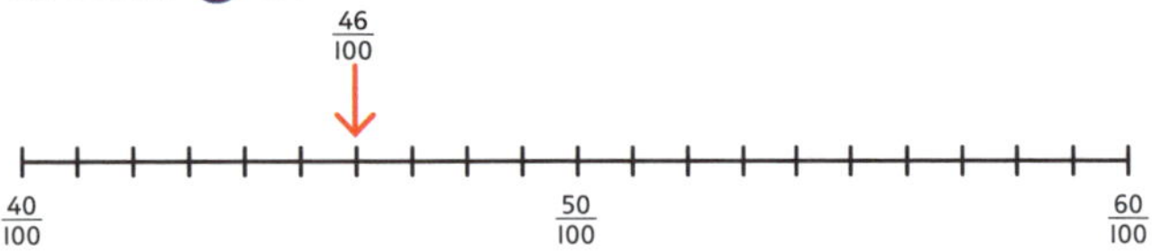

Question **3** b): The fraction of a pound that Emma and Jamie have spent is the same.

Emma's coins are $\frac{4}{10}$ and $\frac{6}{100}$. Jamie's coins are $\frac{46}{100}$.

Question **3** c): Emma and Jamie each have $\frac{54}{100}$ left.

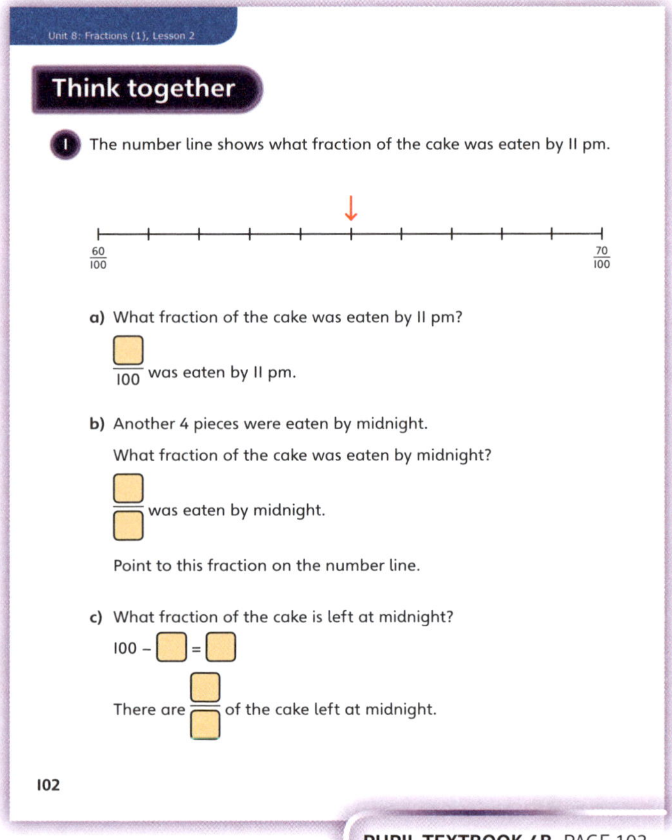

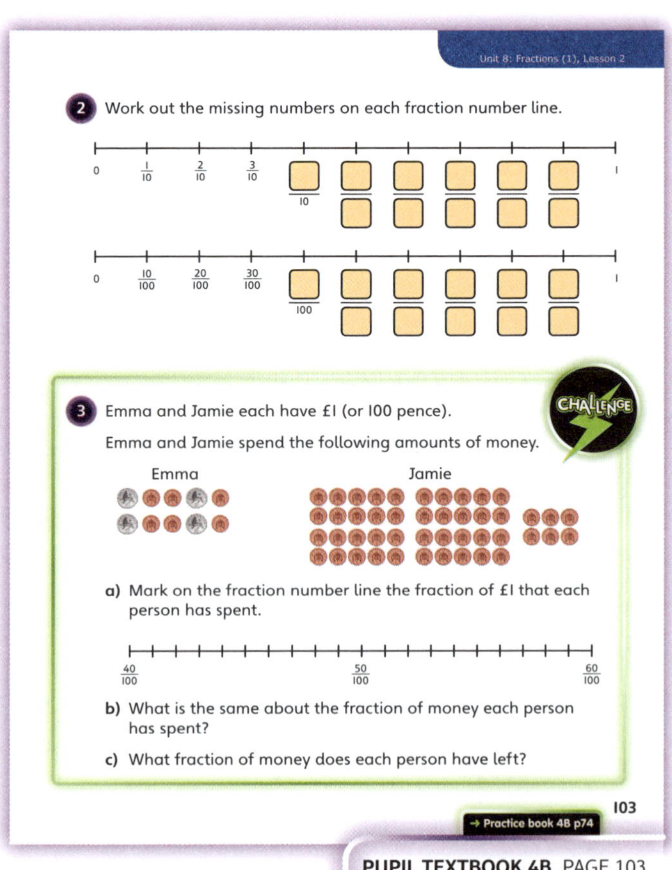

PUPIL TEXTBOOK 4B PAGE 102

PUPIL TEXTBOOK 4B PAGE 103

Practice

WAYS OF WORKING Independent thinking

IN FOCUS Question ❸ requires children to explain their reasoning on the link between hundredths and tenths. Question ❹ uses this understanding in abstract form, without the support of number lines, to convert between fractions in tenths and hundredths.

Question ❻ links the representational forms explored in the previous lesson with the number line work in this lesson. Children need to work out the fraction shown in each grid before marking their position on the number line.

STRENGTHEN If children are struggling to place fractions on the number line, first give them a number line using hundredths grids before moving on to number lines with numbers only.

DEEPEN Ask children to show the link between tenths and hundredths using number lines, hundredths grids and 10 grids.

ASSESSMENT CHECKPOINT Use question ❷ to assess whether children can identify fractions on a number line. Use question ❸ to assess whether they can explain the link between tenths and hundredths. Use question ❺ to check whether they can provide the answers in hundredths and tenths, as well as just hundredths.

ANSWERS Answers for the **Practice** part of the lesson appear in the separate **Practice and Reflect answer guide**.

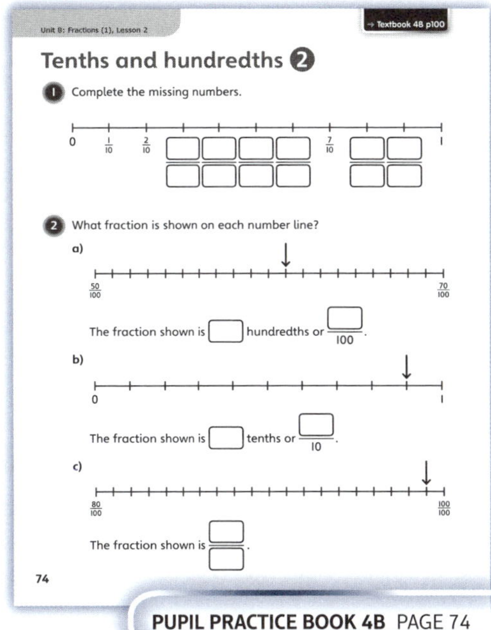

PUPIL PRACTICE BOOK 4B PAGE 74

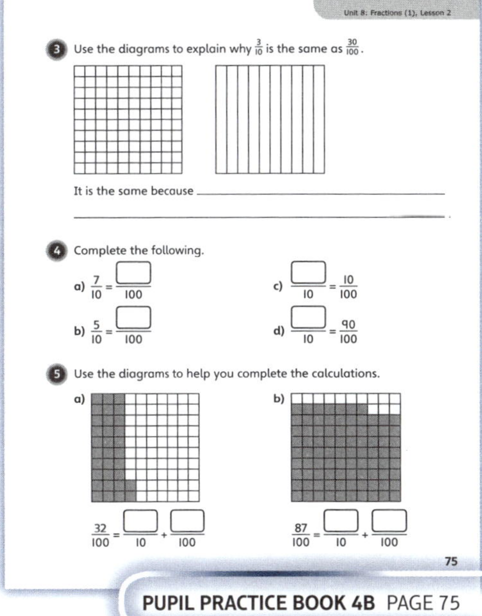

PUPIL PRACTICE BOOK 4B PAGE 75

Reflect

WAYS OF WORKING Pair work

IN FOCUS This activity provides children with the opportunity to articulate the difference between tenths and hundredths. It may also provide discussion about numerators and denominators and the size of fractions.

ASSESSMENT CHECKPOINT Check whether children can see the link between tenths and hundredths, and can articulate the difference between them.

ANSWERS Answers for the **Reflect** part of the lesson appear in the separate **Practice and Reflect answer guide**.

After the lesson 

- Can children identify fractions on a number line?
- Can they show their understanding of tenths and hundredths on number lines?
- Can children represent tenths and hundredths pictorially?

PUPIL PRACTICE BOOK 4B PAGE 76

Equivalent fractions ❶

Learning focus

In this lesson, children will use a fraction wall and fraction strips to identify equivalent fractions.

Small steps

→ Previous step: Tenths and hundredths (2)
→ **This step: Equivalent fractions (1)**
→ Next step: Equivalent fractions (2)

NATIONAL CURRICULUM LINKS

Year 4 Number – Fractions (Including Decimals)

Recognise and show, using diagrams, families of common equivalent fractions.

ASSESSING MASTERY

Children can use a fraction wall to identify and give equivalent fractions. Children begin to see the links between equivalent fractions and multiples.

COMMON MISCONCEPTIONS

Children may not be able to see that some fractions are not equivalent when the numerators are the same, but the denominators are different. Ask:

- *What is the same and what is different about these fractions?*
- *Show me $\frac{2}{3}$ on the fraction wall. Now show me $\frac{2}{4}$. What do you notice about the numerators? What do you notice about the denominators? What does this tell you? (For the same numerator, the greater the denominator, the smaller the number.)*

STRENGTHENING UNDERSTANDING

Provide children with fraction strips so that they can directly compare fractions with the same or different numerators and denominators. Alternatively, children could create their own fraction strips from blank strips, to build conceptual understanding. Using fraction strips should provide a link between fractions and multiples.

GOING DEEPER

Ask children to provide more than one equivalent fraction and to draw their fractions in other ways than using strip diagrams and fraction walls.

KEY LANGUAGE

In lesson: fraction wall, fraction strip, equivalent

Other language to be used by the teacher: fraction, numerator, denominator, greater than (>), less than (<), equal to (=)

STRUCTURES AND REPRESENTATIONS

fraction wall, fraction strips, 2D shapes

RESOURCES

Optional: large fraction wall, fraction strips

 In the eTextbook of this lesson, you will find interactive links to a selection of teaching tools.

Before you teach ❚❚

- Can children identify the numerator and the denominator?
- Can children use <, > and = signs correctly?
- Can children identify fractions from diagrams?

Discover

WAYS OF WORKING Pair work

ASK

- Question ① a): *What is the width of the fraction wall equal to? How many thirds are in the fraction? How many sixths are in the fraction? How can you tell that the fractions are the same? What link can you see between $\frac{1}{3}$ and $\frac{2}{6}$?*
- Question ① b): *What equivalent fractions can you see on the fraction wall?*

IN FOCUS The fraction wall enables children to see which fractions are the same size (equivalent).
In question ① a) they should be able to see that $\frac{1}{3}$ and $\frac{2}{6}$ are equivalent as they are the same size on the fraction wall. In question ① b) they will be able to see that $\frac{4}{6}$ and $\frac{3}{4}$ are not equivalent as they are different sizes on the fraction wall.

PRACTICAL TIPS Provide children with equal lengths of paper so they can make fraction strips and build their own fraction walls. They will then be able to directly compare fractions practically.

ANSWERS

Question ① a): $\frac{1}{3}$ and $\frac{2}{6}$ are the same size on the fraction wall.

$\frac{3}{9}$ is also equivalent to $\frac{1}{3}$.

Question ① b): Max is incorrect. $\frac{3}{4}$ is not equivalent to $\frac{4}{6}$.

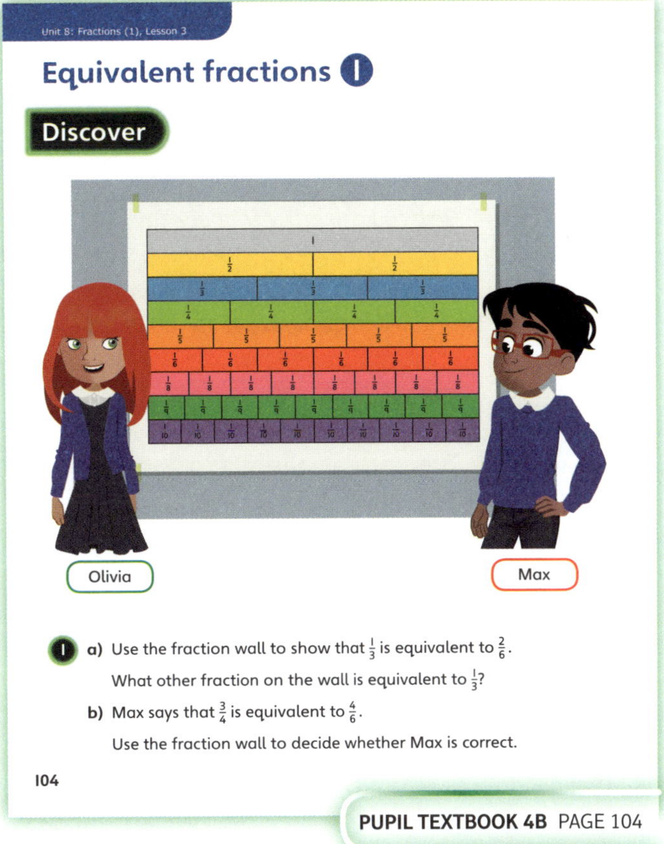

PUPIL TEXTBOOK 4B PAGE 104

Share

WAYS OF WORKING Whole class teacher led

ASK

- Question ① a): *How could you extend the fraction wall to find more equivalent fractions to $\frac{1}{3}$? Do you need to use a fraction wall to find equivalent fractions?*
- Question ① b): *What fraction is equivalent to $\frac{3}{4}$? What fraction is equivalent to $\frac{4}{6}$? How can you use the signs Flo has suggested?*

IN FOCUS Question ① a) enables children to see how to use a fraction wall to find equivalent fractions. Question ① b) starts by establishing whether fractions are equivalent and then introduces children to the concept of using a fraction wall to compare fractions.

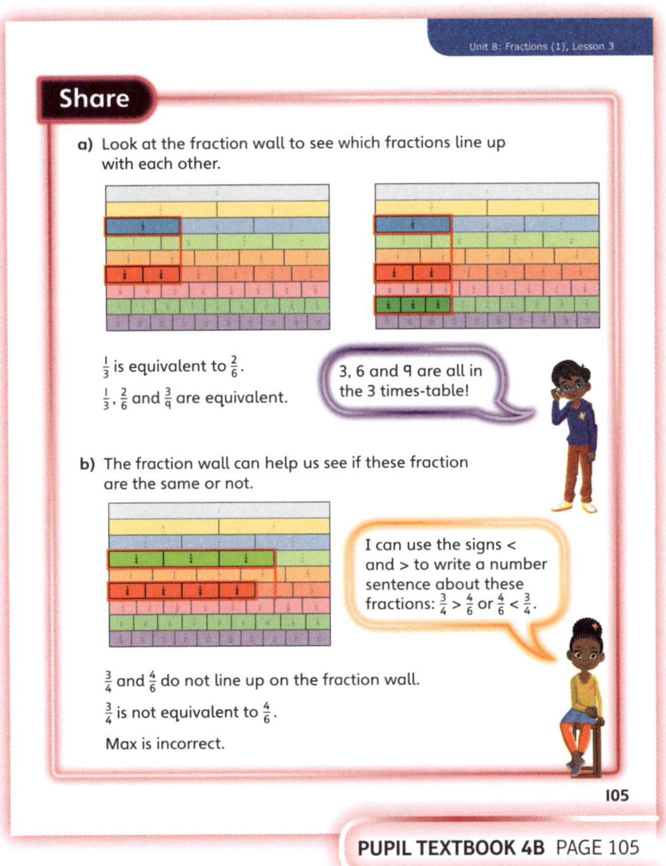

PUPIL TEXTBOOK 4B PAGE 105

Think together

WAYS OF WORKING Whole class teacher led (I do, We do, You do)

ASK

- Question **1** a): *How can you prove that one fraction is greater than the other?*
- Question **1** b): *What is the connection between $\frac{1}{5}$ and $\frac{2}{10}$?*
- Question **3**: *How can a fraction wall help you to compare fractions?*

IN FOCUS Question **1** provides children with the opportunity to offer clear reasons whether or not a statement is true. They can demonstrate conceptual understanding in their answers.

STRENGTHEN Instead of using the fraction wall, ask children to create fraction strips for the relevant fractions. They can then put the strips next to each other to compare fractions more easily.

DEEPEN Ask children to identify pairs of equivalent fractions and then to look for patterns in their fraction pairs.

ASSESSMENT CHECKPOINT Use questions **1** and **2** to assess whether children can identify equivalent fractions. Can children see the link between equivalent fractions and multiples?

ANSWERS

Question **1** a): This is false because $\frac{1}{2}$ does not line up with $\frac{2}{8}$. $\frac{1}{2}$ is equal to $\frac{4}{8}$.

Question **1** b): This is true because $\frac{2}{10}$ and $\frac{1}{5}$ line up. They are equivalent fractions.

Question **1** c): This is false because $\frac{4}{8}$ does not line up with $\frac{4}{10}$, so the fractions are not equivalent. The numerators are the same, but the denominators are different. $\frac{4}{8}$ is greater than $\frac{4}{10}$.

Question **2**: $\frac{2}{4}, \frac{3}{6}, \frac{4}{8}, \frac{5}{10}$

Question **3**: A and D, B and C.

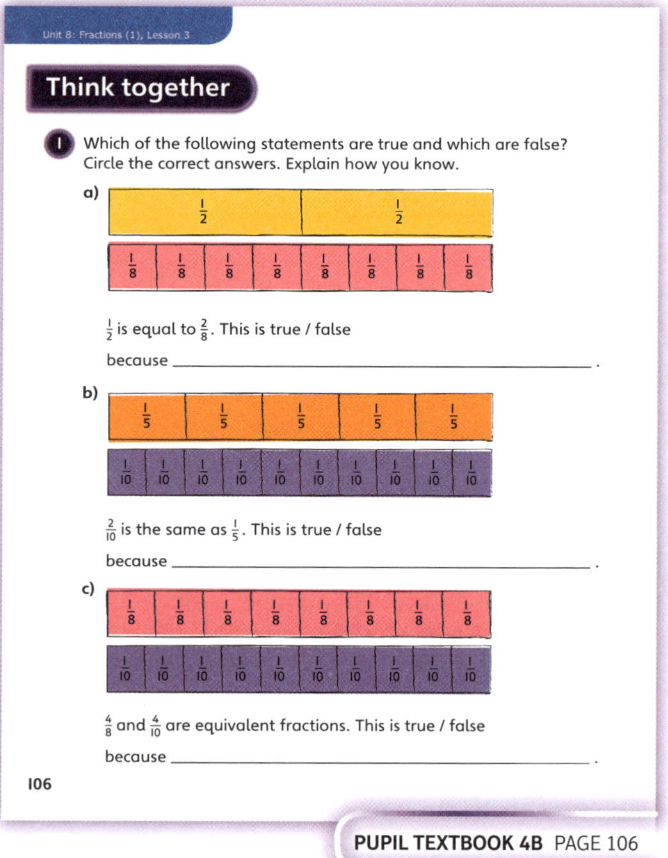

PUPIL TEXTBOOK 4B PAGE 106

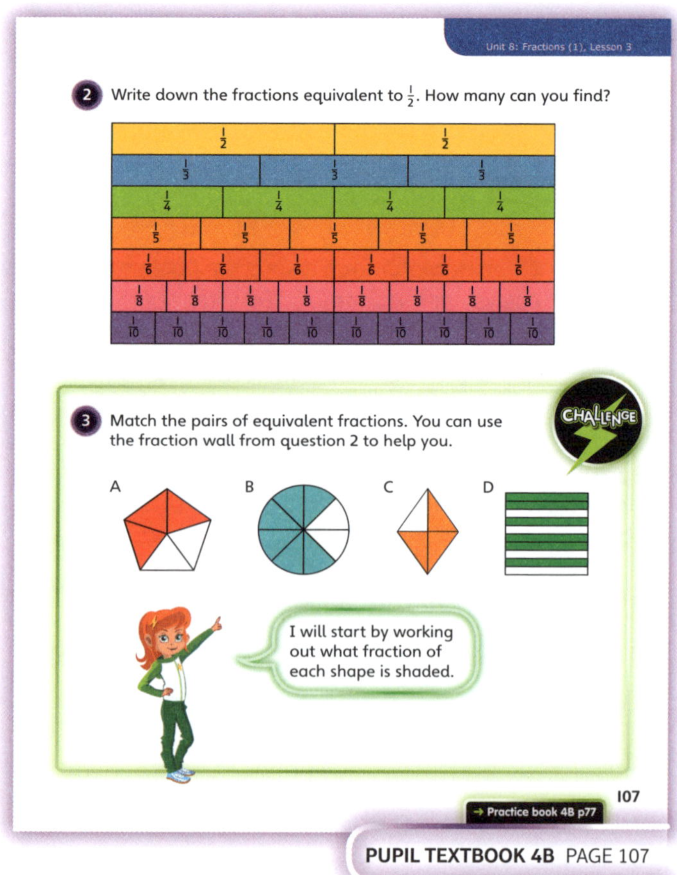

PUPIL TEXTBOOK 4B PAGE 107

Practice

WAYS OF WORKING Independent thinking

IN FOCUS Question **4** addresses the misconception some children have that if the numerators are the same, the fractions are the same size. This should reinforce the concept that when the numerators are the same, the fraction with the smaller denominator will be larger.

STRENGTHEN Give children a square sheet of paper and ask them to fold it in half and shade $\frac{1}{2}$. They can then fold the paper in half again and identify how many quarters make a half. They can continue folding to find other equivalent fractions.

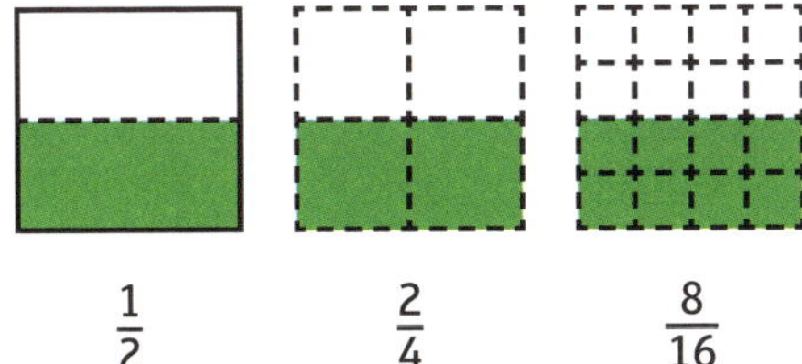

$$\frac{1}{2} \qquad \frac{2}{4} \qquad \frac{8}{16}$$

DEEPEN In question **4**, ask children to articulate a rule for comparing fractions with the same numerator. They could also find as many equivalent fractions to Lee and Zac's fractions as they can.

ASSESSMENT CHECKPOINT Use questions **1** and **2** to assess whether children can find equivalent fractions using fraction strips and whether they can explain why fractions are equivalent.

ANSWERS Answers for the **Practice** part of the lesson appear in the separate **Practice and Reflect answer guide**.

Reflect

WAYS OF WORKING Independent thinking

IN FOCUS This activity allows children to explain what a fraction wall is and how to use one to find equivalent fractions. They may state that $\frac{1}{2}$ is equivalent to $\frac{2}{4}$. 'I can use the fraction wall to show this because …'

ASSESSMENT CHECKPOINT This will show what children know about fraction walls and whether they can explain how they can be used. Can children show their understanding of equivalent fractions or discuss how to compare fractions using a fraction wall and fraction strips?

ANSWERS Answers for the **Reflect** part of the lesson appear in the separate **Practice and Reflect answer guide**.

After the lesson

- Can children use a fraction wall to compare fractions?
- Are children using the language of more than, less than and equal to correctly?
- Can children identify patterns in equivalent fractions?

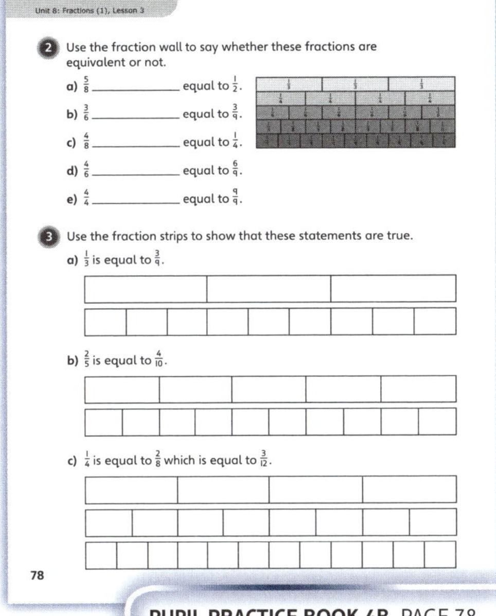

PUPIL PRACTICE BOOK 4B PAGE 77

PUPIL PRACTICE BOOK 4B PAGE 78

PUPIL PRACTICE BOOK 4B PAGE 79

Equivalent fractions ②

Learning focus

In this lesson, children will continue to find equivalent fractions. They will look at the relationship between the numerators and denominators of fractions.

Small steps

→ Previous step: Equivalent fractions (1)
→ **This step: Equivalent fractions (2)**
→ Next step: Simplifying fractions

NATIONAL CURRICULUM LINKS

Year 4 Number – Fractions (Including Decimals)

Recognise and show, using diagrams, families of common equivalent fractions.

ASSESSING MASTERY

Children can explain why two fractions are equivalent using their knowledge of multiplication and division.

COMMON MISCONCEPTIONS

Children may not see that they can multiply or divide fractions by different numbers to find an equivalent fraction, or they may struggle to identify the common multiples. Ask:

• *What times-table do both denominators have in common?*
• *Which times-table has both of these numbers in?*
• *Can you list the multiples of the denominators?*

STRENGTHENING UNDERSTANDING

Provide children with fraction strips or ask them to create fraction strips, so that they can directly compare fractions with the same or different numerators and denominators.

GOING DEEPER

Encourage children to articulate exactly why they know two fractions are equivalent. Ask them to give further examples of equivalent fractions.

KEY LANGUAGE

In lesson: fraction, numerator, denominator, equivalent, fraction wall

Other language to be used by the teacher: fraction strip, multiple

STRUCTURES AND REPRESENTATIONS

fraction cards, shapes split into fractions

RESOURCES

Optional: fraction wall, fraction strips

 In the eTextbook of this lesson, you will find interactive links to a selection of teaching tools.

Before you teach ⏸

• Can children multiply and divide?
• Can children identify common multiples?
• Can children identify if a fraction is larger or smaller than another?

Discover

ASK

- Question **1** a): *How can you sort the fractions? Should you look at the numerators or the denominators? Why?*
- Question **1** b): *How can you find the missing numbers? What information are you given? Can you use your knowledge of multiples to help?*

IN FOCUS Questions **1** a) and b) prompt children to develop the patterns they identified in the previous lesson, to look at the numerators and denominators of the fractions to find connections. It guides them to use their understanding of multiples to find equivalent fractions.

PRACTICAL TIPS Provide children with space to draw the fractions. They may choose to use fractions strips, or represent the fractions within a rectangle or square.

ANSWERS

Question **1** a): $\frac{2}{3}, \frac{4}{6}, \frac{6}{9}, \frac{8}{12}$ and $\frac{1}{2}, \frac{2}{4}, \frac{4}{8}, \frac{5}{10}$

Question **1** b): $\frac{3}{15}, \frac{6}{30}, \frac{1}{5}$

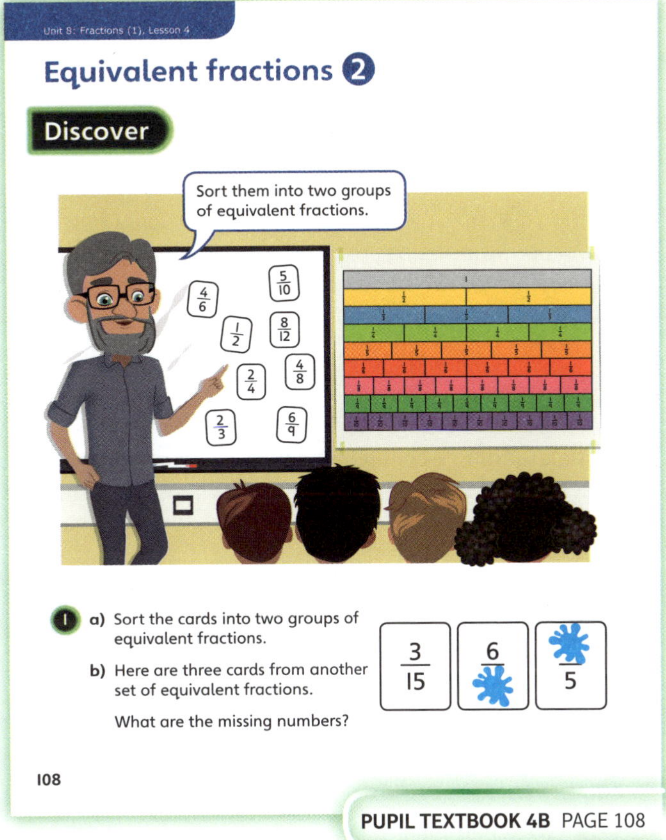

PUPIL TEXTBOOK 4B PAGE 108

Share

ASK

- Question **1** a): *Why is Ash suggesting that you look at the numerators and denominators? What similarities are there between $\frac{4}{6}$ and $\frac{6}{9}$? What is similar about the numerators? What is similar about the denominators?*

IN FOCUS Question **1** a) uses rectangles divided into fractions, giving children an alternative model to the strips used in the previous lesson. Sorting the fractions gives children the opportunity to show their conceptual understanding of equivalent fractions. Question **1** b) helps children to develop the ability to use their understanding of multiples to find equivalent fractions.

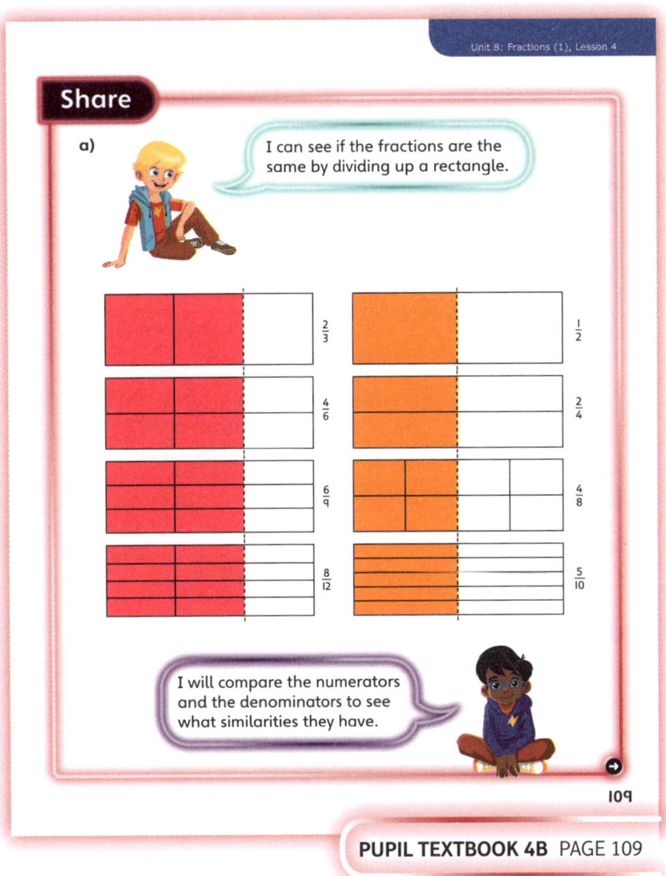

PUPIL TEXTBOOK 4B PAGE 109

Think together

WAYS OF WORKING Whole class teacher led (I do, We do, You do)

ASK

- Question ❶: *How can you compare the fractions? How can fraction strips help you? Which numerators and denominators are in the same times-tables? In your own words, describe what it means for two fractions to be equivalent.*

IN FOCUS Question ❶ gives children practice in identifying equivalent fractions. Question ❷ develops this understanding – children use their knowledge of multiples and times-tables to find unknown numbers in equivalent fractions. Question ❸ allows children to show their understanding by articulating how to find equivalent fractions.

STRENGTHEN Ask children to create fraction strips or draw a diagram so they can see the equivalent fractions. In question ❶, if they are struggling to identify the equivalent fractions, suggest they list multiples of the numerator and denominator of each fraction. Can they see any common multiples?

DEEPEN Give children a fraction, for example $\frac{15}{60}$. Ask them to find the equivalent fraction with the lowest possible numerator/denominator. Challenge them to make an equivalent fraction with as high a numerator as they can. Ask them to explain how they created their fraction.

ASSESSMENT CHECKPOINT Use question ❶ to assess whether children can identify equivalent fractions. Use question ❷ to assess whether they can create equivalent fractions. Ensure they can see that whatever the numerator is multiplied or divided by is the same as the number that the denominator is multiplied or divided by.

ANSWERS

Question ❶: $\frac{1}{3}$ and $\frac{3}{9}$, $\frac{2}{5}$ and $\frac{6}{15}$, $\frac{2}{7}$ and $\frac{10}{35}$

Question ❷: $\frac{1}{4} = \frac{3}{12}$, $\frac{3}{5} = \frac{12}{20}$, $\frac{6}{8} = \frac{12}{16}$

Question ❸ a): Jamie is correct because you can simplify the numerators to 3 and the denominators to 5, by dividing the numerator and denominator of $\frac{6}{10}$ by 2 and the numerator and denominator of $\frac{9}{15}$ by 3.

Question ❸ b): Check children's fractions are equivalent to $\frac{3}{5}$, such as $\frac{12}{20}$, $\frac{15}{25}$.

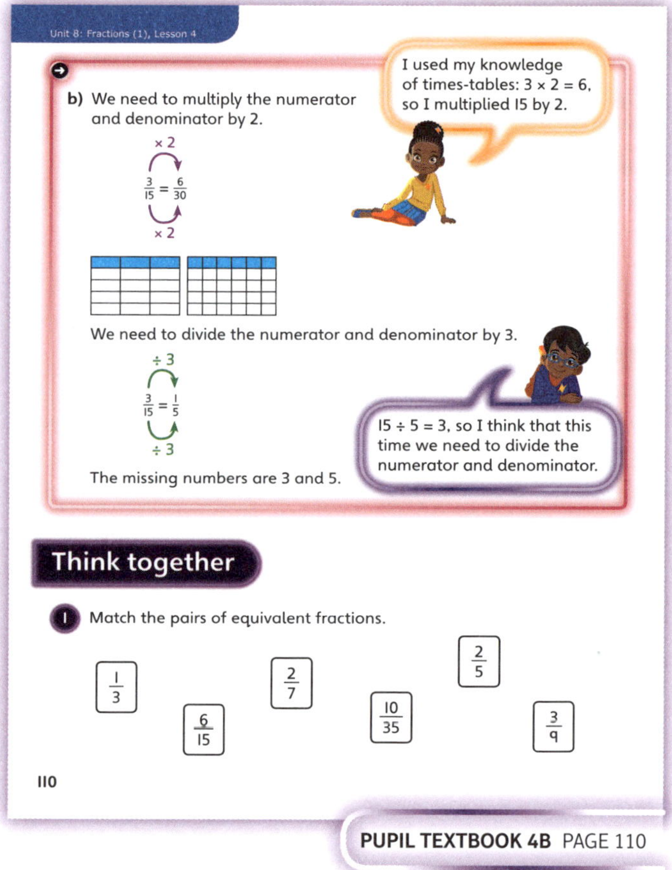

PUPIL TEXTBOOK 4B PAGE 110

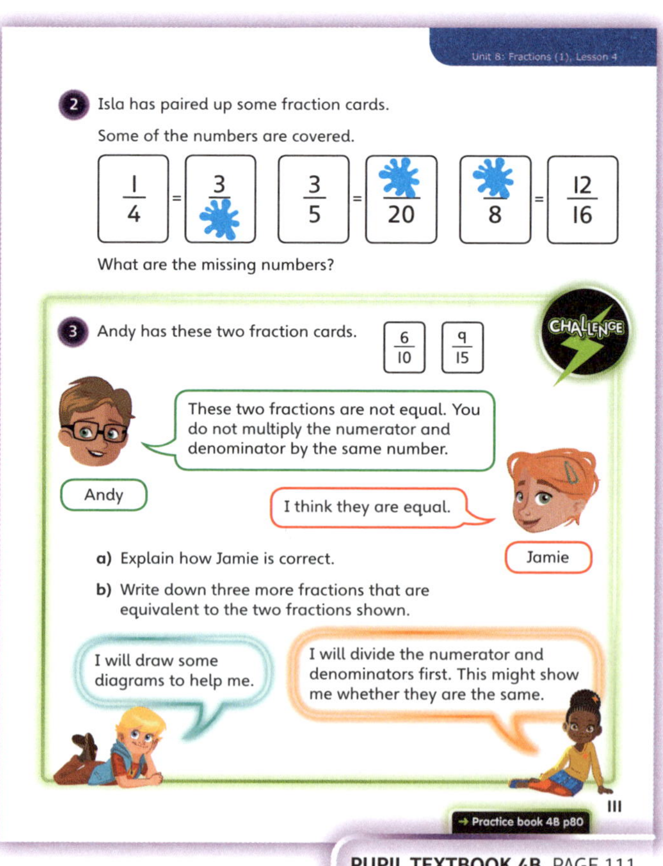

PUPIL TEXTBOOK 4B PAGE 111

Practice

WAYS OF WORKING Independent thinking

IN FOCUS In question ③ children need to use their knowledge of multiples and times-tables to work out equivalent fractions.

STRENGTHEN Provide children with fraction strips or a fraction wall and allow them to draw the fractions, so they have a visual representation. If children lack confidence or familiarity with the times-tables, provide sheets or posters for them to refer to.

DEEPEN Ask children to create similar pairs of fractions to the pair given in question ⑥. They can then give them to a partner to check. Ensure they understand that the multiplier for the numerator/denominator to find one fraction from the other should not be a whole number.

THINK DIFFERENTLY Question ④ is more open-ended, as children can choose any number for the missing number in the first fraction of each pair. Encourage them to investigate different numerators and denominators.

ASSESSMENT CHECKPOINT Use questions ② and ③ to assess whether children can identify and create equivalent fractions. Check whether children can explain why the fractions are equivalent, drawing diagrams if necessary.

ANSWERS Answers for the **Practice** part of the lesson appear in the separate **Practice and Reflect answer guide**.

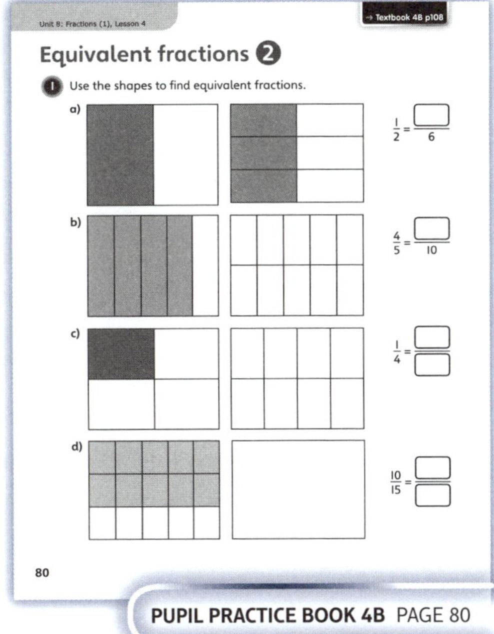

PUPIL PRACTICE BOOK 4B PAGE 80

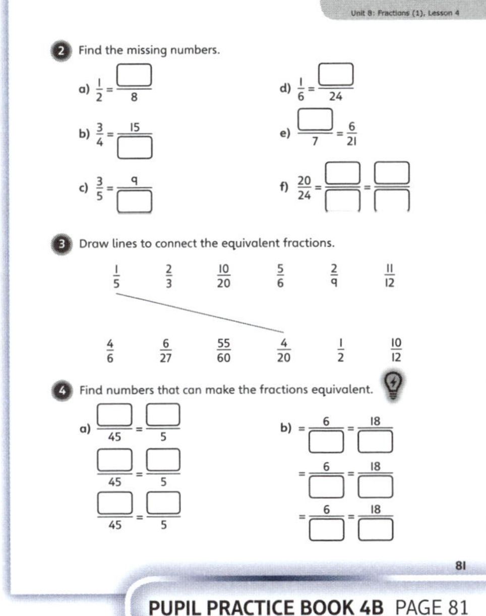

PUPIL PRACTICE BOOK 4B PAGE 81

Reflect

WAYS OF WORKING Independent thinking

IN FOCUS This activity uses a specific fraction to enable children to explain how to find equivalent fractions. They may use a drawing to support their explanation.

ASSESSMENT CHECKPOINT Assess what children know about equivalent fractions. Look at whether and how they use the key words denominator, numerator, multiples and factors.

ANSWERS Answers for the **Reflect** part of the lesson appear in the separate **Practice and Reflect answer guide**.

After the lesson ⏸

- Can children give examples of equivalent fractions?
- Can children use their knowledge of multiplication and division to find equivalent fractions?

PUPIL PRACTICE BOOK 4B PAGE 82

Simplifying fractions

Learning focus

In this lesson, children will simplify fractions, initially with the aid of images and then with abstract fractions.

Small steps

→ Previous step: Equivalent fractions (2)
→ **This step: Simplifying fractions**
→ Next step: Fractions greater than 1 (1)

NATIONAL CURRICULUM LINKS

Year 4 Number – Fractions (Including Decimals)

Recognise and show, using diagrams, families of common equivalent fractions.

ASSESSING MASTERY

Children can simplify fractions, including to their simplest form. Children may also be able to explain the most efficient way to simplify a fraction.

COMMON MISCONCEPTIONS

Children may not simplify a fraction to its simplest form. Ask:
- *How did you simplify this fraction? Can you simplify it any further?*
- *Are there any numbers that will divide into both the numerator and denominator?*

STRENGTHENING UNDERSTANDING

For children who do not have the multiplication and division facts readily available, provide multiplication grids to support them when simplifying fractions (as multiplication facts are not the objective).

Provide children with pictorial representations of the fractions, so that they can see the fraction.

GOING DEEPER

Give children a fraction, such as $\frac{30}{105}$ or $\frac{126}{210}$, and challenge them to find its simplest form. Encourage them to use as few stages as possible.

KEY LANGUAGE

In lesson: fraction, **simplest fraction**, simplest form, **simplify**, numerator, denominator

Other language to be used by the teacher: multiple, common multiple, factor

STRUCTURES AND REPRESENTATIONS

hundredths grids, tenths grids

RESOURCES

Optional: fraction walls

 In the eTextbook of this lesson, you will find interactive links to a selection of teaching tools.

Before you teach 

- Can children draw fractions?
- Can children identify equivalent fractions?
- Can children divide accurately?

Discover

WAYS OF WORKING Pair work

ASK

- Question ❶ a): *What is the numerator? What is the denominator? Is there a common multiple for the numerator and denominator? How can this help you?*
- Question ❶ b): *What is different about Mo's and Lexi's pictures? What is similar?*

IN FOCUS In question ❶ b) children should be able to see the numerator and the denominator clearly from the pictures. This should allow them to see if there is a common number that can be used to divide, so that the fraction can be simplified. It is easier to identify the common number in Mo's picture, as he has shaded in rows. In Lexi's picture, children need to use their knowledge of multiples and times-tables.

PRACTICAL TIPS Give children large drawings of windows to shade different fractions. Suggest that initially they shade rows or columns where possible to aid understanding of the concept.

ANSWERS

Question ❶ a): $\frac{6}{9}$ of Mo's picture is shaded.

$\frac{2}{3}$ of Mo's picture is shaded.

Question ❶ b): $\frac{16}{20}$ of Lexi's picture is shaded.

$\frac{4}{5}$ is the simplest fraction.

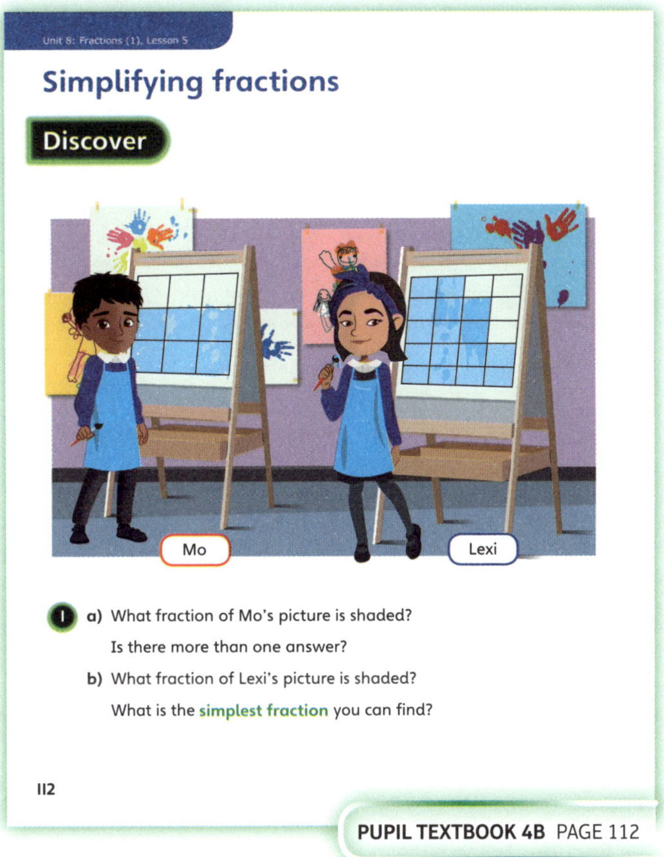

PUPIL TEXTBOOK 4B PAGE 112

Share

WAYS OF WORKING Whole class teacher led

ASK

- Question ❶ a): *In each column how many squares are shaded? What does this show us?*
- Question ❶ b): *How can Astrid's comment help you? Is there a number that goes into both the numerator and denominator?*

IN FOCUS In question ❶ a) children can demonstrate their understanding from the previous lesson by finding as many equivalent fractions as possible. Question ❶ b) introduces children to the concept of the simplest form of a fraction. The fraction $\frac{16}{20}$ can be simplified either in two stages (dividing by 2 twice) or in a single stage (dividing by 4).

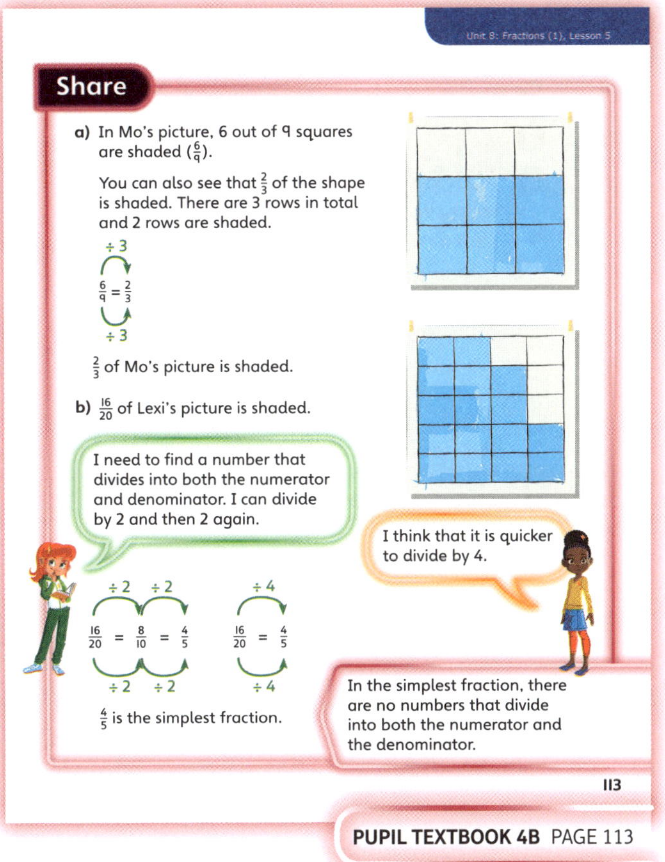

PUPIL TEXTBOOK 4B PAGE 113

Think together

WAYS OF WORKING Whole class teacher led (I do, We do, You do)

ASK

- Question **1**: *What is the whole? How many parts are shaded? What is this as a fraction?*
- Question **2** b): *How many red cars are there? How many cars are there in total? Is there a number that goes into both these numbers? Is that fraction in its simplest form?*

IN FOCUS Question **2** provides a different type of context for children to find fractions, as it does not use regular grids. In both fractions, children only need to divide by a single factor to simplify.

Question **3** encourages children to simplify fractions to compare the proportions of rectangles that are shaded. To simplify Reena's fraction, $\frac{60}{80}$, they can divide by three successive factors (2, 2, 5), two factors (2, 10) or, for the most able children, a single factor (20).

STRENGTHEN In question **2** encourage children to draw each fraction on a grid, so they can see the numerator and denominator clearly.

DEEPEN Challenge children to create a story or context for the fractions being used.

ASSESSMENT CHECKPOINT Use question **2** to assess whether children can simplify fractions. Check that they understand the terms 'simplify' and 'simplest fraction' and that they recognise when a fraction is in its simplest form.

ANSWERS

Question **1**: $\frac{3}{12}$ of Isla's picture is shaded. $\frac{3}{12} = \frac{1}{4}$

Question **2** a): $\frac{5}{15} = \frac{1}{3}$. $\frac{1}{3}$ of the cars are yellow.

Question **2** b): $\frac{6}{15} = \frac{2}{5}$. $\frac{2}{5}$ of the cars are red.

Question **3**: Andy is correct, they have shaded the same fraction.
He has shaded $\frac{9}{12}$, which is $\frac{3}{4}$ when simplified. Reena has shaded $\frac{60}{80}$, which is also $\frac{3}{4}$ when simplified.

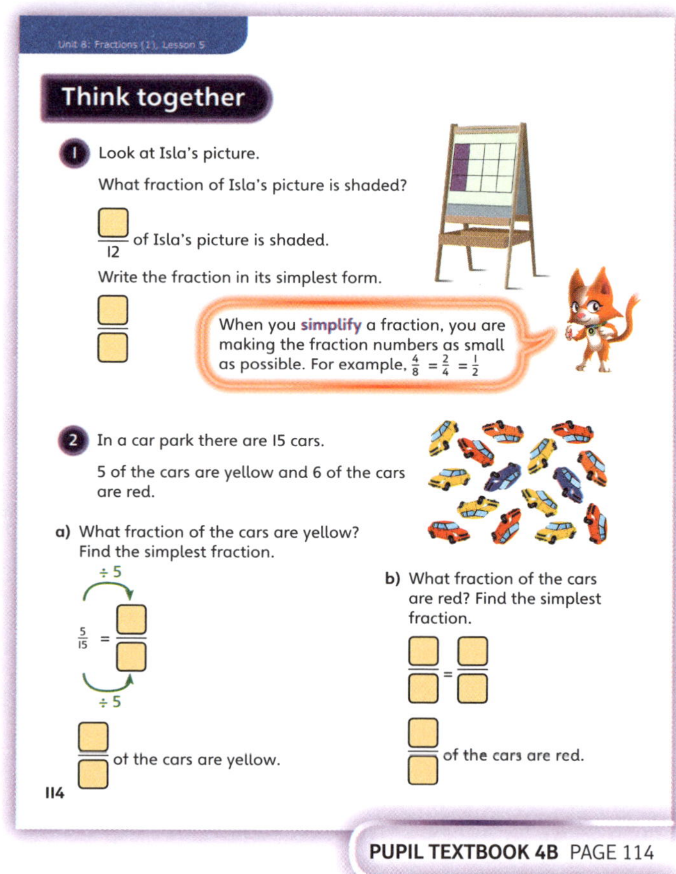

PUPIL TEXTBOOK 4B PAGE 114

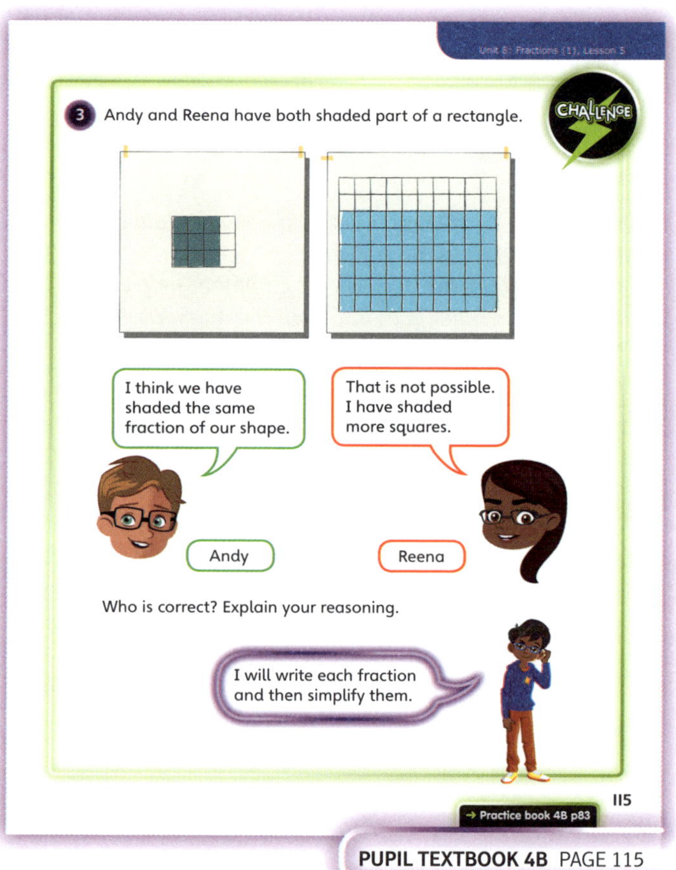

PUPIL TEXTBOOK 4B PAGE 115

Practice

WAYS OF WORKING Independent thinking

IN FOCUS Question **4** is a problem-solving question that requires children to simplify three fractions and then compare them. All the fractions have the same denominator when simplified, so the comparison should be straightforward if they have simplified the fractions correctly.

Question **6** allows children to develop someone else's answer and reinforces the concept of the simplest fraction being one that cannot be simplified any further.

STRENGTHEN A fraction wall may help children to see the simplest form. The highest representation of a particular fraction on the fraction wall is the simplest form. In question **4**, providing diagrams of the fractions will enable children to see them pictorially.

DEEPEN Ask children to find five fractions that all simplify to the same fraction.

THINK DIFFERENTLY Question **5** encourages children to look for the highest times-table in which both the numerator and denominator appear, rather than dividing in successive steps using the smallest factor they can see. They need to be able to articulate how they are finding the most efficient method.

ASSESSMENT CHECKPOINT Use questions **2** and **3** to assess whether children can simplify fractions to their simplest form. Use question **6** to check that they understand when a fraction is in its simplest form.

ANSWERS Answers for the **Practice** part of the lesson appear in the separate **Practice and Reflect answer guide**.

Reflect

WAYS OF WORKING Pair work

IN FOCUS This task requires children to explain how they know they have found the simplest fraction. They need to be able to state that a fraction is in its simplest form when it can no longer be divided. If they cannot do this, they will need to further develop their understanding. This can be done with concrete manipulatives.

ASSESSMENT CHECKPOINT Check whether children understand how to find a fraction's simplest form. Look to see whether they can find the most efficient method.

ANSWERS Answers for the **Reflect** part of the lesson appear in the separate **Practice and Reflect answer guide**.

After the lesson ▮▮

- Can children simplify a fraction?
- Can they identify a fraction that is not in its simplest form?
- Can children show how they know a fraction is in its simplest form?

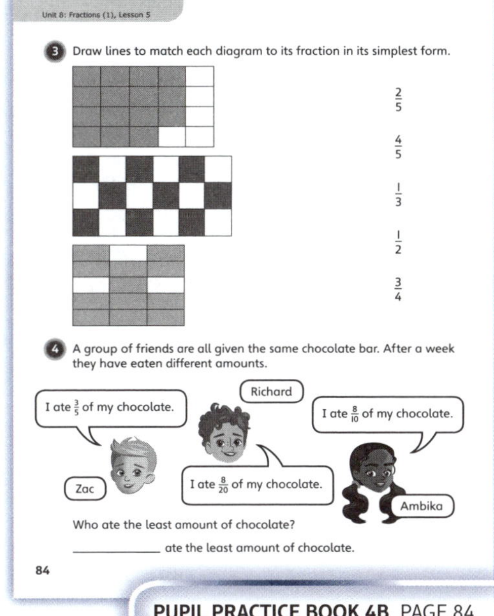

PUPIL PRACTICE BOOK 4B PAGE 83

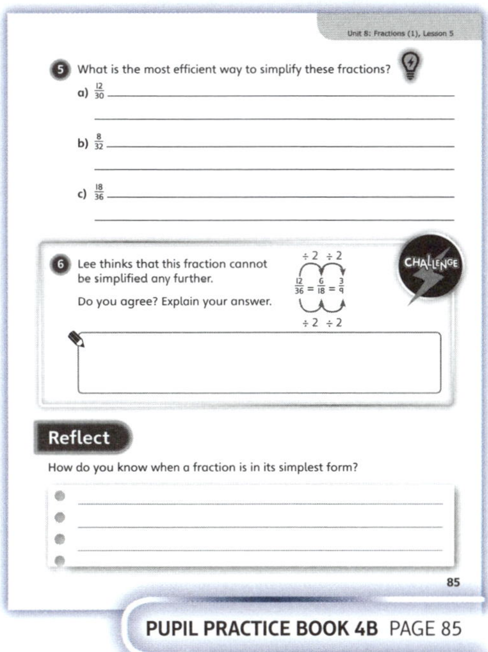

PUPIL PRACTICE BOOK 4B PAGE 84

PUPIL PRACTICE BOOK 4B PAGE 85

Fractions greater than 1 ①

Learning focus

In this lesson, children will identify and draw mixed numbers with the aid of images and sentence stems.

Small steps

➜ Previous step: Simplifying fractions
➜ **This step: Fractions greater than 1 (1)**
➜ Next step: Fractions greater than 1 (2)

NATIONAL CURRICULUM LINKS

Year 4 Number – Fractions (Including Decimals)

Solve problems involving increasingly harder fractions to calculate quantities, and fractions to divide quantities, including non-unit fractions where the answer is a whole number.

ASSESSING MASTERY

Children can identify the wholes and parts of mixed numbers. They can also draw their own fractions greater than 1.

COMMON MISCONCEPTIONS

Children may not see that a number involving a fraction can be more than 1. Ask:
- *How many wholes do you have? How many parts are left over? What is the fraction of a whole for these parts? What is the total?*
- *Can a number with a fraction be more than 1? Why?*

Children may not see how many wholes and parts there are in a mixed number. Ask:
- *How many parts make a whole? How many parts are left over?*

STRENGTHENING UNDERSTANDING

Provide children with concrete shapes so they can create 'wholes' and 'parts' to make the numbers.

GOING DEEPER

Ask children to set problems for each other, making up clues to their number. For example: 'My denominator is less than 8. I have 2 wholes and my numerator is greater than 1. My part is equal to $\frac{1}{3}$.' ($2\frac{2}{6}$)

KEY LANGUAGE

In lesson: fraction, mixed number

Other language to be used by the teacher: numerator, part, whole, denominator, greater than

STRUCTURES AND REPRESENTATIONS

part-whole model

RESOURCES

Optional: shapes split into fractions, concrete shapes

 In the eTextbook of this lesson, you will find interactive links to a selection of teaching tools.

Before you teach 

- Do children know how to show a whole using fractions?
- Can children simplify fractions?

Discover

WAYS OF WORKING Pair work

ASK

- Question **1** a): *How many triangles are there in total? How many triangles will make a complete hexagon (or a whole)? How many triangles are left over?*
- Question **1** b): *How many wholes are there? How many parts remaining? What is the numerator? What is the denominator?*

IN FOCUS Question **1** a) requires children to count how many triangles there are and to work out how many hexagons they can make. They should be able to see that 6 triangles make a hexagon, and that the denominator must be 6. They should also be able to see how many full hexagons can be made and then use the remaining triangles to show a fraction that is less than 1.

PRACTICAL TIPS Provide children with triangles for them to recreate the situation. This will enable them to manipulate the triangles to aid understanding.

ANSWERS

Question **1** a): Jamilla and Richard can make 4 whole hexagons. They will have 5 triangles left over.

Question **1** b): There are $4\frac{5}{6}$ hexagons.

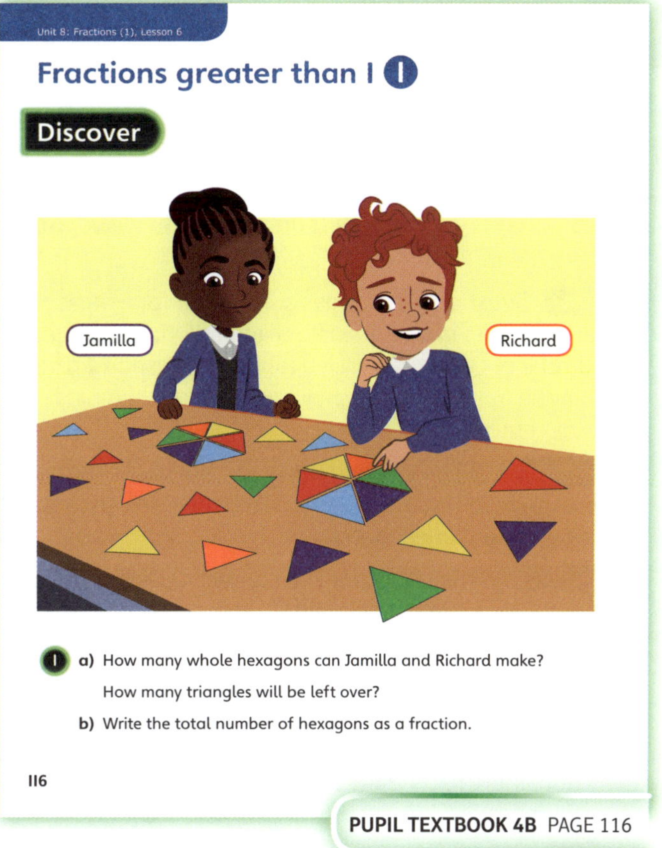

PUPIL TEXTBOOK 4B PAGE 116

Share

WAYS OF WORKING Whole class teacher led

ASK

- Question **1** a): *How many full hexagons can you see? How many triangles are left over?*
- Question **1** b): *How can you write your answer to part a) as a fraction?*

IN FOCUS Question **1** a) prompts children to show their understanding of parts and wholes. Question **1** b) takes the parts to create a fraction and combines this with the whole to write the number as a mixed number. Explain that a mixed number is a number that has a whole and a part (for example, $3\frac{3}{5}$).

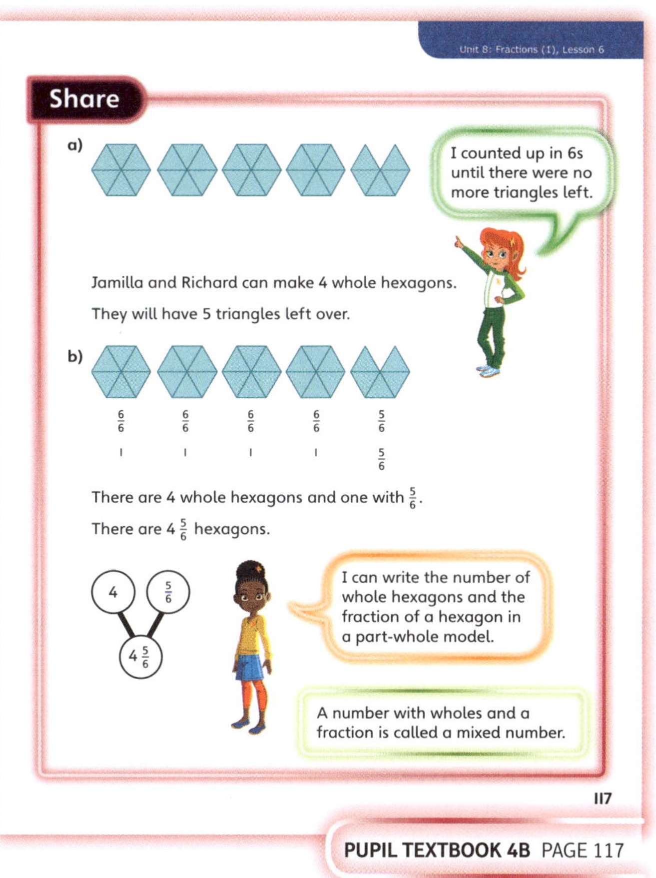

PUPIL TEXTBOOK 4B PAGE 117

Think together

WAYS OF WORKING Whole class teacher led (I do, We do, You do)

ASK

• Question ❶: *How many parts are there in the whole? How many wholes can Max make?*

IN FOCUS Question ❶ provides scaffolding for writing the wholes and parts in a part-whole diagram, prior to writing the number of circles as a mixed number. Question ❸ gives children the opportunity to show their understanding in a different context, using 3D shapes.

STRENGTHEN Provide children with concrete shapes to use to recreate the questions. Alternatively, children can draw the shapes to aid understanding.

DEEPEN Ask children to show their answers using different models, for example in a part-whole diagram and on a number line. Also, ask them to write their answers as whole numbers and remainders, as well as mixed numbers.

ASSESSMENT CHECKPOINT Use questions ❶ and ❷ to assess whether children can identify how many wholes and how many remaining parts there are, and write the total as a mixed number.

ANSWERS

Question ❶ a): Max can make 5 complete circles with 3 pieces left over.

Question ❶ b): Max can make $5\frac{3}{5}$ circles.

Question ❷: Lexi made 3 complete rectangles with 3 squares left over. She made $3\frac{3}{6}$ rectangles. Amelia made 4 complete rectangles with 2 squares left over. She made $4\frac{2}{6}$ rectangles.

Question ❸ a): 3 boxes

Question ❸ b): 1 will be left over.

Question ❸ c): There will be $3\frac{1}{8}$ boxes of cubes.

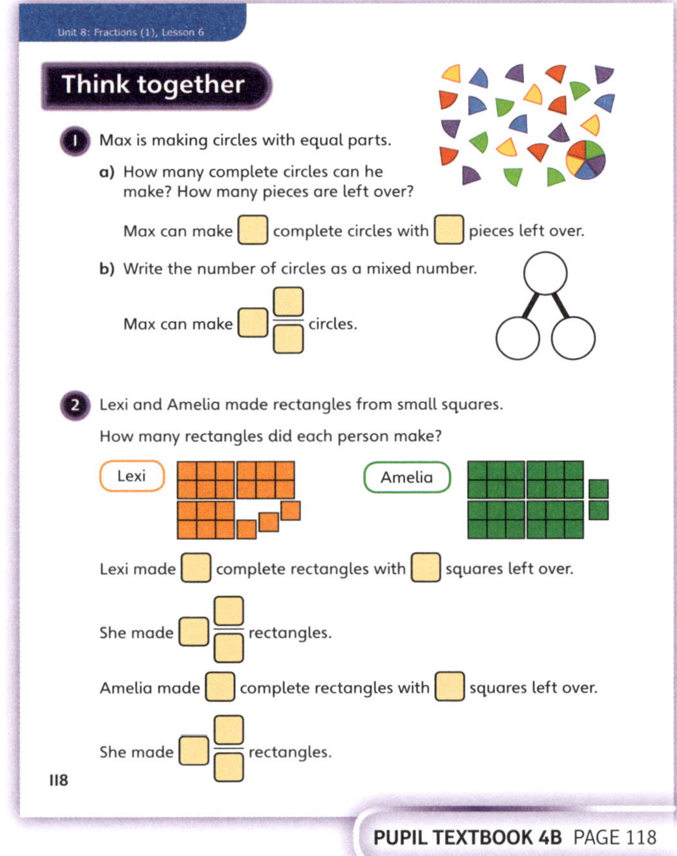

PUPIL TEXTBOOK 4B PAGE 118

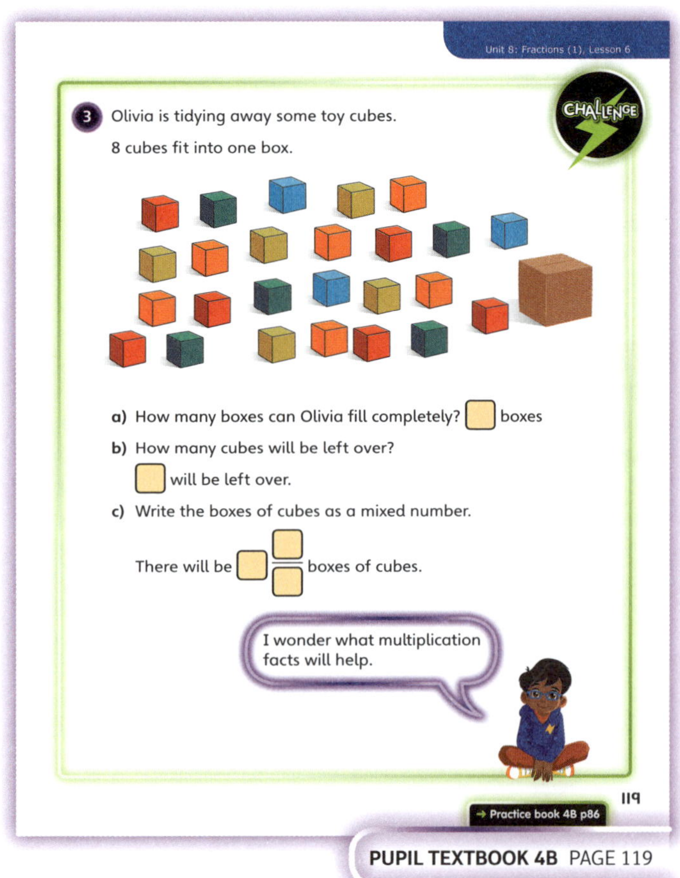

PUPIL TEXTBOOK 4B PAGE 119

150

Practice

WAYS OF WORKING Independent thinking or pair work

IN FOCUS Questions ❶ and ❷ require children to work out the mixed number from a diagram, which will show their understanding of finding fractions greater than 1. In question ❸ children need to shade a diagram to show their understanding of mixed numbers.

STRENGTHEN Provide children with concrete shapes to use (or images of the shapes) or allow children to draw the shapes to aid understanding.

DEEPEN Ask children to write down a mixed number. Their partner can then ask questions with yes or no answers to guess the number. (For example, is the numerator a multiple of 4?)

ASSESSMENT CHECKPOINT Use questions ❶, ❷ and ❹ to assess whether children can identify wholes and parts and express the total as a mixed number. Check that children can use the vocabulary wholes and parts correctly. Use question ❸ to assess whether children can shade diagrams to represent mixed numbers.

ANSWERS Answers for the **Practice** part of the lesson appear in the separate **Practice and Reflect answer guide**.

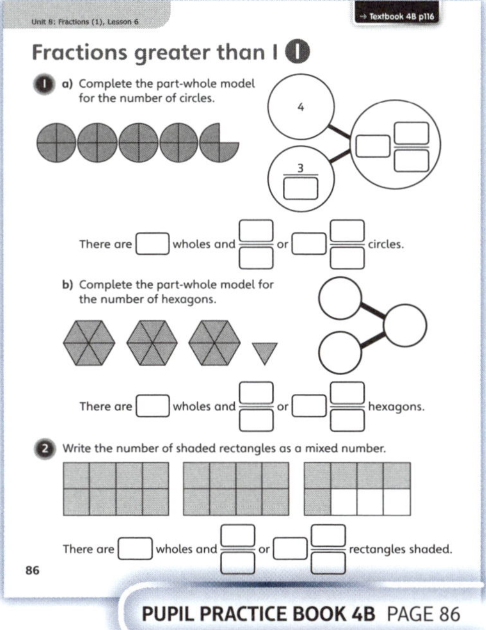

PUPIL PRACTICE BOOK 4B PAGE 86

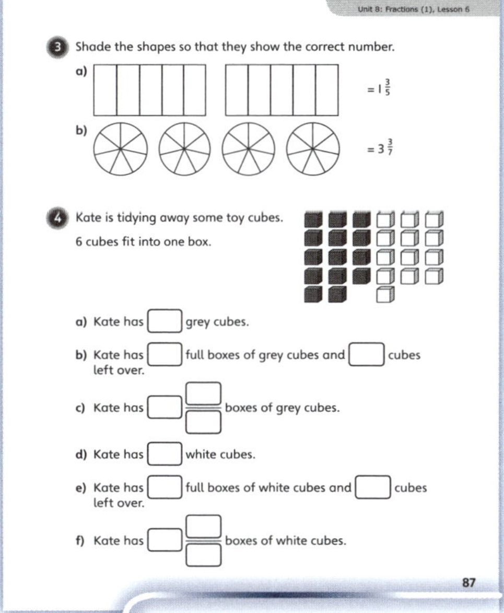

PUPIL PRACTICE BOOK 4B PAGE 87

Reflect

WAYS OF WORKING Pair work

IN FOCUS Children draw a diagram for a mixed number and compare their drawing with a partner's. They can discuss the similarities and differences between the two drawings, giving them the opportunity to practise using the vocabulary associated with mixed numbers.

ASSESSMENT CHECKPOINT Assess whether children can draw a diagram to represent a mixed number. Ask them to identify the wholes and parts.

ANSWERS Answers for the **Reflect** part of the lesson appear in the separate **Practice and Reflect answer guide**.

After the lesson ⏸

• Can children identify wholes and parts of a mixed number?
• Can children draw a diagram that shows a mixed number?

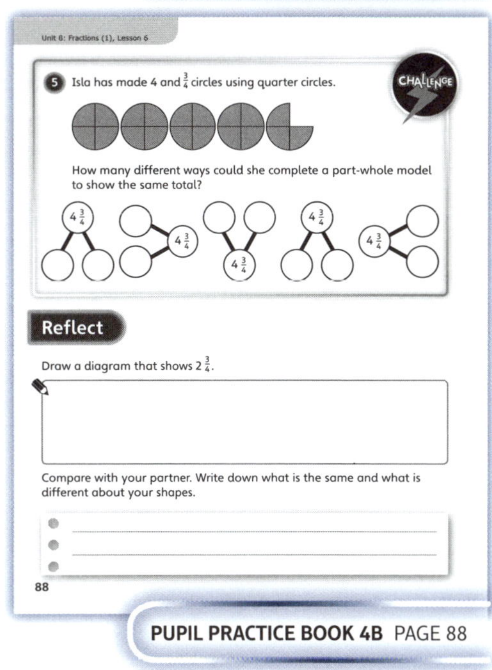

PUPIL PRACTICE BOOK 4B PAGE 88

Fractions greater than I ❷

Learning focus

In this lesson, children will count up in fractions beyond 1 using a number line. Children understand what an improper fraction is.

Small steps

➜ Previous step: Fractions greater than 1 (1)
➜ **This step: Fractions greater than 1 (2)**
➜ Next step: Adding fractions

NATIONAL CURRICULUM LINKS

Year 4 Number – Fractions (Including Decimals)

Solve problems involving increasingly harder fractions to calculate quantities, and fractions to divide quantities, including non-unit fractions where the answer is a whole number.

ASSESSING MASTERY

Children can identify and mark a fraction greater than 1 on a number line. They can state the similarities and differences between mixed numbers and improper fractions.

COMMON MISCONCEPTIONS

Children may fail to use a number line correctly. Ask:
- *How many jumps do you need to make? Where do you start to jump? How do you know when you have one/two whole(s)?*

Children may not understand that a fraction can be greater than 1. Ask:
- *How many parts make a whole? Can you colour in a whole on the number line? How many wholes are in your fraction? How many parts are left over?*

Children may not see the links between mixed numbers and improper fractions. Ask:
- *Is 1 and $\frac{2}{3}$ the same as $\frac{5}{3}$? How do you know? Can you show me on two number lines?*

STRENGTHENING UNDERSTANDING

Provide children with number lines that are blank or have pre-drawn increments. The number lines with increments will allow children to see how many fifths or eighths are in a whole.

GOING DEEPER

Ask children to provide equivalent improper fractions. Can they use a drawing to show their conceptual understanding?

KEY LANGUAGE

In lesson: fraction, numerator, denominator, wholes, parts, mixed number, **improper fraction**

Other language to be used by the teacher: greater than, less than, equal to

STRUCTURES AND REPRESENTATIONS

number line

RESOURCES

Optional: number lines

 In the eTextbook of this lesson, you will find interactive links to a selection of teaching tools.

Before you teach ⏸

- Can children write mixed numbers?
- Can they identify how many wholes and how many parts there are in a mixed number?
- Can children use a number line?

Discover

 Pair work

ASK

- Question ❶ a): *How many bottles are in a pack? How many bottles are left over? What is the denominator if you write this as a fraction?*
- Question ❶ b): *How many wholes are there? How many parts? How many parts are there in one whole?*

IN FOCUS Question ❶ b) shows that the same number can be written as a mixed number and as an improper fraction. These definitions should be discussed with children, and examples given, to ensure they understand the vocabulary.

PRACTICAL TIPS Allow children to use concrete equipment to recreate the image so they have a better conceptual understanding of what is shown. For example, children could use cubes to represent the bottles of water.

ANSWERS

Question ❶ a): 1 whole pack was used. $\frac{2}{5}$ of the next pack was needed.

Question ❶ b): $1\frac{2}{5}$ packs are needed.
$\frac{7}{5}$ packs are needed.

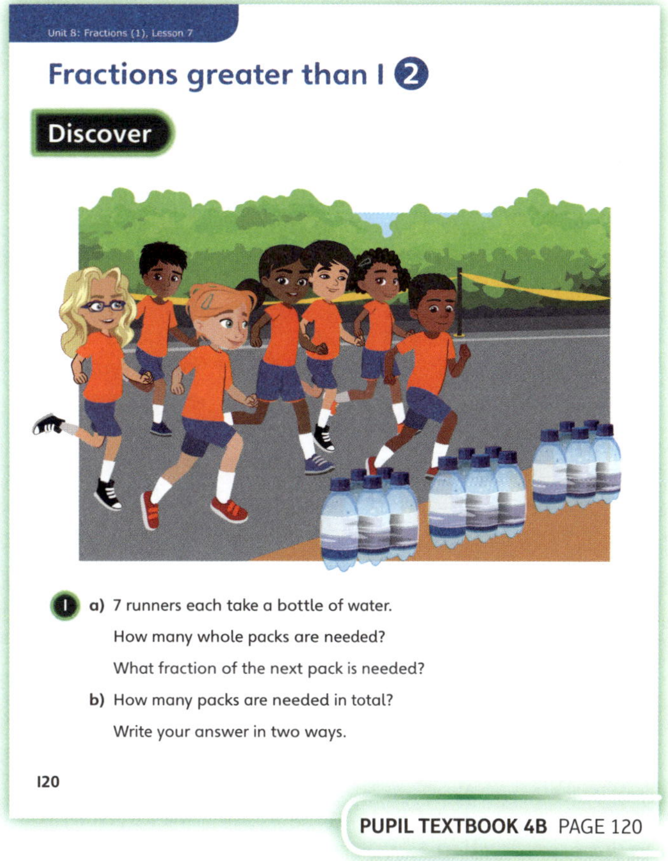

PUPIL TEXTBOOK 4B PAGE 120

Share

 Whole class teacher led

ASK

- Question ❶: *What do the number lines show?*
- Question ❶ b): *What happens when this number line goes over 1? How many fifths are in 1 whole and 2 fifths? Look at what Sparks has said. What is the same about an improper fraction and a mixed number? What is different?*

IN FOCUS Question ❶ a) shows children how a fraction greater than 1 can be shown as a mixed number on a number line. Question ❶ b) shows the same number on a number line using improper fractions, enabling children to see how a mixed number can be written as an improper fraction. Explain that an improper fraction is a fraction that has a numerator equal to or greater than the denominator (such as $\frac{6}{4}$).

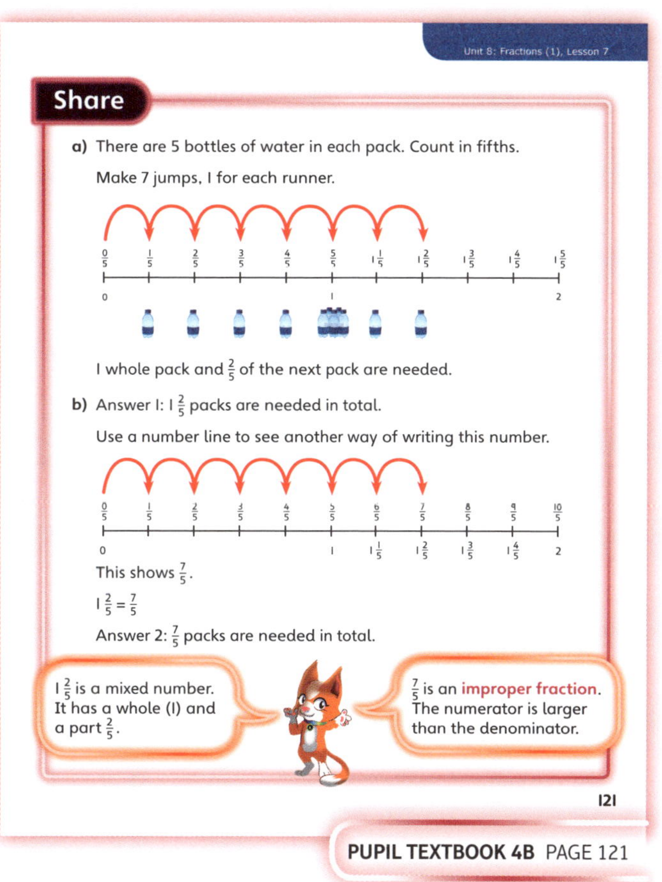

PUPIL TEXTBOOK 4B PAGE 121

Think together

WAYS OF WORKING Whole class teacher led (I do, We do, You do)

ASK

- Question **1**: *How many energy bars are in a pack? How many whole packs are used? How many parts of another pack are used? How do the answer sentences help you?*

IN FOCUS Question **3** a) requires children to use a number line to show or write a mixed number. Children may assume that all number lines provided are the same and therefore place the fraction incorrectly. It is important for them to recognise that each number line uses different increments. Question **3** b) gives children the opportunity to combine their knowledge of equivalent fractions with what they have learnt about mixed numbers and improper fractions.

STRENGTHEN Provide children with both blank and pre-drawn number lines, so they can show the jumps that take place.

DEEPEN Ask children to investigate how to convert between mixed numbers and improper fractions.

ASSESSMENT CHECKPOINT Use questions **1** and **2** to assess whether children can use a number line to show an improper fraction. Check that they understand that an improper fraction is always greater than 1.

ANSWERS

Question **1**: $1\frac{4}{5}$ packs were eaten.

$\frac{9}{5}$ packs were eaten.

Question **2**: Kate walks $2\frac{3}{4}$ km.

Kate walks $\frac{11}{4}$ km.

Question **3** a)

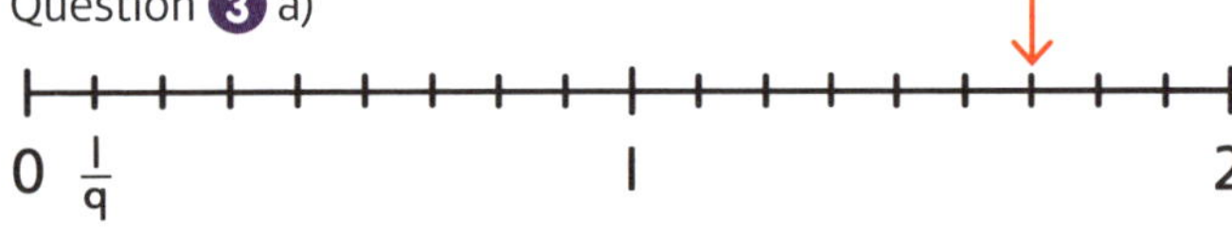

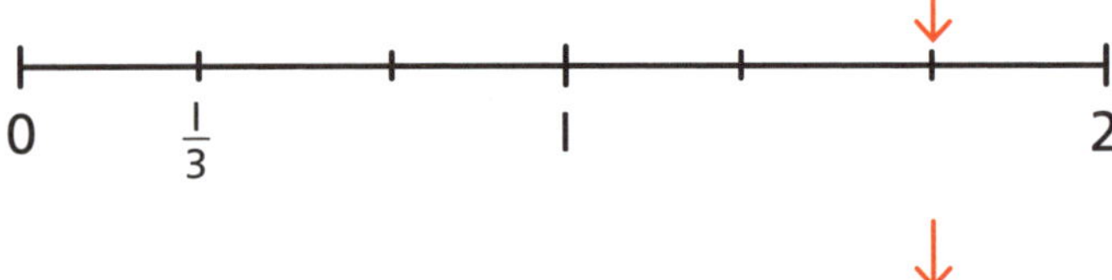

Question **3** b): Children's numbers equivalent to $1\frac{6}{9}$, for example, $\frac{15}{9}, \frac{5}{3}, \frac{10}{6}, 1\frac{2}{3}, 1\frac{4}{6}$.

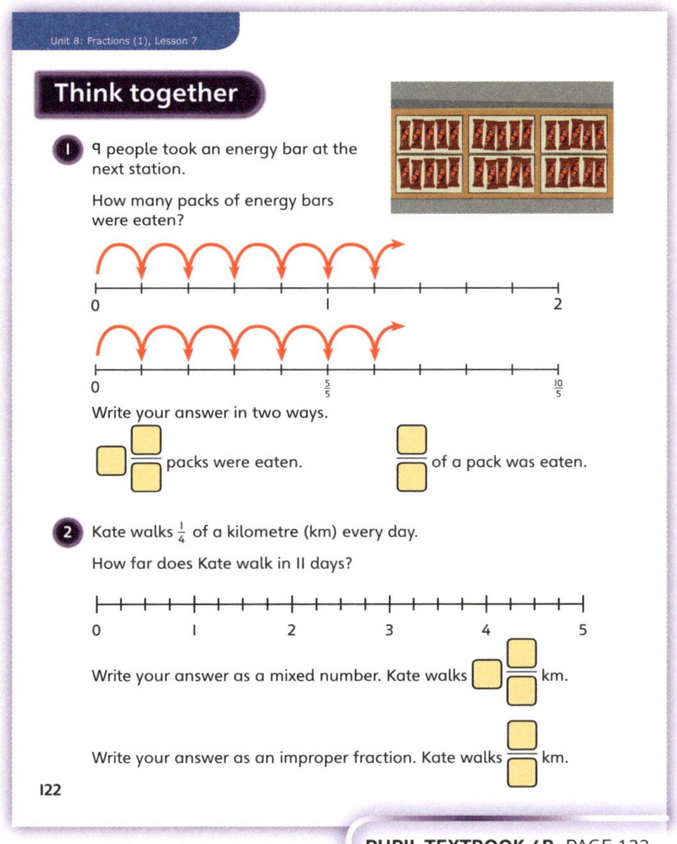

PUPIL TEXTBOOK 4B PAGE 122

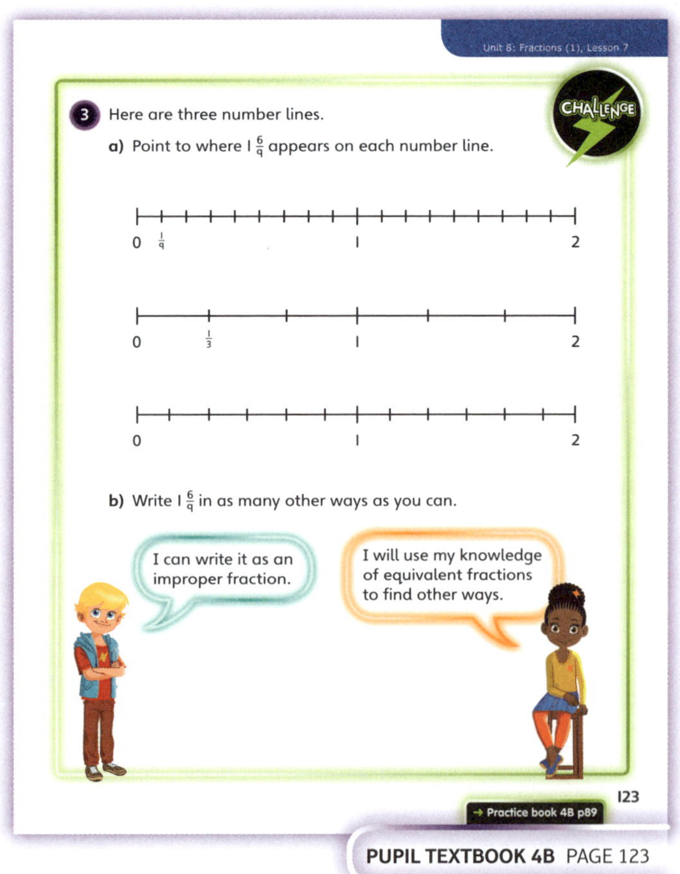

PUPIL TEXTBOOK 4B PAGE 123

Practice

WAYS OF WORKING Independent thinking

IN FOCUS Question ❷ asks children to identify what fraction is being shown on a number line.

Question ❸ gives children the opportunity to place mixed numbers and improper fractions on a number line.

Question ❻ lets children explain their reasoning and explain how fractions with different denominators can be equivalent.

STRENGTHEN Provide children with a blank number line or a number line with the increments required by a question. For question ❹ b) ask them to indicate where the whole numbers 1 and 2 would lie on the number line.

DEEPEN Can children find fractions that are equivalent to the ones provided or shown on a number line?

ASSESSMENT CHECKPOINT Use questions ❶ and ❷ to assess whether children can identify mixed numbers and improper fractions on a number line. Use question ❸ to assess whether children can mark mixed numbers and improper fractions on a number line. Check that children can provide answers both as mixed numbers and as improper fractions.

ANSWERS Answers for the **Practice** part of the lesson appear in the separate **Practice and Reflect answer guide**.

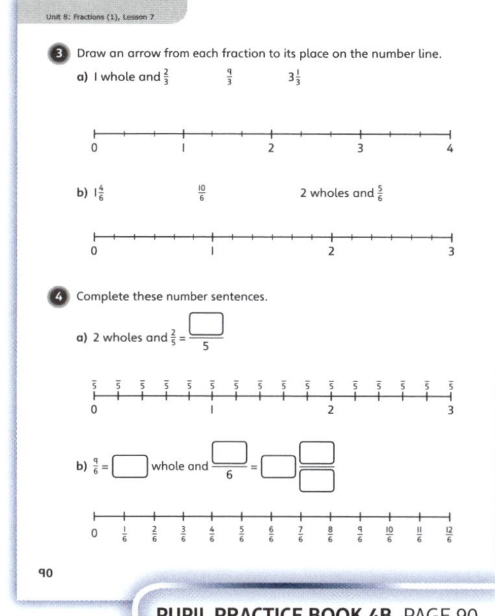

PUPIL PRACTICE BOOK 4B PAGE 89

PUPIL PRACTICE BOOK 4B PAGE 90

Reflect

WAYS OF WORKING Pair work

IN FOCUS This activity requires children to articulate how to find mixed numbers and improper fractions and which they prefer to find.

ASSESSMENT CHECKPOINT Assess whether children can explain the similarities and differences between mixed numbers and improper fractions.

ANSWERS Answers for the **Reflect** part of the lesson appear in the separate **Practice and Reflect answer guide**.

PUPIL PRACTICE BOOK 4B PAGE 91

After the lesson 

- Can children identify mixed numbers and improper fractions?
- Do children know that if the numerator is greater than the denominator, the fraction is greater than 1?
- Can children show mixed numbers and improper fractions on a number line?

End of unit check

Don't forget the *Power Maths* unit assessment grid on p26.

WAYS OF WORKING Group work adult led

IN FOCUS Questions **1** and **3** provide an opportunity to see if children can apply their understanding of parts and wholes to simplify fractions, tenths and hundredths.

Question **4** allows children to show their understanding of writing a fraction in its simplest form, while providing other options that are not in the simplest form.

Question **6** is a SATs-style question that provides children with an opportunity to explain their reasoning.

ANSWERS AND COMMENTARY Children who have mastered the concepts in this unit will recognise the link between tenths and hundredths (for example, that 1 tenth is equal to 10 hundredths). They will be able to use strip diagrams to show equivalent fractions and will be able to simplify fractions to their simplest form. Children will also be able to count beyond 1 using a number line and understand that fractions greater than 1 can be written in different ways.

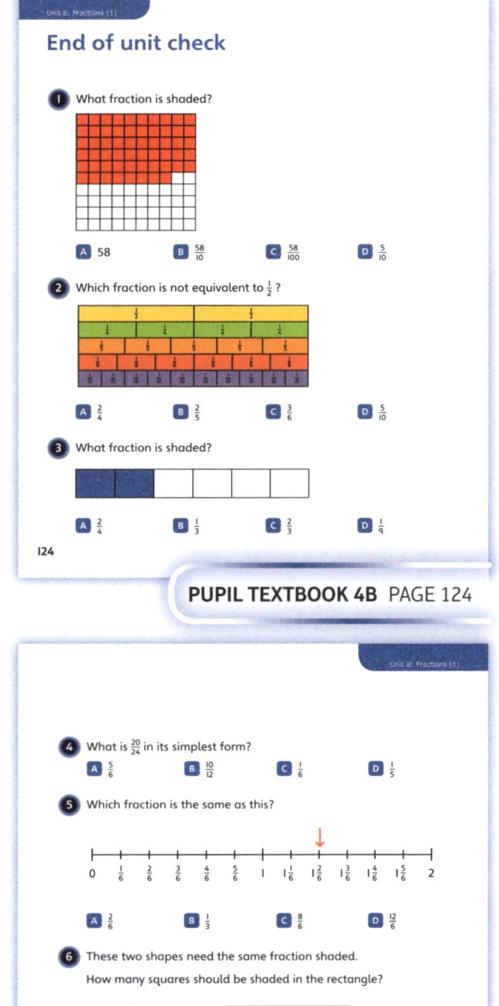

PUPIL TEXTBOOK 4B PAGE 124

PUPIL TEXTBOOK 4B PAGE 125

Q	A	WRONG ANSWERS AND MISCONCEPTIONS	STRENGTHENING UNDERSTANDING
1	C	A suggests children do not understand that each small square is $\frac{1}{100}$.	To help children gain fluency in identifying shaded fractions, ask: • *How many parts are shaded?* • *What is the whole?* • *What fraction is shaded?*
2	B	A, C or D suggests children have made a mistake lining up the fractions on the fraction wall.	To help children identify equivalent fractions, ask: • *What does equivalent mean?*
3	B	A indicates that children have confused the numerator and denominator. C or D suggests they have made a mistake when simplifying.	To help children gain fluency in simplifying fractions, ask: • *What do you do to the numerator and denominator to simplify?* • *Can you divide them again?*
4	A	D indicates that they have subtracted the same number from the numerator and denominator.	
5	C	A or B indicates children have used only the fraction part of the mixed number.	
6	15 squares	Some children may shade the same number of squares as segments in the circle (6).	

My journal

WAYS OF WORKING Independent thinking

ANSWERS AND COMMENTARY Possible answers include:

$\frac{1}{4}$ $\frac{2}{8}$ $\frac{3}{12}$ $\frac{4}{16}$ $\frac{5}{20}$ $\frac{6}{24}$ $\frac{7}{28}$ $\frac{8}{32}$ $\frac{9}{36}$ $\frac{10}{40}$ $\frac{11}{44}$ $\frac{12}{48}$ $\frac{13}{52}$ $\frac{14}{56}$ $\frac{15}{60}$ $\frac{16}{64}$ $\frac{17}{68}$ $\frac{18}{72}$ $\frac{19}{76}$ $\frac{20}{80}$

$\frac{1}{3}$ $\frac{2}{6}$ $\frac{3}{9}$ $\frac{4}{12}$ $\frac{5}{15}$ $\frac{6}{18}$ $\frac{7}{21}$ $\frac{8}{24}$ $\frac{9}{27}$ $\frac{10}{30}$ $\frac{11}{33}$ $\frac{12}{36}$ $\frac{13}{39}$ $\frac{14}{42}$ $\frac{15}{45}$ $\frac{16}{48}$ $\frac{17}{51}$ $\frac{18}{54}$ $\frac{19}{57}$ $\frac{20}{60}$

$\frac{11}{20}$ $\frac{22}{40}$ $\frac{33}{60}$ $\frac{44}{80}$ $\frac{55}{100}$ $\frac{66}{120}$ $\frac{77}{140}$ $\frac{88}{160}$ $\frac{99}{180}$ $\frac{110}{200}$ $\frac{121}{220}$ $\frac{132}{240}$ $\frac{143}{260}$ $\frac{154}{280}$ $\frac{165}{300}$ $\frac{176}{320}$ $\frac{187}{340}$

$\frac{198}{360}$ $\frac{209}{380}$ $\frac{220}{400}$

The simplest form is shown in red.

Children may find multiple answers for each fraction and need to justify their responses (i.e. I multiplied or divided the numerators and denominators by *x*). They need to explain how they know the simplest form for each fraction (i.e. the numerator and denominator cannot be divided by the same number).

Power check

WAYS OF WORKING Independent thinking

ASK

- *What can you do now that you could not do at the start of the unit?*
- *What do you know now that you did not know at the start of the unit?*
- *Can you write down how your learning has developed?*

Power play

WAYS OF WORKING Pair work or small groups

IN FOCUS This **Power play** gives children the opportunity to apply the skills taught in the unit. As they move around the board they need to draw or simplify fractions, explain if fractions can be simplified or provide equivalent fractions. They can also create their own game on the board provided.

ANSWERS AND COMMENTARY Children may choose to use a number line, hundredths grid or shapes to help them with the game.

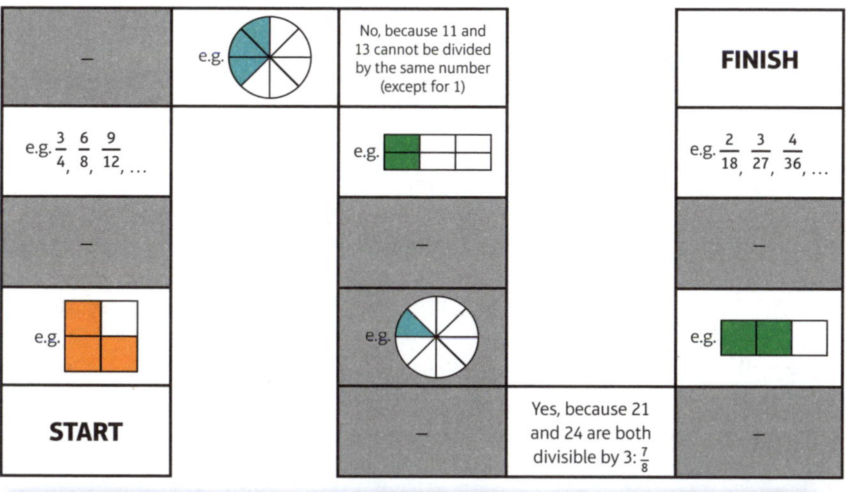

After the unit

- Can children confidently simplify fractions and explain what mixed numbers and improper fractions are?
- Do children know how many hundredths make a tenth?

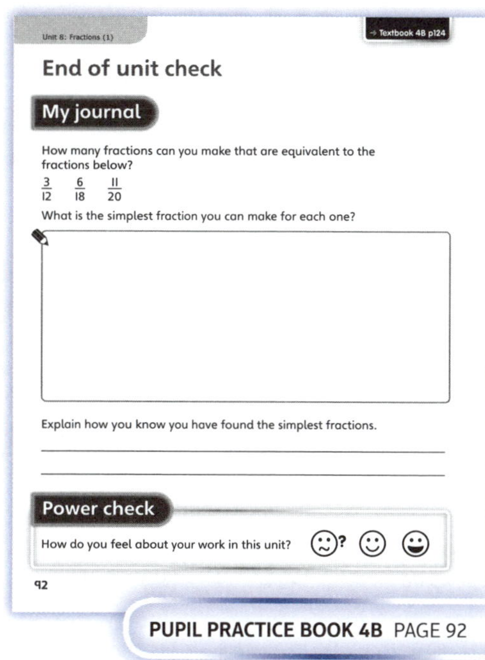

PUPIL PRACTICE BOOK 4B PAGE 92

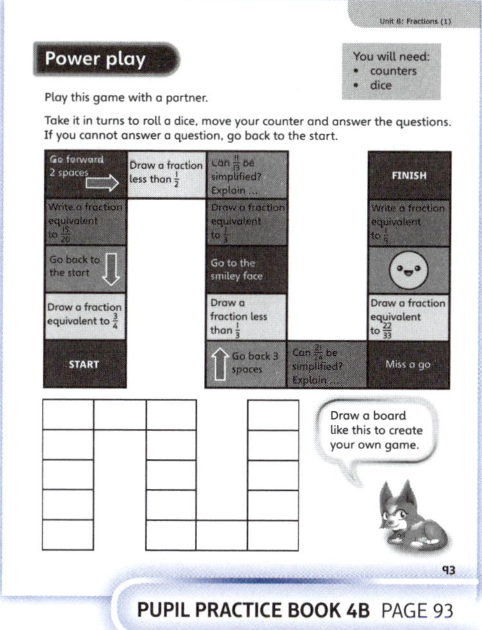

PUPIL PRACTICE BOOK 4B PAGE 93

Strengthen and **Deepen** activities for this unit can be found in the *Power Maths* online subscription.

Unit 9
Fractions ❷

Don't forget to watch the Unit 9 video!

WHY THIS UNIT IS IMPORTANT

This unit is important as children look to build on their fraction work from Year 3 and the previous unit, by extending their knowledge of adding and subtracting fractions to fractions where the answers are greater than 1. Children realise that, as long as the denominators are equal, they can add the numerators. Children start to explore subtracting a fraction from a whole number, which will support their more in-depth exploration of fractions in Year 5. Finally, children continue to find a fraction of an amount, working with divisions within times-tables they have learnt. It is vital that children see the connection with division and are able to use visual representations, such as fraction strips, to represent a given problem.

WHERE THIS UNIT FITS

→ Unit 8: Fractions (1)

→ **Unit 9: Fractions (2)**

→ Unit 10: Decimals (1)

This unit builds on children's work in Year 3 when they added and subtracted fractions with the same denominator. They deepen their understanding of finding a fraction of an amount using both unit and non-unit fractions. Children see the link between fractions and the work they have done on multiplication and division and they should now be able to deal with any times-table facts.

Before they start this unit, it is expected that children:
- know how to add and subtract two fractions with the same denominator
- know when two fractions add up to 1
- know how to subtract a fraction from 1.

ASSESSING MASTERY

Children will be able to confidently add and subtract fractions with the same denominator. They will be able to write their answers as either mixed numbers or improper fractions. Children will be able to subtract a fraction from a whole number. They will be able to find a fraction of an amount and also find the whole when given a fraction of an amount.

COMMON MISCONCEPTIONS	STRENGTHENING UNDERSTANDING	GOING DEEPER
Children add or subtract the denominators when adding or subtracting fractions.	Use a fraction strip above a number line to show how to add fractions, and cross out parts when subtracting.	Challenge children to add or subtract three or more numbers and find three fractions that total 1, for example.
Children may find it difficult to work with improper fractions and mixed numbers.	Use a number line to help children by counting aloud above 1 to help them see the connection. Try to interchange between improper fractions and mixed numbers regularly to get children used to both.	Children apply their knowledge of subtracting fractions from a whole number by applying it to questions such as $3 - 1\frac{1}{3}$. Children realise they can subtract the whole first and then subtract $\frac{1}{3}$ from 2.
When finding the whole amount, children just find a fraction of an amount.	When finding fractions of an amount, or when you are given a fraction of an amount as a whole number, ask children to draw a fraction strip to support their understanding.	Children start to solve more complicated problems involving finding a fraction of an amount. For example, find $\frac{2}{3}$ of $\frac{1}{4}$ of 24.

WAYS OF WORKING

Use these pages to introduce the unit focus to children as a whole class. Ensure they understand how the fraction strips represent the calculation. Talk through the keywords to see if there are any they do not recognise.

STRUCTURES AND REPRESENTATIONS

Fraction number line: Fraction number lines are used in this unit to show how whole numbers can be split into fractions. Fractions can be written on number lines either as proper and improper fractions (shown here above the line) or as mixed numbers (shown below the line).

Fraction strips: Fraction strips can be used to find the fraction of an amount of the whole. Braces above and below help to calculate parts.

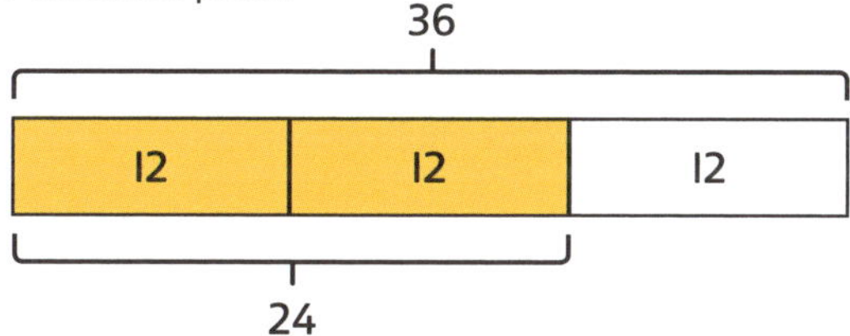

Fraction strips with number line below: Fraction strips are used in this unit to represent fractions of a whole. The fraction number line underneath allows children to add or subtract fractions by adding together on a number line or crossing out. For example:

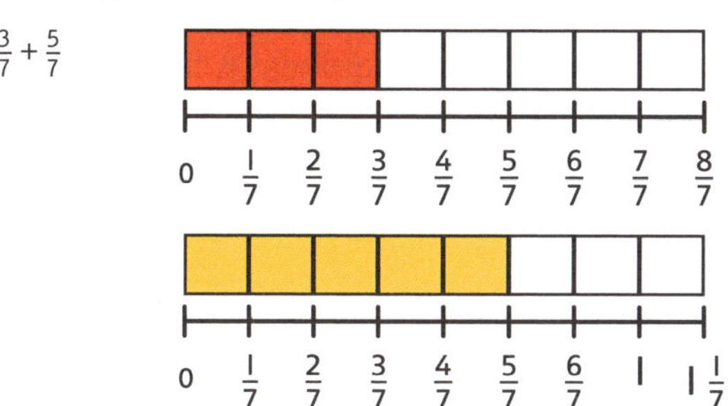

$$\frac{3}{7} + \frac{5}{7} = \frac{8}{7}$$

KEY LANGUAGE

There is some key language that children will need to know as a part of the learning in this unit:

→ numerator, denominator

→ fraction, whole number, mixed number, proper fraction, improper fraction

→ add (+), subtract (−), multiply (×), divide (÷), sign, greater than (>), less than (<)

→ whole, part, find … of …

→ fraction strip, represent, number line, diagram, problem solving

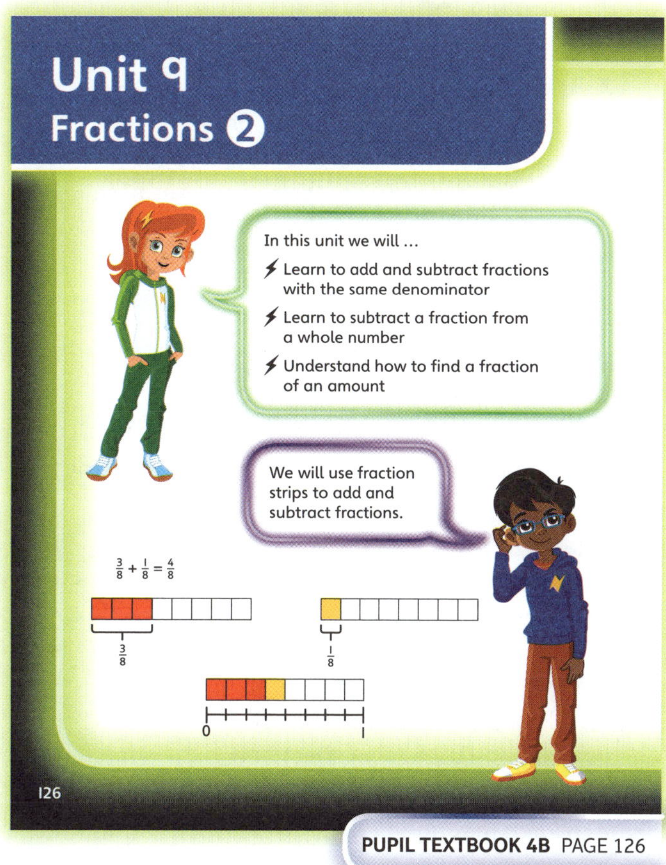

PUPIL TEXTBOOK 4B PAGE 126

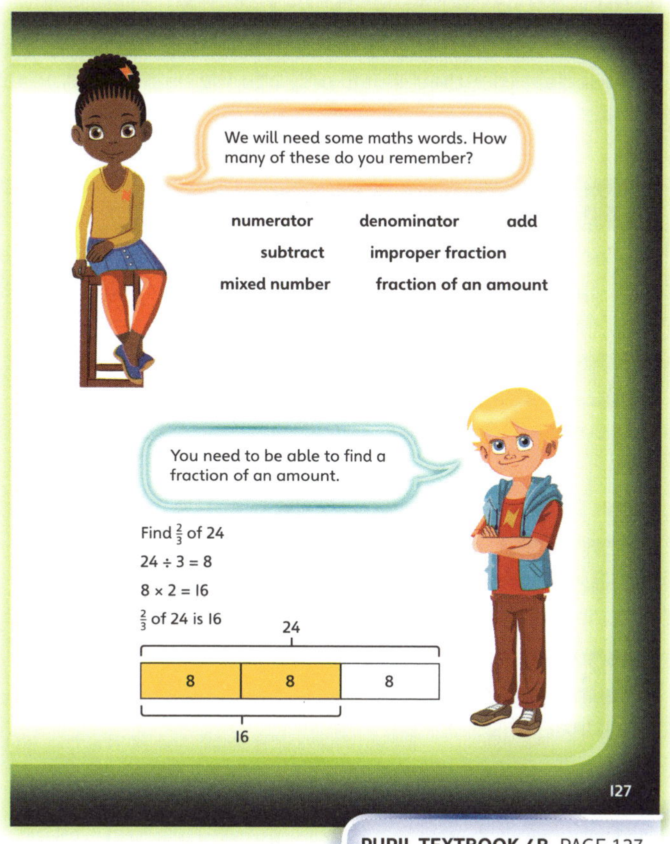

PUPIL TEXTBOOK 4B PAGE 127

Adding fractions

Learning focus

In this lesson, children will add fractions with the same denominator where their answer is greater than one. Children will write their answers as both improper fractions and mixed numbers, using visual aids such as fraction strips and number lines.

Small steps

→ Previous step: Fractions greater than 1 (2)
→ **This step: Adding fractions**
→ Next step: Subtracting fractions (1)

NATIONAL CURRICULUM LINKS

Year 4 Number – Fractions (Including Decimals)

Add and subtract fractions with the same denominator.

ASSESSING MASTERY

Children can add fractions with the same denominator where the answer is greater than 1. They can express their answers as either improper fractions or mixed numbers, using a number line to help them. Children can answer abstract questions or those presented in a context, and draw diagrams to explain their thinking.

COMMON MISCONCEPTIONS

Children may add both the numerator and the denominator. Ask:

• *What is the numerator? What is the denominator? What do these represent? Which should we add to solve the question?*

Some children may misinterpret the question and subtract the fractions rather than adding them. Ask:

• *What word(s) in the question will tell you to add or subtract?*

STRENGTHENING UNDERSTANDING

Strengthen children's understanding by asking them to use fraction strips, bar models or a similar pictorial representation to make it clear that they should only add the numerators. To take a step back, use concrete objects such as base 10 equipment, counters or even everyday objects such as coloured pencils. For example, children can use different coloured cubes to show that $\frac{2}{3} + \frac{2}{3} = \frac{4}{3}$.

GOING DEEPER

Challenge children to write answers as both mixed numbers and improper fractions and to answer questions with more than one possible answer. For example, ask: *How many ways can you make an answer of $\frac{11}{9}$?* You could also incorporate prior learning of inequalities into questions, such as $\frac{4}{7} + \frac{6}{7} < \frac{5}{7} + ?$

KEY LANGUAGE

In lesson: fraction, proper fraction, improper fraction, mixed number, whole, equal parts, numerator, denominator, add, sum, total

Other language to be used by the teacher: represent, kilometre (km), litre (l)

STRUCTURES AND REPRESENTATIONS

fraction strips, number line, bar model

RESOURCES

Mandatory: fraction cards

Optional: base 10 equipment

 In the eTextbook of this lesson, you will find interactive links to a selection of teaching tools.

Before you teach 

• Do children understand what a fraction is?
• Can children draw a diagram or use cubes to represent a fraction?
• Do children understand the concepts of total, altogether and sum?

Discover

 Pair work

ASK

- Question ❶ a): *How many pieces is each pizza cut into? What fraction of their pizza has each child eaten?*
- Question ❶ b): *Why is this question different?*

IN FOCUS Question ❶ a) will highlight the main misconception – whether children add both the numerator and the denominator or just the numerator. Question ❶ b) asks children to work slightly differently, linking the fractions with a unit of measure. Children need to realise that this does not make the question any more difficult than part a).

PRACTICAL TIPS Consider allowing children to make/cook their own pizzas and set up a party-style atmosphere for them so they feel fully engrossed in the lesson. Children could bring in empty litre bottles and fill them with water to re-create the juice activity.

ANSWERS

Question ❶ a): In total, $\frac{7}{5}$ were eaten. This is the same as $1\frac{2}{5}$ pizzas.

Question ❶ b): Kate's and Luis's friends drink $1\frac{6}{10}$ litres in total. This is the same as $\frac{16}{10}$ litres.

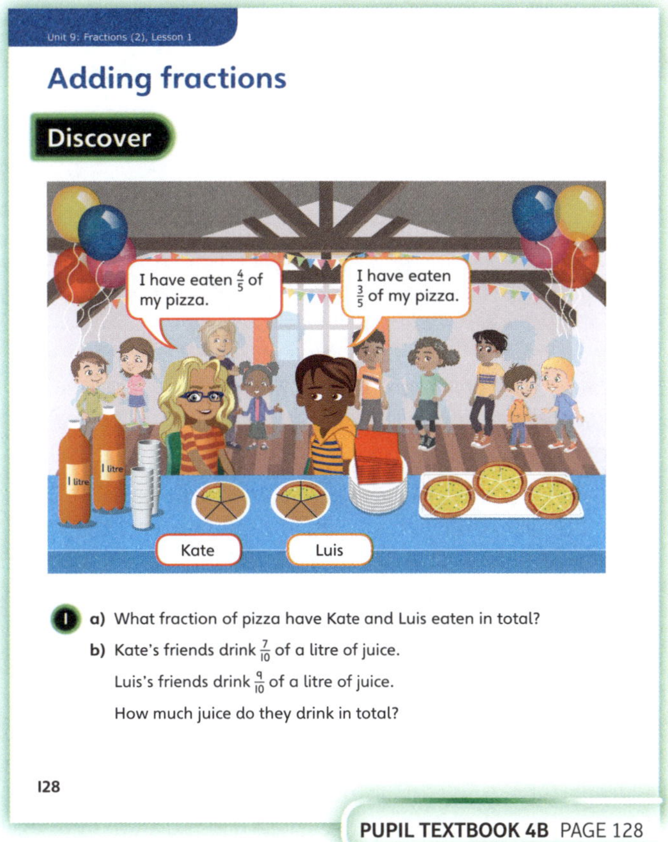

PUPIL TEXTBOOK 4B PAGE 128

Share

  Whole class teacher led

ASK

- Question ❶ a): *To answer this question do you add or subtract? What part of the question tells you this? Can you shade a fraction strip to show how much of each pizza has been eaten? How many fifths have been eaten in total? How many whole pizzas and how many fifths is this? Does this mean that $\frac{7}{5}$ and $1\frac{2}{5}$ are the same? Can you prove this by using number lines?*
- Question ❶ b): *Can you estimate the answer? Will it be bigger or smaller than 1? Why is the denominator not 20? Can you draw a diagram to show why? What should you include with your answer? Do you remember what is meant by 'improper fraction' and by 'mixed number'?*

IN FOCUS For question ❶ a) children should be able to articulate why they have added the numerators but not the denominators. They should be able to use fraction strips to represent their answers and be able to express their answers as an improper fraction and as a mixed number. It is important at this point that they do not formally convert between improper fractions and mixed numbers; instead, they should use a number line to count up and see the equivalence between the representations.

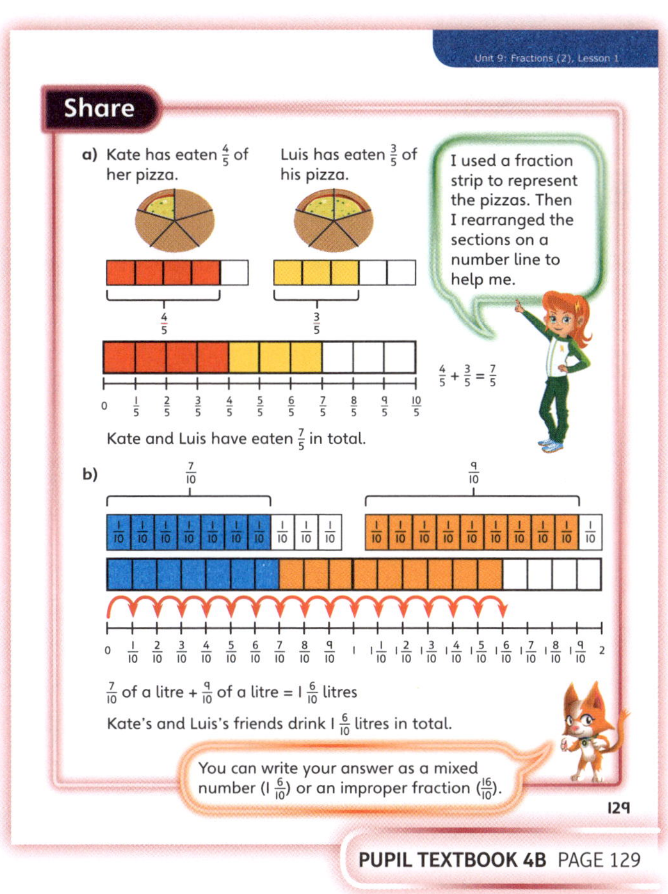

PUPIL TEXTBOOK 4B PAGE 129

Think together

WAYS OF WORKING Whole class teacher led (I do, We do, You do)

ASK

- Question **1**: *What is the same and what is different about this question and the pizza question in the **Discover** section?*
- Question **2**: *Do the units make this question any harder? What must you include with your answer?*
- Question **3**: *What strategy have you used? Does each question have only one answer or is there more than one correct option?*

IN FOCUS Question **3** encourages children to reason about finding different ways to make the same amount. Pay attention to children's strategies, in particular, how they find the sum of all five cards. Do they work from left to right or do they find pairs that make one whole and then add the final fraction? For example, they should recognise that $\frac{1}{8} + \frac{7}{8} = 1$ whole and $\frac{3}{8} + \frac{5}{8} = 1$ whole, with $\frac{9}{8}$ left over; therefore the total is 3 whole numbers and $\frac{1}{8}$ ($3\frac{1}{8}$) or $\frac{25}{8}$.

STRENGTHEN To support understanding, ensure children have access to base 10 equipment so they can physically make each fraction and rearrange them if they wish. Ideally, it should be possible to arrange these cubes over a number line. Perhaps you could have a large number line accessible that fits the cubes exactly. This will really help children to understand why they are adding the numerators only.

DEEPEN Encourage children to explore all the possible ways of making an amount. Can they work systematically to check they have found all possible combinations of fractions that make a specific total? Ask: *Can you come up with your own question for a partner to try?* Children could also explore more open-ended questions such as: find five fractions that have a sum of $1\frac{6}{11}$.

ASSESSMENT CHECKPOINT Question **3** will demonstrate whether children can explain how to add two fractions with a total that is greater than 1. It will show whether they can draw a diagram to support their explanation and write their answers both as improper fractions and as mixed numbers.

ANSWERS

Question **1**: $\frac{9}{7}$ or $1\frac{2}{7}$ of cake is eaten altogether.

Question **2**: Max walks $\frac{15}{9}$ or $1\frac{6}{9}$ km altogether.

Question **3** a): $\frac{9}{8}$ and $\frac{1}{8}$, or $\frac{3}{8}$ and $\frac{7}{8}$

Question **3** b): $\frac{9}{8}$, $\frac{7}{8}$ and $\frac{5}{8}$

Question **3** c): $\frac{1}{8}$ and $\frac{7}{8}$, or $\frac{3}{8}$ and $\frac{5}{8}$

Question **3** d): $\frac{25}{8}$ or $3\frac{1}{8}$

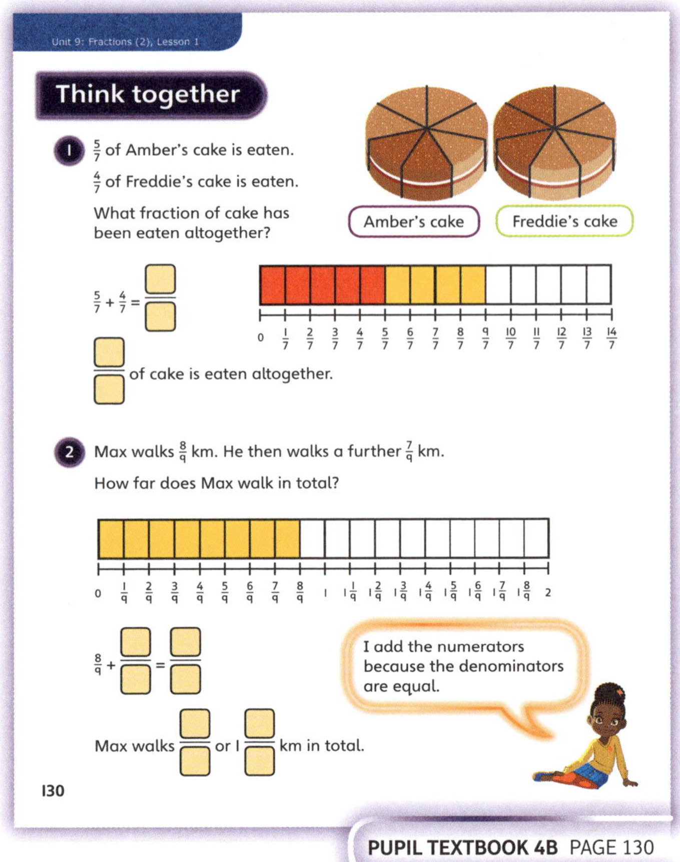

PUPIL TEXTBOOK 4B PAGE 130

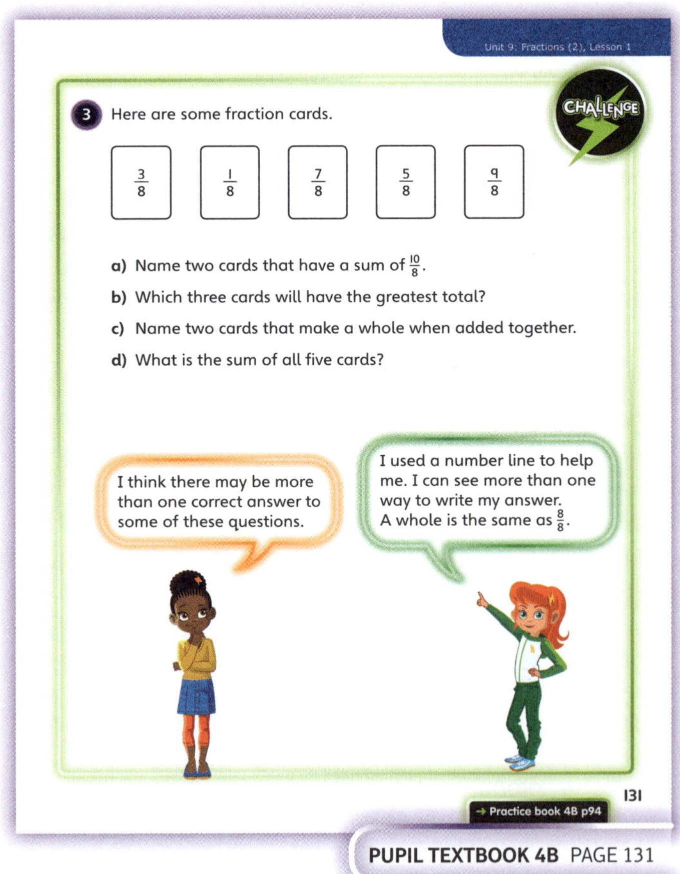

PUPIL TEXTBOOK 4B PAGE 131

Practice

WAYS OF WORKING Independent work

IN FOCUS Questions ❶ and ❷ consolidate children's understanding of adding fractions. Question ❶ expresses the answer as a mixed number whereas question ❷ uses improper fractions. This is to remind children that either answer is acceptable and that they should be able to use both. Question ❸ is purposefully abstract and varies the way in which the question is presented; this is to ensure that children do not always expect fraction questions to be presented in the same way. Children who are not secure in their understanding may struggle with question ❺, which is set up to highlight the misconception that denominators should be added. Question ❻ should be a challenge and will require a deep understanding not only of adding fractions but also of prior learning on addition and subtraction.

STRENGTHEN Ensure children have access to concrete materials such as base 10 equipment and number lines throughout the exercise. Encourage them to draw bar models to represent their answers, as doing this will consolidate their understanding.

DEEPEN Question ❻ allows children to deepen their understanding by finding missing fractions. They should realise that some of the answers are equal and explain why (for example, because $\frac{7}{5}$ and $1\frac{2}{5}$ are equal). In part c), children extend their understanding to adding three fractions and realise that, once again, they can add the numerators as the denominators are all the same. Encourage children to show this works using a bar model.

ASSESSMENT CHECKPOINT Questions ❹ and ❺ highlight the main misconception in this lesson; children should be able to explain clearly why they do not add the denominators when adding two or more fractions.

ANSWERS Answers for the **Practice** part of the lesson appear in the separate **Practice and Reflect answer guide**.

Reflect

WAYS OF WORKING Pair work

IN FOCUS This simple activity will check that children thoroughly understand the lesson and that they have not completed the **Practice** section by following a process. If they cannot draw a diagram explaining this calculation then their understanding is unlikely to be deep enough to move on.

ASSESSMENT CHECKPOINT Ideally children should draw both fraction strips and a number line to prove that their answer is $\frac{8}{5}$. This will give you the opportunity to check that they are confident counting fractions and understand that they are made up of equal parts.

ANSWERS Answers for the **Reflect** part of the lesson appear in the separate **Practice and Reflect answer guide**.

After the lesson ⏸

- Can children count in fractions?
- Can children add two fractions with an answer greater than one?
- Can children express their answer as a mixed number and as an improper fraction?
- Can children explain, using a diagram, why they only add the numerators when adding fractions?

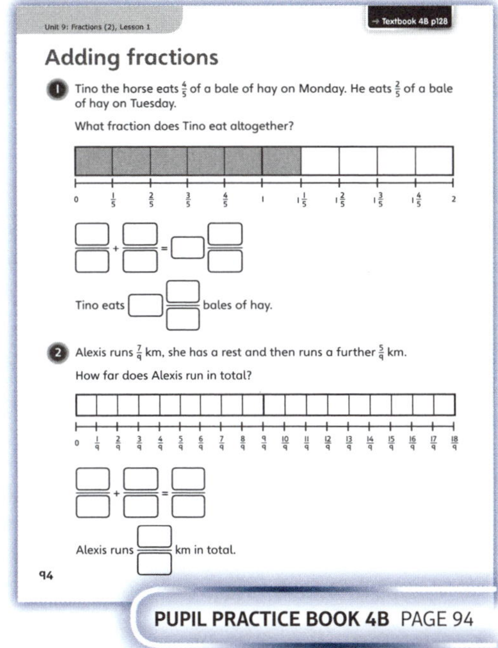

PUPIL PRACTICE BOOK 4B PAGE 94

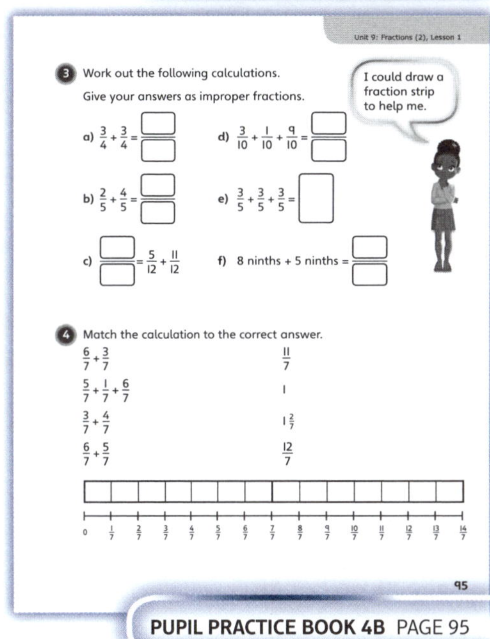

PUPIL PRACTICE BOOK 4B PAGE 95

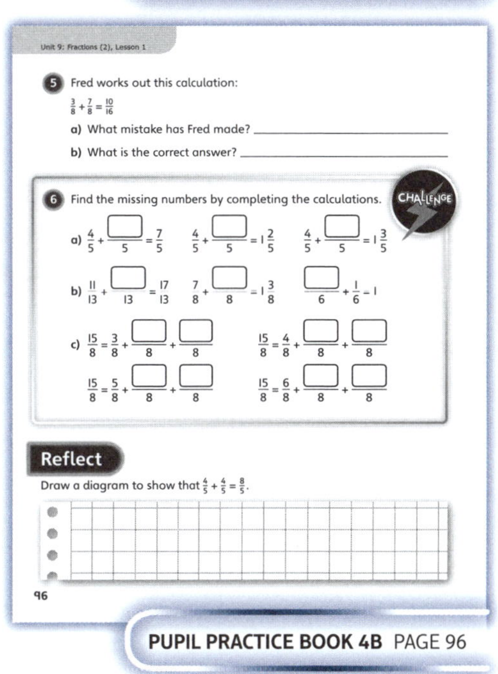

PUPIL PRACTICE BOOK 4B PAGE 96

Subtracting fractions ①

Learning focus

In this lesson, children will subtract proper fractions from mixed numbers with the same denominator. They will use fraction strips and number lines to help them to visualise what is happening.

Small steps

→ Previous step: Adding fractions
→ **This step: Subtracting fractions (1)**
→ Next step: Subtracting fractions (2)

NATIONAL CURRICULUM LINKS

Year 4 Number – Fractions (Including Decimals)
- Add and subtract fractions with the same denominator.
- Solve problems involving increasingly harder fractions to calculate quantities, and fractions to divide quantities, including non-unit fractions where the answer is a whole number.

ASSESSING MASTERY

Children can subtract proper fractions from mixed numbers with the same denominator. They can express their answers as either improper fractions or mixed numbers, using a number line to help them. Children can answer abstract questions or those presented in a context and draw diagrams to explain their thinking.

COMMON MISCONCEPTIONS

Children may treat the whole number as a fraction. Ask:
- *Can you identify which are whole numbers and which are fractions?*

Some children may be inclined to try to convert the mixed number into an improper fraction before completing the subtraction, but in this lesson they will use counting back to calculate their answers. Ask:
- *Will it be easier to complete the subtraction with a mixed number or an improper fraction?*

STRENGTHENING UNDERSTANDING

The lesson includes the use of fraction strips and fraction number lines to support children's understanding of a fraction as part of the whole. To strengthen understanding, children can use base 10 equipment or counters to physically subtract an amount. Cubes work nicely alongside fraction strips to help children see the link between the concrete and the pictorial.

GOING DEEPER

Children can complete answers as both mixed numbers and improper fractions. They should be encouraged to spot patterns in their answers and thus to make reasoned judgements about answers to other calculations without working them out.

KEY LANGUAGE

In lesson: fraction, proper fraction, improper fraction, mixed number, whole, equal parts, numerator, denominator, subtract, difference

Other language to be used by the teacher: represent, pattern, metre (m), kilogram (kg), litre (l)

STRUCTURES AND REPRESENTATIONS

fraction strips, fraction number lines

RESOURCES

Optional: base 10 equipment

 In the eTextbook of this lesson, you will find interactive links to a selection of teaching tools.

Before you teach 

- Are children confident with whole number subtraction?
- Can children draw a diagram or use cubes to represent a fraction?
- Can children count forwards and backwards in fractions?

Discover

WAYS OF WORKING **WAYS OF WORKING** Pair work

ASK

- Question ❶ a): *Do you think this question is about addition or subtraction? Do you think your answer will start with a 2 or a 1?*
- Question ❶ b): *Can you answer this question without doing a calculation? How do you know you are correct?*

IN FOCUS Question ❶ a) highlights the fact that using something up will result in a subtraction. Pick up on key language to help children make this link, such as the word 'left'. Question ❶ b) checks that children can understand the size of fractions and recognise that $\frac{3}{4}$ is smaller than $1\frac{2}{4}$ so Olivia must have enough left. They can complete another subtraction to check.

PRACTICAL TIPS Consider setting up the **Discover** activity practically using a bag of dried spaghetti. This will give children a clear sense of what is going on and what $2\frac{1}{4}$ kg of spaghetti looks like. The spaghetti could even be measured out on scales to check there is exactly $2\frac{1}{4}$ kg.

ANSWERS

Question ❶ a): There are $1\frac{2}{4}$ kg of spaghetti left.

Question ❶ b): Yes, Olivia has enough spaghetti to make the same meal again.

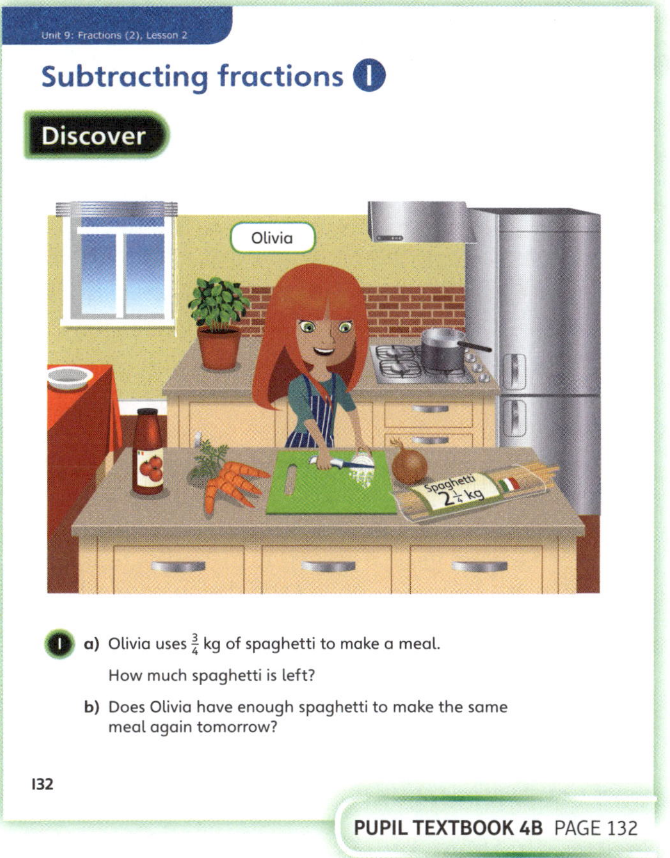

PUPIL TEXTBOOK 4B PAGE 132

Share

WAYS OF WORKING **WAYS OF WORKING** Whole class teacher led

ASK

- Question ❶ a): *To answer this question, do you need to add or subtract? What part of the question tells you this? Can you draw a fraction strip to show how much spaghetti is left? How can you show the subtraction? Will a number line help you to find your answer?*
- Question ❶ b): *Do you need to do a calculation to find the answer? Will Olivia have enough to make the meal a third time? Will there be any spaghetti left over?*

IN FOCUS In question ❶ a), children should be able to articulate why they have performed a subtraction and not an addition. Children will use fraction strips and number lines to prove their answers are correct and to help them see how they can work out answers by counting back in fractions.

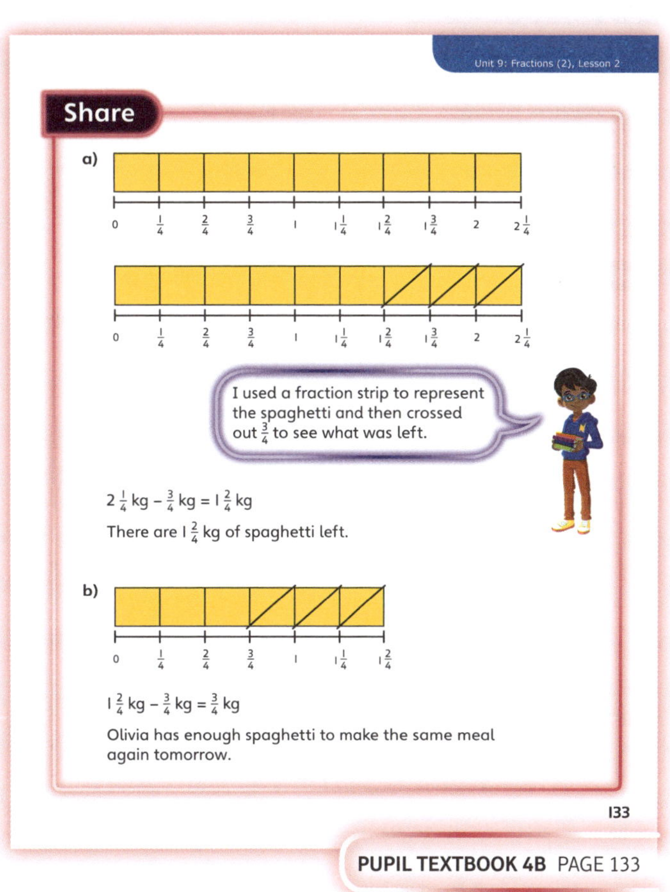

PUPIL TEXTBOOK 4B PAGE 133

Think together

WAYS OF WORKING   Whole class teacher led (I do, We do, You do)

ASK

- Question **3** a): *Which calculation did you find easier to calculate? How could this idea help you in other questions?*
- Question **3** b): *Are your answers getting bigger or smaller? What would you have to subtract to get an answer of $\frac{1}{9}$?*

IN FOCUS   Question **3** introduces the idea of subtracting from mixed numbers and improper fractions. Children are likely to find the improper fraction question easier to calculate. This may encourage them to change other mixed numbers into improper fractions before calculating. This is fine as long as they understand, using a number line, how to do this and are not given a process to follow (for example, multiply by the denominator and add the numerator).

STRENGTHEN   To support understanding, ensure children have access to base 10 equipment so they can physically make each fraction and rearrange them if they wish. Ideally, they should be able to arrange the cubes over a number line. Perhaps have a large number line accessible that fits the cubes exactly and is stuck onto a desk.

DEEPEN   Challenge children to create their own subtraction word problems. Ask: *What language will you use to ensure it is a subtraction? Can you use a variety of different words?* Challenge children to make up subtraction word problems that include crossing the whole.

ASSESSMENT CHECKPOINT   Question **3** will assess whether children have understood how to subtract fractions. Consider whether they can use a fraction strip and a number line to explain their answers. Check if some children are beginning to calculate mentally. If so, ask them to describe their thought process to the rest of the group. Ask: *What pictures are they making in their head? How are they checking the calculation?*

ANSWERS

Question **1**: $3\frac{1}{5} - \frac{4}{5} = 2\frac{2}{5}$. There are $2\frac{2}{5}$ litres of water left.

Question **2**: $2\frac{3}{8} - \frac{5}{8} = 1\frac{6}{8}$ km. Mawusi has to walk $1\frac{6}{8}$ km further.

Question **3** a): They both give the same answer of $\frac{4}{7}$. One of them includes a number written as a mixed number and the other includes a number written as an improper fraction.

Question **3** b): $1\frac{1}{9} - \frac{5}{9} = \frac{5}{9}$, $1\frac{1}{9} - \frac{6}{9} = \frac{4}{9}$, $1\frac{1}{9} - \frac{7}{9} = \frac{3}{9}$, $1\frac{1}{9} - \frac{8}{9} = \frac{2}{9}$. Each successive answer goes down by a ninth. Each calculation involves subtracting an extra ninth.

Question **3** c): $3\frac{5}{6} - \frac{10}{6} = 2\frac{1}{6}$

$3\frac{4}{6} - \frac{9}{6} = 2\frac{1}{6}$

$3\frac{3}{6} - \frac{8}{6} = 2\frac{1}{6}$

$3\frac{2}{6} - \frac{7}{6} = 2\frac{1}{6}$

$3\frac{1}{6} - \frac{6}{6} = 2\frac{1}{6}$

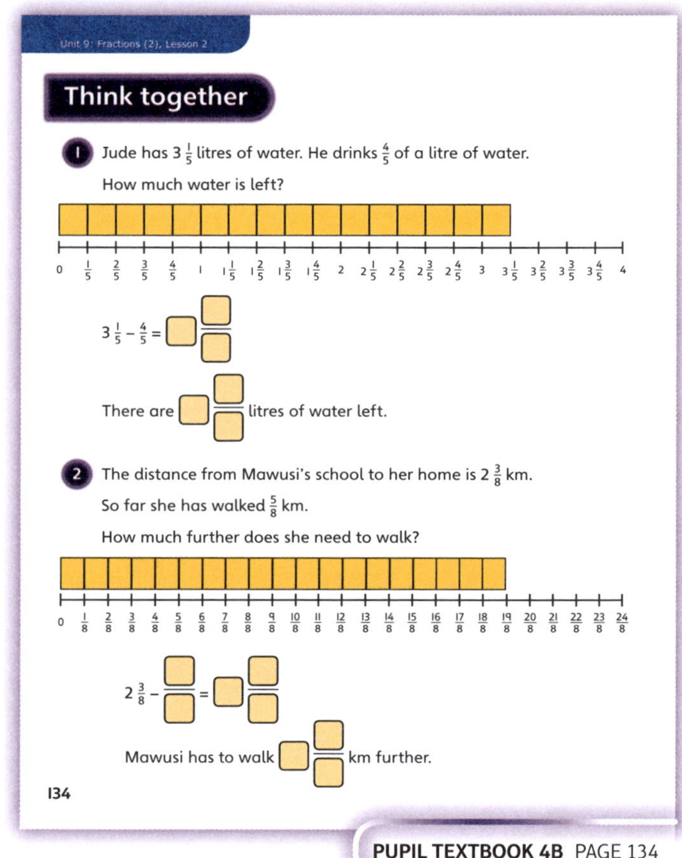

PUPIL TEXTBOOK 4B PAGE 134

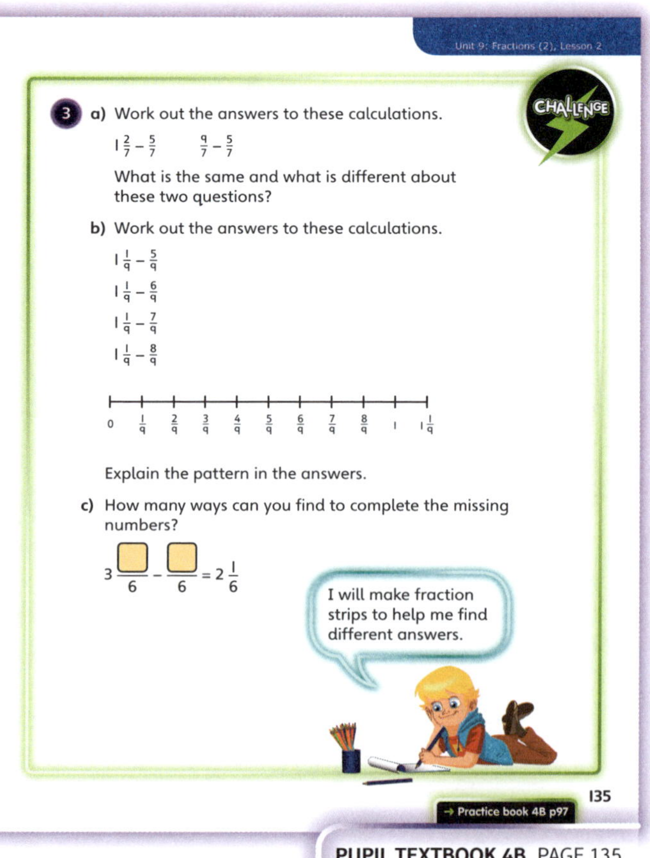

PUPIL TEXTBOOK 4B PAGE 135

Practice

WAYS OF WORKING Independent work

IN FOCUS Question **2** is an abstract question; however, children are given the fraction strip and number line so they can use crossing out to find the answer. Question **3** is purposefully varied to ensure that children are exposed to calculations set out in different ways; part c) links back to other areas of maths where children are encouraged to find efficient ways of working out calculations rather than simply working from left to right.

STRENGTHEN Encourage children to continuously explain their thought process. This will help them to build up a picture in their head and give them a variety of strategies to use. As children use concrete materials, ensure they are thinking about how this links to abstract concepts. For example, they should consider, when they remove cubes, what part of the calculation this represents.

DEEPEN Challenge children to discuss how they tackled question **6**. Ask: *Is there only one way to approach it? How many different approaches can you think of? Which is most efficient?*

ASSESSMENT CHECKPOINT Question **5** will indicate whether children have understood the lesson or not. Subtraction questions are presented in a variety of ways with a variety of different fractions. Some children may need some more time to consolidate their understanding using concrete materials and to talk about mental strategies they could use.

ANSWERS Answers for the **Practice** part of the lesson appear in the separate **Practice and Reflect answer guide**.

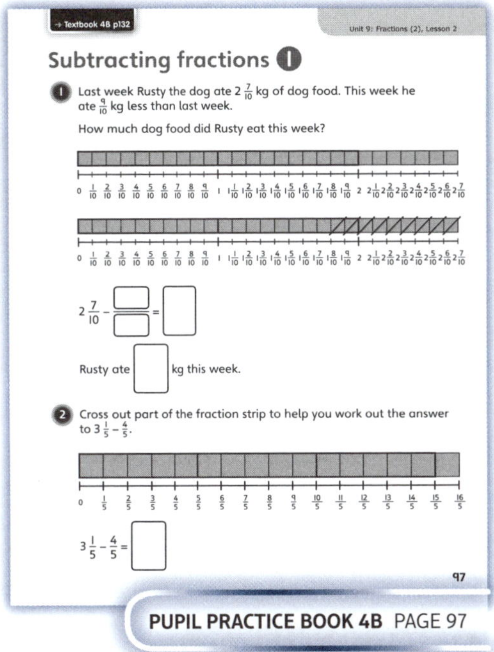

PUPIL PRACTICE BOOK 4B PAGE 97

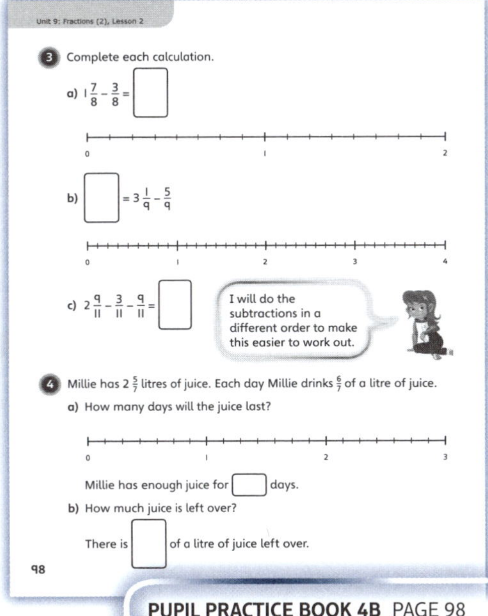

PUPIL PRACTICE BOOK 4B PAGE 98

Reflect

WAYS OF WORKING Pair work

IN FOCUS By drawing a fraction strip and a number line, children are proving that they understand the concept of subtracting fractions. Through this activity it is also possible to check that they understand what a fraction is and that they can set up a number line going up in fractions.

ASSESSMENT CHECKPOINT Check that children are able to explain what is happening when they subtract fractions. They should use phrases like 'take away' and 'left over' to explain their answers. Check that children can recognise different methods and be flexible in the way they approach questions like this.

ANSWERS Answers for the **Reflect** part of the lesson appear in the separate **Practice and Reflect answer guide**.

After the lesson ⏸

- Can children count backwards in fractions?
- Can they subtract a proper fraction from a mixed number?
- Do children have a mental strategy to use?

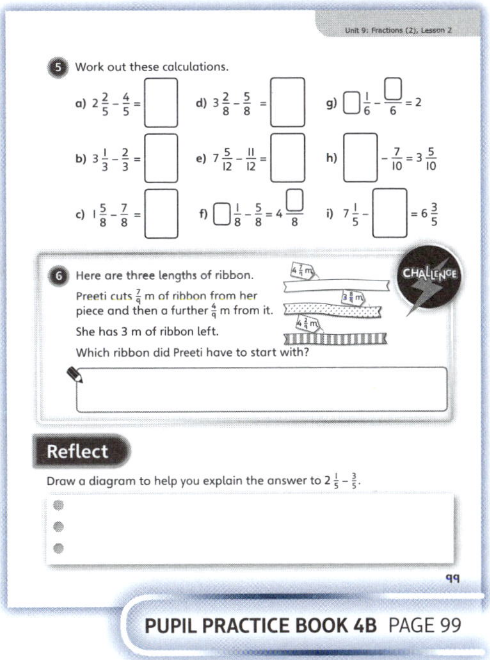

PUPIL PRACTICE BOOK 4B PAGE 99

Subtracting fractions ❷

Learning focus

In this lesson, children will subtract fractions from a whole number and explore different methods.

Small steps

→ Previous step: Subtracting fractions (1)
→ **This step: Subtracting fractions (2)**
→ Next step: Problem solving – adding and subtracting fractions (1)

NATIONAL CURRICULUM LINKS

Year 4 Number – Fractions (Including Decimals)
- Add and subtract fractions with the same denominator.
- Solve problems involving increasingly harder fractions to calculate quantities, and fractions to divide quantities, including non-unit fractions where the answer is a whole number.

ASSESSING MASTERY

Children can use and explain different methods for subtracting a fraction from a whole. Children can decide which method is the most efficient within a context.

COMMON MISCONCEPTIONS

Children may subtract the numerator from the whole without making the connection between the denominator and the whole. For example, they subtract $\frac{2}{6}$ from 4 and get $\frac{2}{6}$. Children need to understand the importance of the denominator representing how many equal parts a whole has been divided into. Ask:
- *What is the whole divided into? How many parts has the whole been divided into? How does the denominator represent the whole?*

STRENGTHENING UNDERSTANDING

If children need more support subtracting fractions, they should use concrete representations. Fraction strips may also support children who cannot see the connection between the whole and the denominator. Encourage children to partition the whole when the whole is more than 1, and then subtract the fraction from 1.

GOING DEEPER

Challenge children to explore and explain different methods for subtracting fractions. Encourage them to decide which method is more efficient, depending on the context. Challenge children to compare the two methods presented and show how they are the same. Children discuss that $2 - \frac{2}{6}$ is the same as converting one of the wholes into $\frac{6}{6}$ and then subtracting $\frac{2}{6}$ from $\frac{6}{6}$. Some children should be able to do this without conversion. Ask them to consider which is more efficient.

KEY LANGUAGE

In lesson: fraction, proper fraction, improper fraction, mixed number, whole, equal parts, numerator, denominator, subtract, difference

Other language to be used by the teacher: represent, partition, remaining, equivalent, kilometre (km), minutes

STRUCTURES AND REPRESENTATIONS

fraction strips, fraction number lines

RESOURCES

Optional: fraction strips, base 10 equipment, real objects to represent fractions of a whole

 In the eTextbook of this lesson, you will find interactive links to a selection of teaching tools.

Before you teach 

- Do children know that the whole can be worth more than 1?
- Do children understand what the denominator and numerator represent?
- Can children make the link between the denominator and the whole?

Discover

 Pair work

ASK

- Question **1** a): *How many slices in total have the 2 cheese and tomato pizzas been cut into? How can you find out how many slices are left? What could you use to represent the pizzas and the slices?*
- Question **1** b): *Can you represent the slices pictorially? How will this help you to calculate how many slices Sofia has sold? How is this question different to the first question?*

IN FOCUS Question **1** a) is used to get children to visualise how many equal parts the whole has been split into. Encourage children to subtract 2 slices from 10 slices. Children can see this subtraction as taking away.

Question **1** b) encourages children to find the difference between the whole and a part. Instead of taking away, children can count up to find the difference.

PRACTICAL TIPS Suggest to children that they can represent the pizzas cut into sixths or 6 equal slices by using circular pieces of paper or paper plates.

ANSWERS

Question **1** a): Sofia has $1\frac{4}{6}$ cheese and tomato pizzas left.

Question **1** b): Sofia sold 5 slices of pizza.

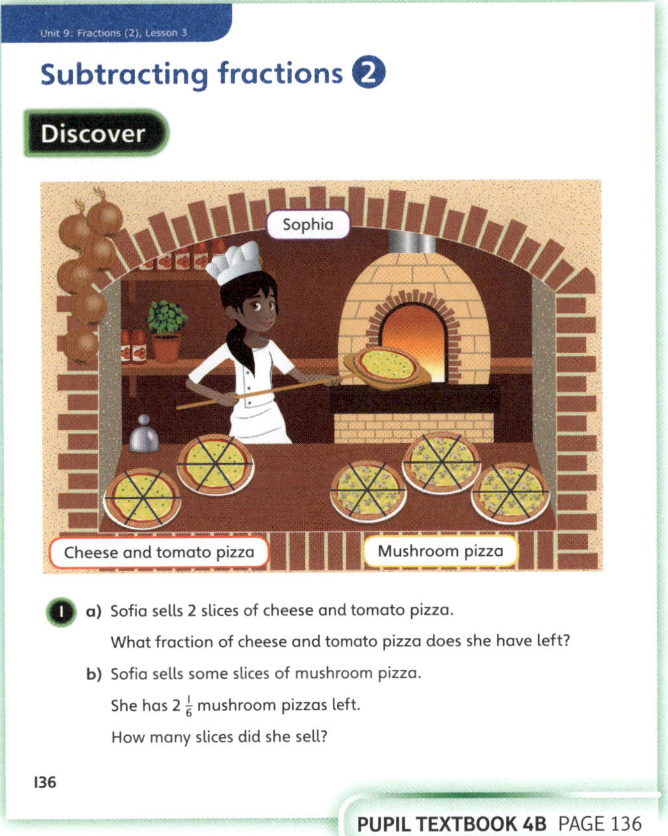

PUPIL TEXTBOOK 4B PAGE 136

Share

 Whole class teacher led

ASK

- Question **1** a): *How many slices have the 2 pizzas been cut into altogether? Can you subtract $\frac{2}{6}$ from 2 easily? Can you partition 2 to help you subtract? Why is 2 the same as 1 and $\frac{6}{6}$?*
- Question **1** b): *How can you calculate what fraction of pizza Sofia sold? How does Flo help us to do it differently? Can you still partition the whole?*

IN FOCUS For question **1** a), encourage children to identify the whole and make connections with the denominator. Show children fraction strips and number lines to represent how the wholes can be partitioned in different ways, where one is made into an improper fraction. For example, one pizza is split into $\frac{6}{6}$ and the rest remain as whole numbers. Can children explain why partitioning in this way is the same and how it helps us to subtract? By the end of the **Share** section, children should have an understanding of how to subtract fractions from a whole amount by using different methods.

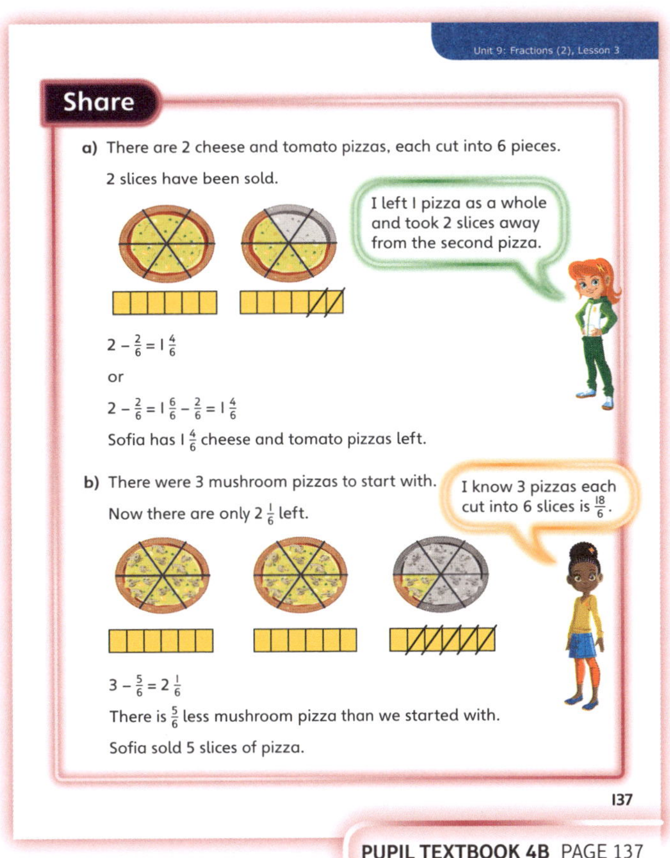

PUPIL TEXTBOOK 4B PAGE 137

Think together

WAYS OF WORKING Whole class teacher led (I do, We do, You do)

ASK

- Question ❶: *How can you partition the whole to help you subtract? What fraction is left?*
- Question ❷: *How many equal parts have the pizzas been divided into? What is the fraction you are subtracting? How can you partition the whole? What fraction is left?*

IN FOCUS Question ❶ looks at partitioning the whole amount into 1 as an improper fraction and the remaining part as a whole number, for example $2 = 1\frac{6}{6}$. Encourage children to subtract the fraction from the improper fraction by using their understanding of number bonds and fractions as a whole. In question ❸ a), children need to solve the subtraction by counting on.

STRENGTHEN To support children's understanding, the questions could be represented using fraction strips. Other concrete representations such as paper plates or circles could also be used.

DEEPEN For question ❷, ask children to work out what fraction would be left if Abdul had eaten $1\frac{1}{5}$ of the pizzas instead. Ask children to work out what fraction would be left if Abdul ate 6 slices.

ASSESSMENT CHECKPOINT By completing question ❸, children are demonstrating that they can partition numbers to help them subtract. Consider whether children recognise how to turn 1 into an improper fraction and whether they can choose the correct denominator based on the context.

ANSWERS

Question ❶: $3 - \frac{2}{3} = 2\frac{3}{3} - \frac{2}{3}$
$= 2\frac{1}{3}$
Holly has $2\frac{1}{3}$ pizzas left for her family.

Question ❷: $8 - \frac{3}{5} = 7\frac{5}{5} - \frac{3}{5}$
$= 7\frac{2}{5}$
There are $7\frac{2}{5}$ pizzas left for Abdul's friends.

Question ❸ a): $5 - 4\frac{2}{7} = 4\frac{7}{7} - 4\frac{2}{7} = \frac{5}{7}$.
$\frac{5}{7}$ of a cheese and tomato pizza has been eaten.

Question ❸ b): As in question ❸ a), $\frac{5}{7}$ of a pizza has been eaten. This is the same as 5 slices, which can be shared between 2 people as: 1 and 4, 2 and 3, or 5 and 0.

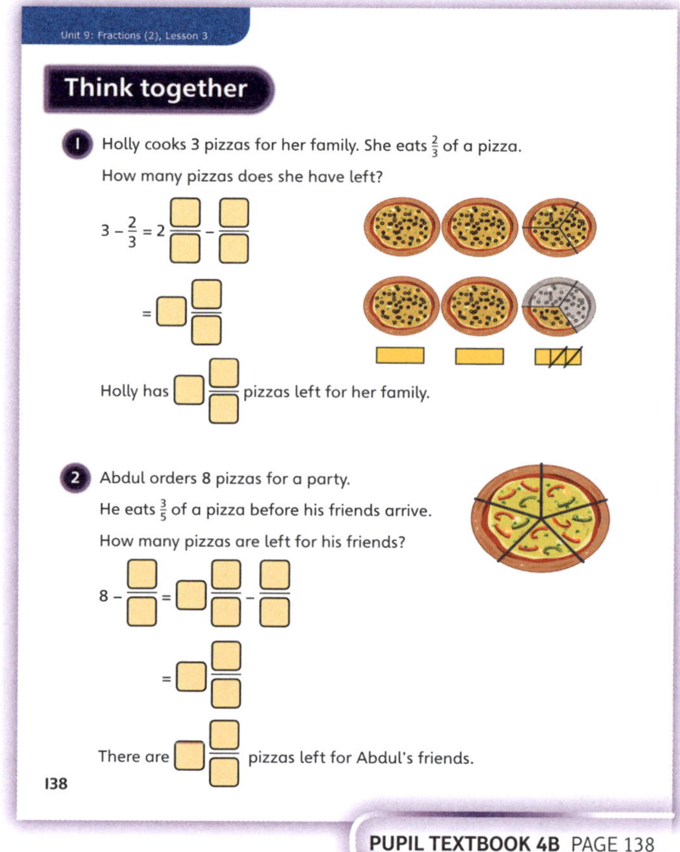

PUPIL TEXTBOOK 4B PAGE 138

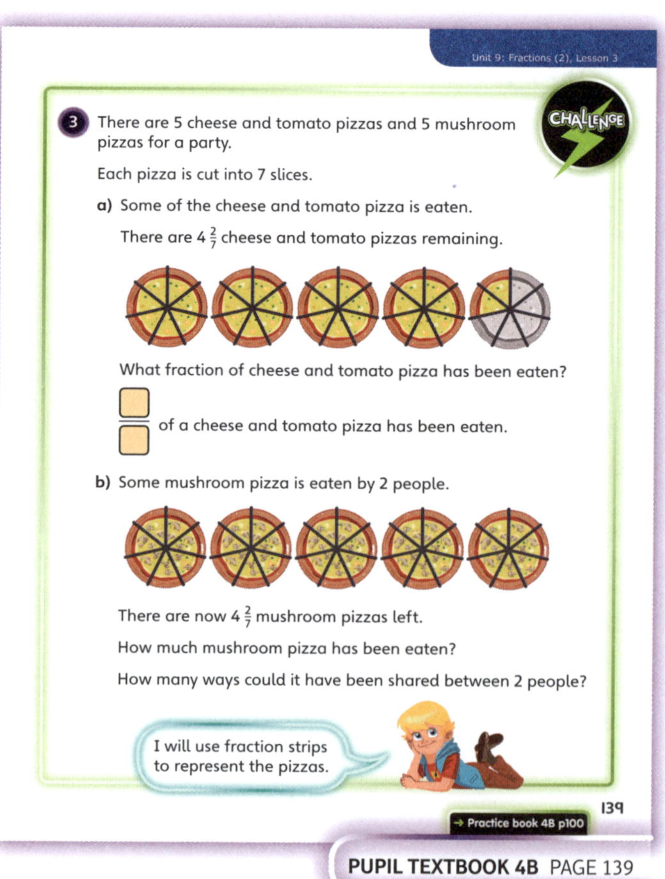

PUPIL TEXTBOOK 4B PAGE 139

Practice

WAYS OF WORKING Independent work

IN FOCUS In question ❶, children realise that they have 2 wholes and they need to subtract $\frac{3}{8}$. They do this by considering the wholes as $1\frac{8}{8}$ and then subtracting $\frac{3}{8}$. The structure is provided in full for children alongside the diagrams.

Question ❸ b) addresses the misconception where children simply subtract the numerator from the whole.

Questions ❹ and ❺ provide abstract practice for subtraction. The questions vary slightly to allow children to try and spot connections and patterns and help them gain a deeper understanding.

STRENGTHEN Children could use fraction strips and number lines to help them with the subtractions from a whole. Children may need support moving between mixed numbers and improper fractions. In Year 4 they do not have to convert, but they need to work with both numbers. A double number line showing the comparison should help children to see the equivalence of fractions.

DEEPEN In question ❻, children have to work out that there are 40 eighths in 5 km, and there are 8 lots of five-eighths in 40 eighths; so she takes 8 lots of 10 minutes.

ASSESSMENT CHECKPOINT Questions ❹ and ❺ will show the level of children's confidence when subtracting a fraction from a whole.

ANSWERS Answers for the **Practice** part of the lesson appear in the separate **Practice and Reflect answer guide**.

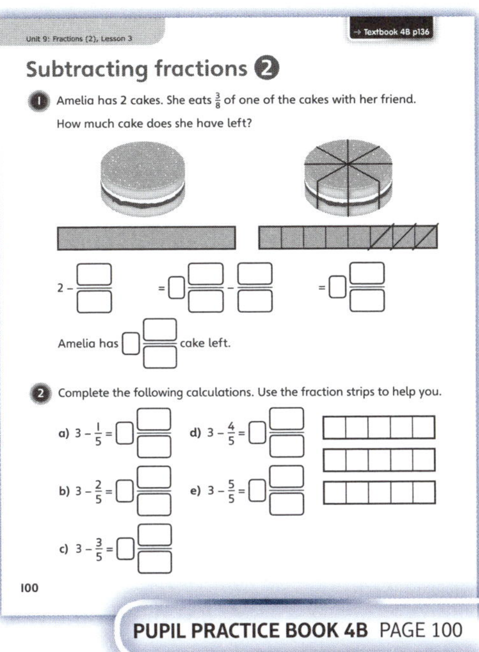

PUPIL PRACTICE BOOK 4B PAGE 100

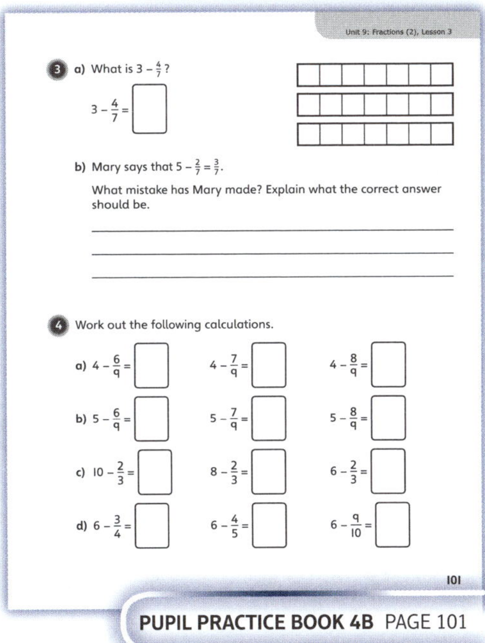

PUPIL PRACTICE BOOK 4B PAGE 101

Reflect

WAYS OF WORKING Pair work

IN FOCUS This **Reflect** question addresses the common misconception of just subtracting the numerator from the whole. Children should explain using diagrams why the answer is not correct. They may explain the mistake that has been made, too.

ASSESSMENT CHECKPOINT Children should be able to work out the correct answer to the question and to explain, through diagrams and other means, why the answer given is not correct.

ANSWERS Answers for the **Reflect** part of the lesson appear in the separate **Practice and Reflect answer guide**.

After the lesson ▮▮

• Can children subtract a fraction from a whole number?
• Can children explain a common mistake that is made?

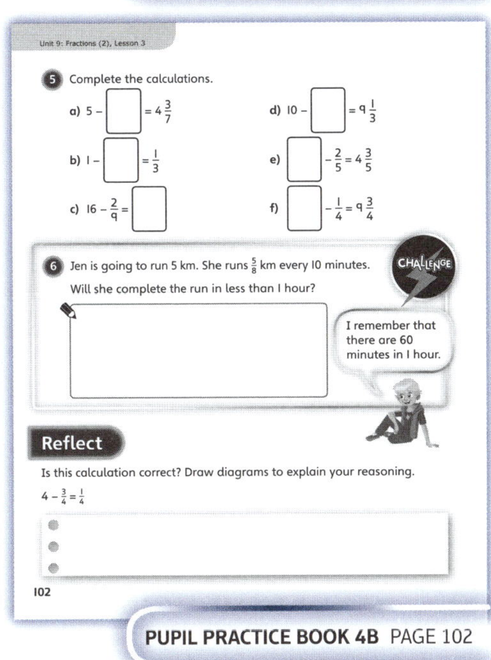

PUPIL PRACTICE BOOK 4B PAGE 102

Problem solving – adding and subtracting fractions ❶

Learning focus

In this lesson, children will apply their understanding of adding and subtracting fractions to solve problems.

Small steps

→ Previous step: Subtracting fractions (2)
→ **This step: Problem solving – adding and subtracting fractions (1)**
→ Next step: Problem solving – adding and subtracting fractions (2)

NATIONAL CURRICULUM LINKS

Year 4 Number – Fractions (Including Decimals)

Solve problems involving increasingly harder fractions to calculate quantities, and fractions to divide quantities, including non-unit fractions where the answer is a whole number.

ASSESSING MASTERY

Children can correctly apply their understanding of adding and subtracting fractions. They can select the correct operation and choose the most efficient method to solve the problem.

COMMON MISCONCEPTIONS

Children may misinterpret which operation to use, and subtract when they should add or vice versa. Ask:
• *Do you need to add or subtract a fraction here?*

Children may think that both the numerator and the denominator should be added/subtracted. Ask:
• *Should you add/subtract the numerator or the denominator?*

STRENGTHENING UNDERSTANDING

Support children with adding and subtracting fractions by providing concrete objects they can manipulate, such as base 10 equipment, to represent problems visually. Encourage children to create fraction strips to help them solve problems.

GOING DEEPER

Children could create two problems where the answer is the same, for example $\frac{5}{8}$, but where the operation is different: one is an addition and the other is a subtraction. Children could also explore solving problems and writing answers both as mixed numbers and as improper fractions.

KEY LANGUAGE

In lesson: problem solving, fraction, proper fraction, improper fraction, mixed number, whole, equal parts, numerator, denominator, add, subtract, partition

Other language to be used by the teacher: addition, subtraction, sum, difference, equivalent, total, remaining, kilometre (km), metre (m), kilogram (kg), acre

STRUCTURES AND REPRESENTATIONS

fraction strips, number line

RESOURCES

Optional: base 10 cubes, fraction wheels, real objects to represent fractions of a whole

 In the eTextbook of this lesson, you will find interactive links to a selection of teaching tools.

Before you teach 

• Do children know that the denominator has to be the same in order to add and to subtract fractions?
• Can children use efficient methods to add and subtract fractions?

Discover

 Pair work

- Question ❶ a): *Show where Lee walks from and to. How many kilometres is it from his home to school? How many kilometres is it from the school to the library? How can you write this as a calculation?*
- Question ❶ b): *How far has Lee walked so far? How far does he walk in total? Do you know how far the harbour is from the library? How can you work it out?*

 Question ❶ a) encourages children to add two fractions that will total an improper fraction. Encourage them to convert the improper fraction to a mixed number.

Question ❶ b) encourages children to use the total from question ❶ a) to find the missing distance. They may take away the given amount from the whole or find the difference to answer the question.

 Children can create their own version of the map, using cubes or counters between the locations to represent each $\frac{1}{8}$ km. This will help them to visualise the distances and the numbers they need to add.

Question ❶ a): Lee walks $1\frac{2}{8}$ km in total.

Question ❶ b): It is $\frac{6}{8}$ km from the library to the harbour.

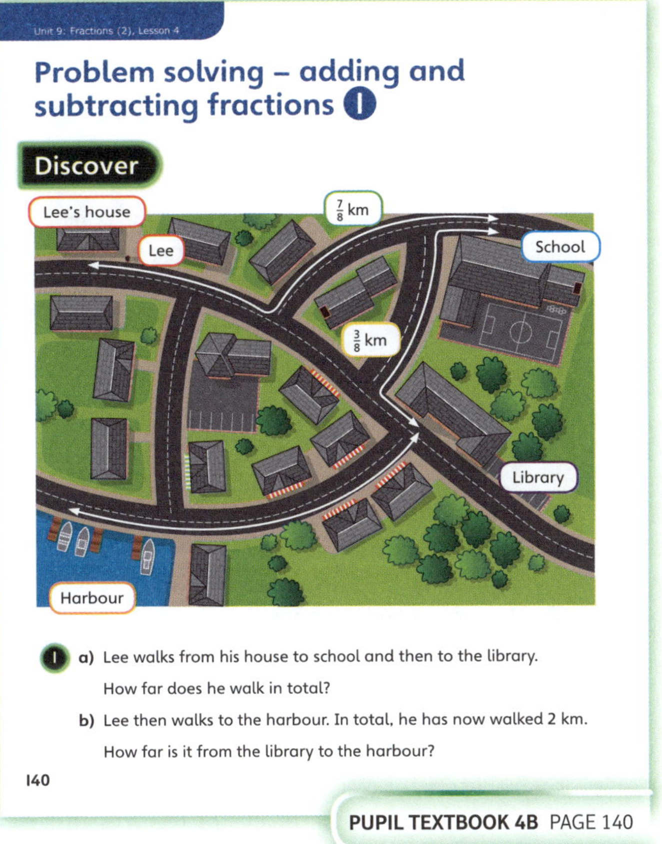

PUPIL TEXTBOOK 4B PAGE 140

Share

 Whole class teacher led

- Question ❶ a): *What distances do you already know? Can you say the answer in a sentence? How can you write this as a mixed number?*
- Question ❶ b): *Which operation will you use to work out how far it is from the library to the harbour? How can you write this as a calculation?*

 In question ❶ a) children will identify the two parts of the journey and add them together. Encourage them to write their answer as a mixed number rather than an improper fraction. Children should explain how they know the improper fraction will convert to that mixed number. Discuss with them the same process for question ❶ b). Children identify what the numbers represent and could link this to their understanding of parts and part-whole models. They could also use a fraction strip to represent the problem.

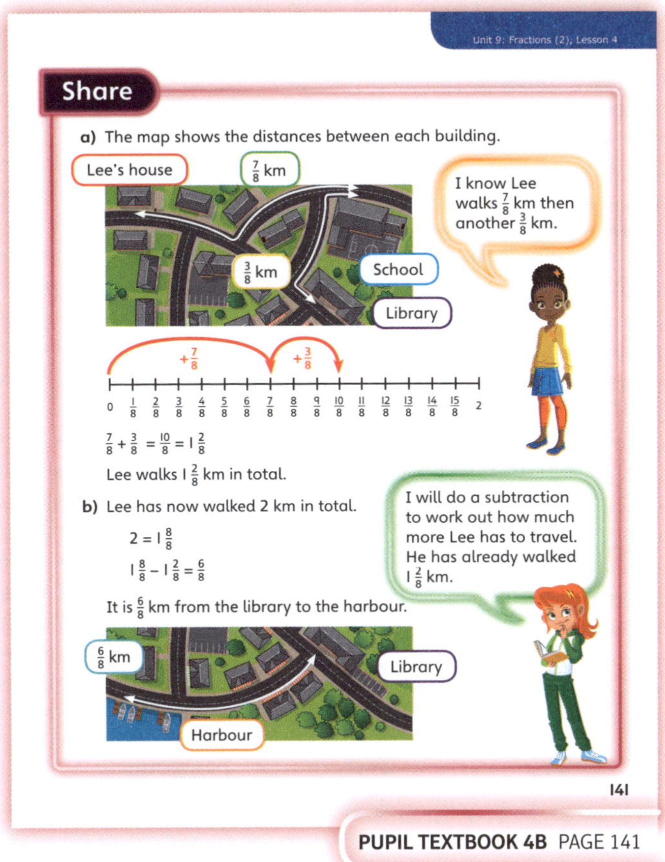

PUPIL TEXTBOOK 4B PAGE 141

Think together

WAYS OF WORKING Whole class teacher led (I do, We do, You do)

ASK

- Question ❶: *What fractions will you add to represent the problem? How far has Ogenna travelled so far? What missing numbers and fractions are needed to complete the subtraction?*
- Question ❷: *What is the whole worth? Can we represent this as a fraction? How can you make the same whole or total with the pieces Aaron has? Are there any pieces of ribbon Aaron cannot use?*
- Question ❸: *How can you calculate how far Max, Lexi and Amelia ran? What strategies can you use to add these fractions? How far did Jamie and Danny run together? How can you work out how far each of them ran?*

IN FOCUS Question ❷ encourages children to apply their skills in a different context. Children could use subtraction to work out the missing parts, or use their understanding of number bonds to work out the missing numbers. Encourage children to link this to a part-whole model.

Question ❸ encourages children to be more flexible in their approach and to interpret the abstract with no other representations. Encourage children to work out how far Max, Lexi and Amelia ran and to write this as a mixed number. Discuss how to find how far Jamie and Danny ran and link this to finding the difference.

STRENGTHEN To support understanding, the problems could be represented with strips of paper to make fraction strips.

DEEPEN For question ❷, ask children to explain and identify if addition could be used to solve all parts of the question and if subtraction could be used to solve all parts. Children could explain why four of the pieces of ribbon could not make 1 m.

ASSESSMENT CHECKPOINT Questions ❶ and ❸ will assess whether children can both add and subtract fractions. Can children explain which operation to use for each problem? Can they explain why in some problems you could use either operation?

ANSWERS

Question ❶: $\frac{1}{7} + \frac{2}{7} = \frac{3}{7}$
$1 - \frac{3}{7} = \frac{4}{7}$
The park is $\frac{4}{7}$ km away from Aisha's house.

Question ❷ a): 1 m (Use the red ribbon.)

Question ❷ b): $\frac{5}{11} + \frac{6}{11} = 1$ (Use the green and purple ribbons.)

Question ❷ c): $\frac{2}{11} + \frac{4}{11} + \frac{5}{11} = 1$ (Use the yellow, blue and green ribbons.)

Question ❸: Danny and Jamie ran $\frac{2}{5}$ km altogether, so $\frac{1}{5}$ km each.

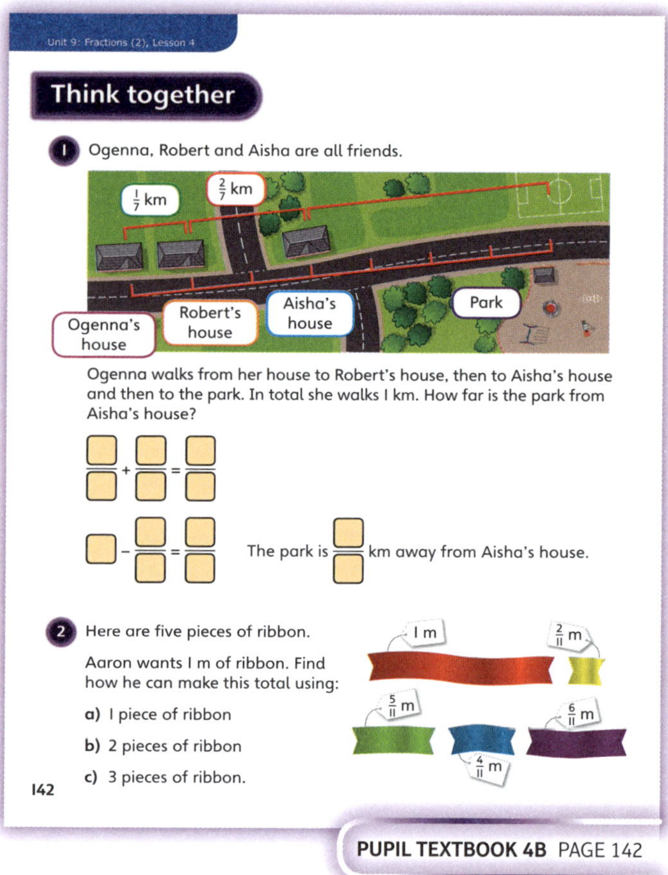

PUPIL TEXTBOOK 4B PAGE 142

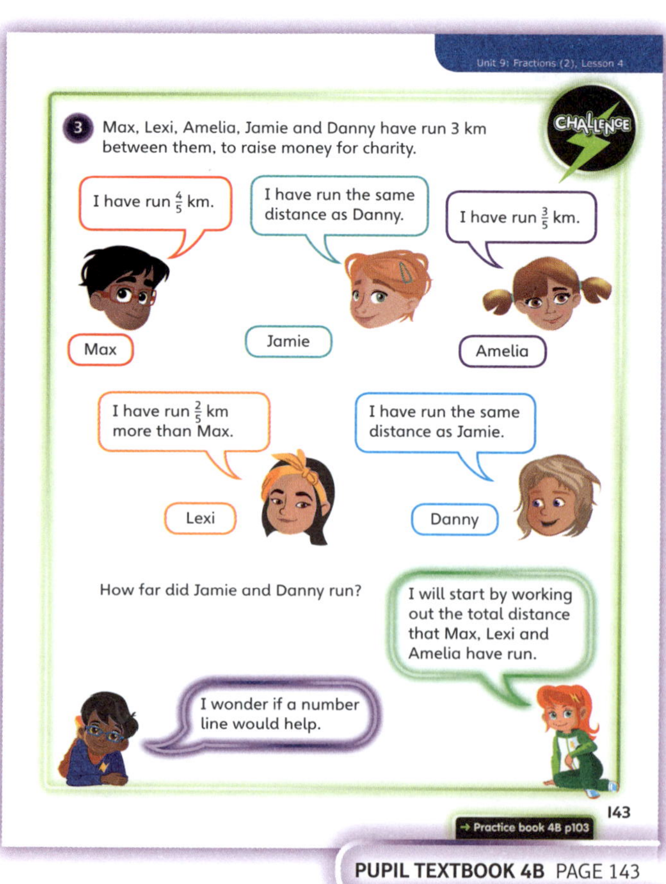

PUPIL TEXTBOOK 4B PAGE 143

Practice

WAYS OF WORKING Independent work

IN FOCUS Questions **1** and **2** consolidate children's understanding of solving word problems involving adding and subtracting fractions. Children will need to extract the relevant information from the questions to decide whether they need to add or subtract. Diagrams are provided to try to help children reason as to why they need to add or subtract.

In question **3**, children need to be careful when interpreting the word 'more' as they tend to just add the values together. In fact, children first need to work out how much Phoebe drinks and then add again, before subtracting from 1.

STRENGTHEN Use fraction strips or fraction wheels to support children's understanding of adding and subtracting fractions. Look for those children who still add and subtract the denominators. Explain to children why they should not do this using concrete and pictorial methods. Children might find it useful if you act out some of the questions, to practise this concept.

DEEPEN Question **5** is deceptively easy as children are adding fractions and then subtracting from 1. However the denominators of the fractions are different and so some children may think they are unable to solve this question. Ensure children realise that they make a whole first and remind them that they know how to subtract a fraction from a whole. You might want to give them other, similar questions, to practise this concept.

ASSESSMENT CHECKPOINT Children should be able to solve simple word problems involving adding and/or subtracting fractions. They should be able to extract the relevant information, draw a supporting diagram and solve it, as in question **5**.

ANSWERS Answers for the **Practice** part of the lesson appear in the separate **Practice and Reflect answer guide**.

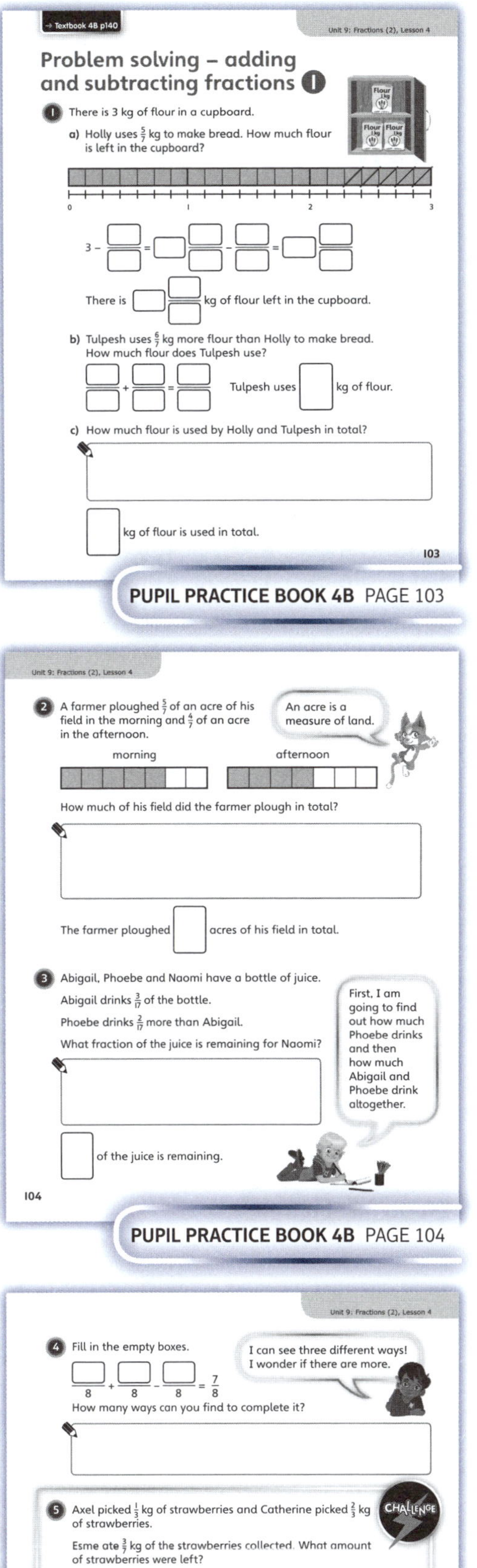

PUPIL PRACTICE BOOK 4B PAGE 103

PUPIL PRACTICE BOOK 4B PAGE 104

Reflect

WAYS OF WORKING Pair work

IN FOCUS In this **Reflect** section, children have to make up their own story that either adds or subtracts two fractions. Look for children using words such as 'sum', 'total', 'numerator', 'denominator', and so on. Share children's responses with the class and discuss the similarities and differences between them.

ASSESSMENT CHECKPOINT Children are able to compose a simple addition or subtraction problem involving fractions.

ANSWERS Answers for the **Reflect** part of the lesson appear in the separate **Practice and Reflect answer guide**.

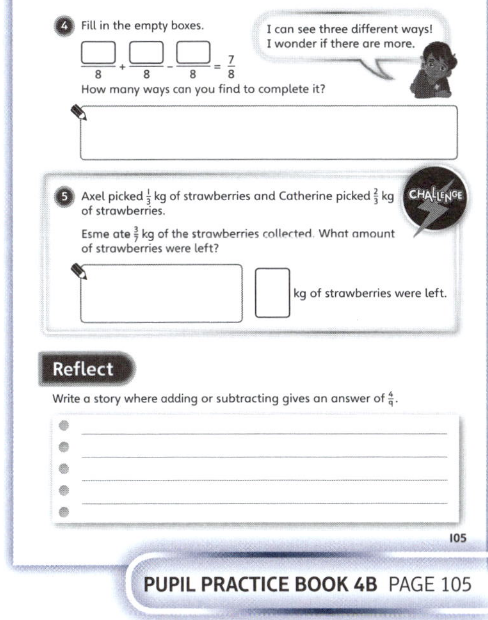

PUPIL PRACTICE BOOK 4B PAGE 105

After the lesson ⏸

- Can children solve simple addition and subtraction problems involving fractions?
- Can they draw supporting diagrams to explain their answers?
- Are they able to write their own addition and subtraction problems for other children to solve?

Problem solving – adding and subtracting fractions ❷

Learning focus

In this lesson, children will solve problems involving adding and subtracting fractions.

Small steps

→ Previous step: Problem solving – adding and subtracting fractions (1)
→ **This step: Problem solving – adding and subtracting fractions (2)**
→ Next step: Calculating fractions of a quantity

NATIONAL CURRICULUM LINKS

Year 4 Number – Fractions (Including Decimals)

Solve problems involving increasingly harder fractions to calculate quantities, and fractions to divide quantities, including non-unit fractions where the answer is a whole number.

ASSESSING MASTERY

Children can determine whether a problem involving fractions requires them to add or subtract, and then solve the problem using their knowledge from previous lessons. Children can solve simple 2-step problems that involve a mix of addition and subtraction of fractions.

COMMON MISCONCEPTIONS

Children may be unsure whether they need to do an addition or subtraction in a given problem. Encourage them to look for keywords and draw a fraction strip to represent the problem. Ask:
• *What is this question asking you to do? What keywords will help you know whether to add or subtract?*

STRENGTHENING UNDERSTANDING

Support children who need help with adding or subtracting fractions by supplying concrete representations such as plates or shapes. Strips of paper used as fraction strips may also support children who cannot see the connection between the whole and the denominator.

GOING DEEPER

Challenge children to make up a word problem that has an answer of $\frac{7}{10}$. Ask: *Can you write one that requires you to add? What about one that involves subtracting? Can you write a 2-step problem?*

KEY LANGUAGE

In lesson: problem solving, fraction, proper fraction, improper fraction, mixed number, whole, equal parts, numerator, denominator, add, subtract, partition

Other language to be used by the teacher: sum, difference, remaining, total, length, metre (m), kilometre (km), kilogram (kg), more than (>), less than (<)

STRUCTURES AND REPRESENTATIONS

fraction strips, number line

RESOURCES

Optional: base 10 equipment, strips of paper

 In the eTextbook of this lesson, you will find interactive links to a selection of teaching tools.

Before you teach 

• Can children add and subtract two fractions with the same denominator?
• Can children subtract a fraction from a whole number?

Discover

  Pair work

ASK

- Question **1** a): *How much ribbon does each person use? What do you need to do to find the total?*
- Question **1** b): *What information do you have now? How much more is there to subtract?*

IN FOCUS In question **1** b), children can solve the problem in two ways. They can add together the total ribbon and then subtract from 2 m, or they can subtract $\frac{4}{7}$ from the answer to **1** a). Children will have to subtract a mixed number from a whole, so review subtracting methods, reminding children to subtract the whole amount first and then the fraction part. Children may find a number line helpful when trying to find the answer.

PRACTICAL TIPS Provide string or fraction strips (divided into 14 sections) and boxes for children to use to represent the ribbon and the presents.

ANSWERS

Question **1** a): They have $1\frac{2}{7}$ metres of ribbon left.

Question **1** b): There is $\frac{5}{7}$ of a metre left now.

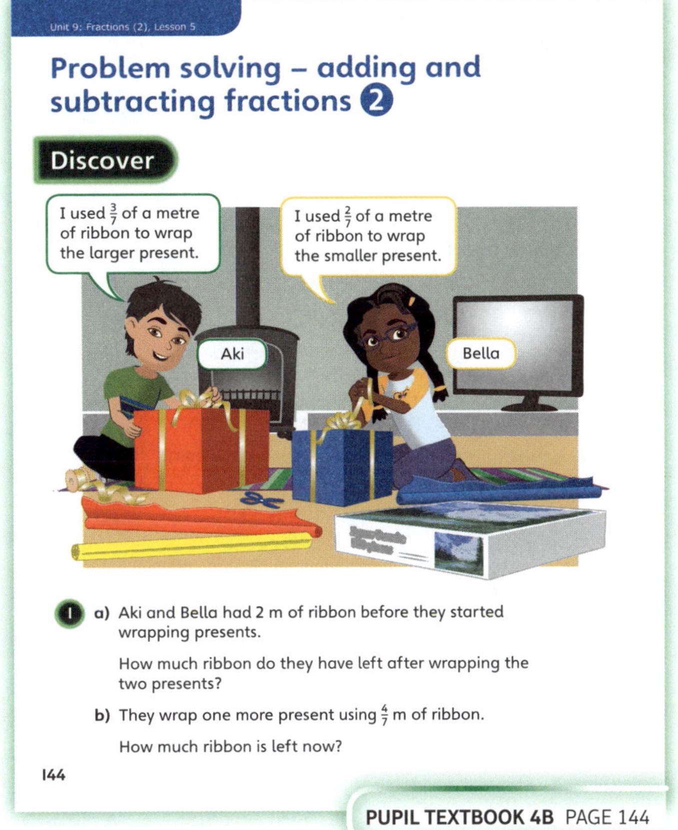

PUPIL TEXTBOOK 4B PAGE 144

Share

  Whole class teacher led

ASK

- Question **1** a): *How do the fraction strip and number line represent the problem?*
- Question **1** b): *How did Flo approach the problem? Is there another way? How can you use your answer from **1** a) to help you?*

IN FOCUS For question **1** a), discuss Astrid's method of adding together the two lengths and then subtracting this from the whole. Ask children about other approaches, such as subtracting the lengths one at a time from the total. Would this result in the same answer? Use the fraction strip and number line to help explain the addition and subtraction and how the two methods are the same.

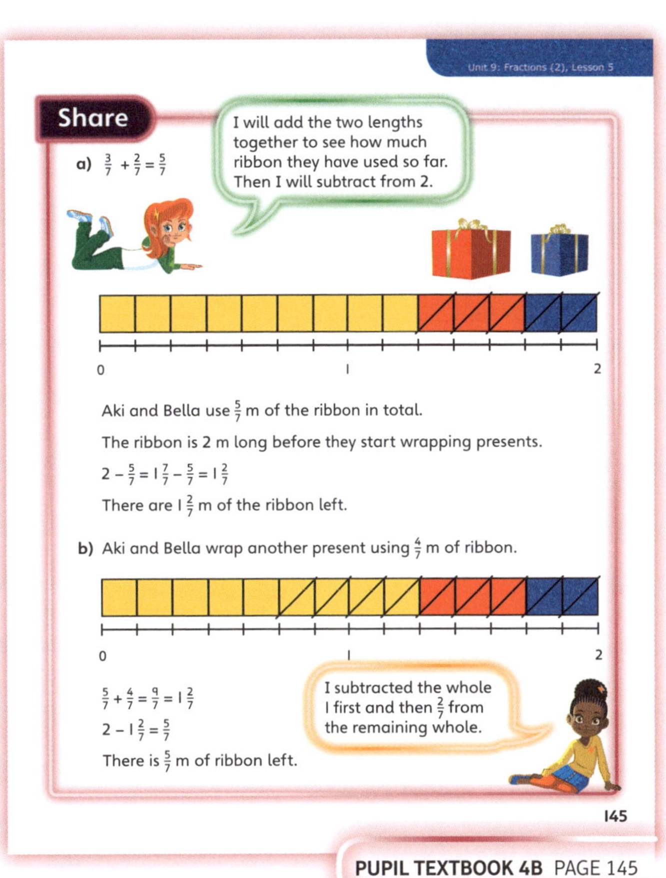

PUPIL TEXTBOOK 4B PAGE 145

Think together

WAYS OF WORKING Whole class teacher led (I do, We do, You do)

ASK

- Question ❶: *How much juice is in each jug? How do you know if it will fit into one jug?*
- Question ❷: *How is this question similar to the last one? What do you have to do differently?*
- Question ❸: *Which fractions have something in common? Will putting the fractions in a different order help?*

IN FOCUS In question ❸, children are presented with fractions where the denominators are different. They may approach this problem by adding the numerators *and* the denominators, so remind them that they can only add numerators when the denominators are the same. Encourage children to pair up the fractions which have the same denominator. Each pair of fractions adds together to make a whole and so they can all be added together to make 2.

STRENGTHEN Strengthen understanding of questions ❶ and ❷ by encouraging children to perform the problem practically with water and measuring equipment marked off in tenths. This should help to cement understanding of the various fractional amounts.

DEEPEN For question ❸, challenge children to write their own fraction problems which involve matching up pairs of fractions with different denominators. They could then ask a partner to solve their problem.

ASSESSMENT CHECKPOINT Questions ❶ and ❷ assess whether children can solve a variety of problems involving addition and subtraction of fractions. Do they know when they need to add and when they need to subtract?

ANSWERS

Question ❶: $\frac{4}{10} + \frac{3}{10} + \frac{2}{10} = \frac{9}{10}$

$\frac{9}{10} < 1$

The juice will all fit into 1 jug.

Question ❷: $\frac{5}{8} + \frac{7}{8} = \frac{12}{8} = 1\frac{4}{8}$

$2 - 1\frac{4}{8} = 1\frac{8}{8} - 1\frac{4}{8} = \frac{4}{8}$

The third jug is $\frac{4}{8}$ filled with juice.

Question ❸: Max could rearrange the fractions and add up each pair that makes a whole.

$= \frac{1}{5} + \frac{4}{5} + \frac{3}{4} + \frac{1}{4}$

$= \frac{5}{5} + \frac{4}{4}$

$= 1 + 1$

$= 2$

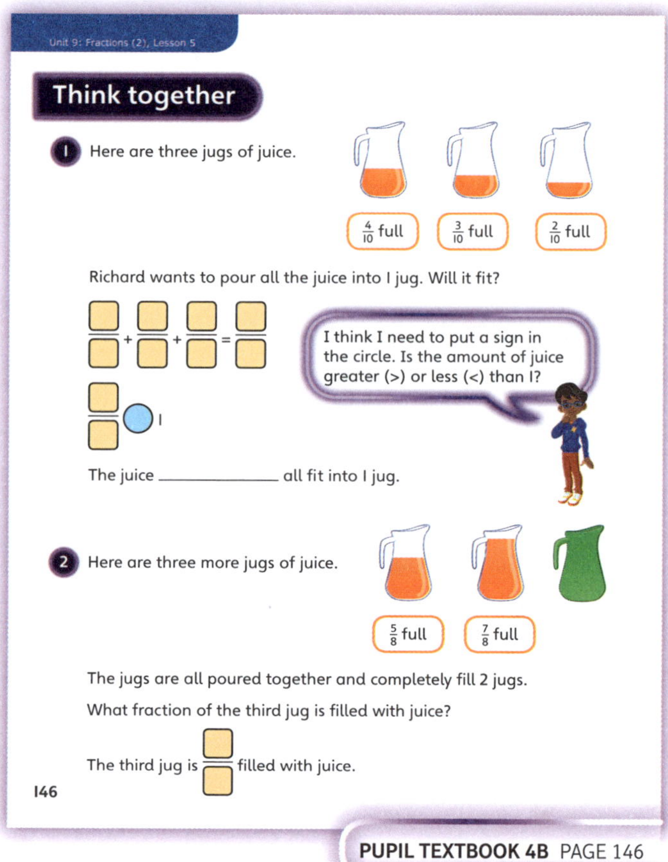

PUPIL TEXTBOOK 4B PAGE 146

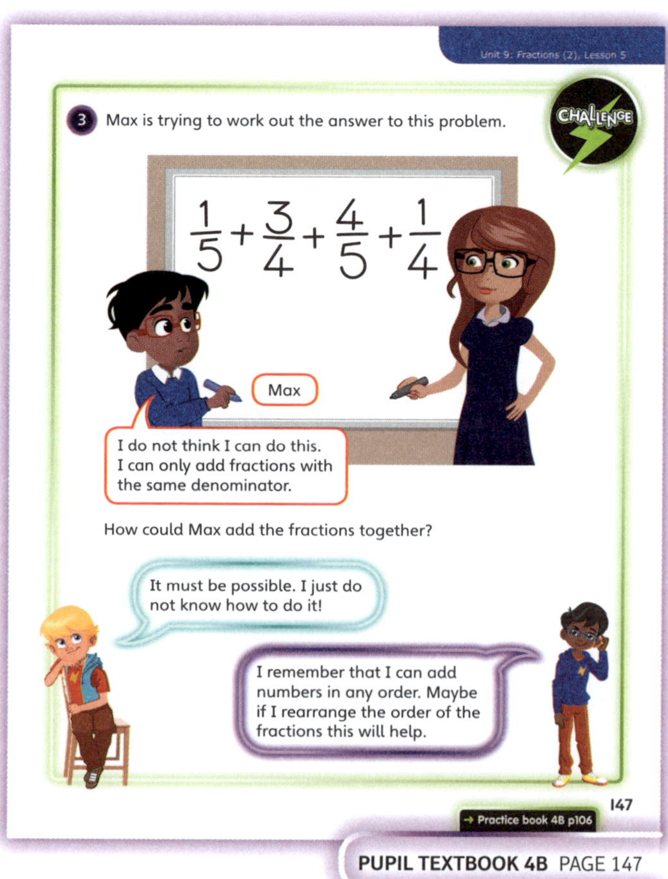

PUPIL TEXTBOOK 4B PAGE 147

Practice

WAYS OF WORKING Independent work

IN FOCUS In question ②, children find missing numbers that will make an answer of $\frac{7}{5}$. There are many possible answers, using mixed numbers and improper fractions. Children should be able to see the connection between these types of number and use both. For example, there is $\frac{10}{5} - \_ = \frac{7}{5}$ and $2 - \_ = \frac{7}{5}$. Children should be able to see that these are the same.

STRENGTHEN Children may need support moving between mixed numbers and improper fractions. In Year 4 they do not have to convert, but they need to work with both numbers. A double number line showing the comparison should help children to see the equivalence of fractions.

DEEPEN For question ②, ask: *Can you make up your own problems that give an answer of $\frac{7}{5}$? Can you turn the questions into word problems?* Challenge children to write other questions that all have the same answer. For question ⑤, deepen understanding by asking: *If he moves $\frac{3}{4}$ km per hour, how long will it take Kofi to travel 3 km?* Children could approach this by adding on $\frac{3}{4}$ until they get an answer equal to or greater than 3. Question ⑥ challenges children to work with different denominators in the same calculation.

ASSESSMENT CHECKPOINT Questions ① and ⑤ will demonstrate whether children are confident when solving word problems involving adding and subtracting fractions.

ANSWERS Answers for the **Practice** part of the lesson appear in the separate **Practice and Reflect answer guide**.

Reflect

WAYS OF WORKING Pair work

IN FOCUS In this activity, children are given an answer and they have to come up with a question. Ask them to write an addition problem and a subtraction problem. Some children may create a 2-step problem that involves an addition and a subtraction. Ask them to get their partner to check their answer.

ASSESSMENT CHECKPOINT Can children write an addition, a subtraction and a word problem?

ANSWERS Answers for the **Reflect** part of the lesson appear in the separate **Practice and Reflect answer guide**.

After the lesson 

- Are children confident adding and subtracting fractions with the same denominator?
- Can they subtract a fraction from a given number of wholes?
- Can children apply this knowledge to solving simple word problems and other problem solving questions?

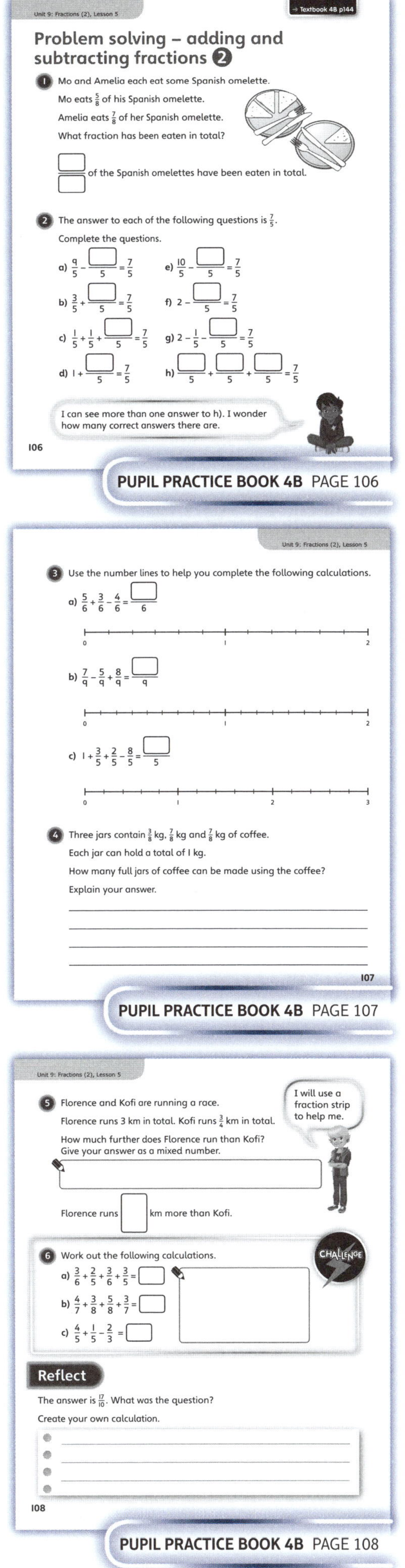

PUPIL PRACTICE BOOK 4B PAGE 106

PUPIL PRACTICE BOOK 4B PAGE 107

PUPIL PRACTICE BOOK 4B PAGE 108

Calculating fractions of a quantity

Learning focus

In this lesson, children will calculate a fraction of a quantity. They will use fraction strips to help them visualise the concept and then use their knowledge of finding a unit fraction of an amount to find non-unit fractions of an amount.

Small steps

→ Previous step: Problem solving – adding and subtracting fractions (2)
→ **This step: Calculating fractions of a quantity**
→ Next step: Problem solving – fraction of a quantity (1)

NATIONAL CURRICULUM LINKS

Year 4 Number – Fractions (Including Decimals)

Solve problems involving increasingly harder fractions to calculate quantities, and fractions to divide quantities, including non-unit fractions where the answer is a whole number.

ASSESSING MASTERY

Children can find a fraction of an amount using unit fractions and non-unit fractions. They can apply understanding to a variety of situations and use a variety of different units, such as measures and money.

COMMON MISCONCEPTIONS

Children may try to memorise a process rather than understand the concept of finding a fraction of an amount. They could therefore get confused and start doing things like dividing by the numerator and multiplying by the denominator. Ask:
• *How do you find a fraction of a quantity/amount? Do you need to divide or multiply?*

Children may find it difficult to split their fraction strip into the appropriate number of parts. For example, when they see fifths they may want to draw five lines on their fraction strip when in fact they only need four lines. Ask:
• *How many parts do you need to split your fraction strip into? How many parts is the whole split into?*

STRENGTHENING UNDERSTANDING

The fraction strip is a powerful tool for the concept learnt in this lesson. Children's understanding could be strengthened by using counters and a large fraction strip drawn onto A3 paper. They can then clearly see what is happening. For example, they could find $\frac{1}{5}$ of 20 counters by sharing their counters into 5 equal parts.

GOING DEEPER

Challenge children by asking them to complete questions that require the inverse calculation; that is finding the whole given a fraction. They have the opportunity to do this in question **5** in the **Practice Book**.

KEY LANGUAGE

In lesson: problem solving, fraction, proper fraction, improper fraction, mixed number, whole, equal parts, numerator, denominator, add, subtract, partition

Other language to be used by the teacher: calculate, sum, difference, multiply, divide, small, smaller, smallest, large, larger, largest, middle, medium, height, tall, long, length, diagram, quantity, centimetre (cm), metre (m), kilogram (kg), money, pound (£)

STRUCTURES AND REPRESENTATIONS

fraction strips, number line

RESOURCES

Optional: base 10 equipment, counters, real objects to represent fractions

 In the eTextbook of this lesson, you will find interactive links to a selection of teaching tools.

Before you teach 

• Do children understand what a fraction is?
• Can children draw a diagram or use cubes to represent a fraction?
• Can children divide and multiply using the times-table facts they know?

Discover

 Pair work

ASK

- Question **1** a): *If the largest doll is 24 cm and you have 8 cubes, how many centimetres does each cube represent? How many cubes stand for $\frac{1}{8}$?*
- Question **1** b): *If you know what $\frac{1}{8}$ is, how can you find out what $\frac{3}{8}$ is?*

IN FOCUS Question **1** a) introduces children to finding a fraction of an amount. Some children may realise that 1 out of the 8 cubes in their doll must represent $\frac{1}{8}$. Encourage discussion around why this is the case. Question **1** b) highlights how children can use the unit fraction of an amount to find the non-unit fraction.

PRACTICAL TIPS Consider allowing children to make their own Russian dolls using cubes. Ensure they use either 8 cubes or a multiple of 8 for the height of the doll so they can find the fractions of an amount easily.

ANSWERS

Question **1** a): The smallest doll is 3 cm tall.

Question **1** b): The middle doll is 9 cm tall.

PUPIL TEXTBOOK 4B PAGE 148

Share

 Whole class teacher led

ASK

- Question **1** a): *If this fraction strip represents the largest doll, how long is it? What part of the fraction tells you how many sections to split your strip into? How many sections are you interested in if you want to know what $\frac{1}{8}$ is? What part of the fraction tells you this?*
- Question **1** b): *How can you use your answer to part a) to help you with part b)? What calculation can you do to find your answer? Which part of the fraction tells you to look at 3 sections of the fraction strip?*

IN FOCUS Children should be able to explain, step by step, which part of the question represents which part of the diagram. The fraction strip diagram helps to reinforce which calculation they should perform and when. For example, the act of splitting up the strip indicates division.

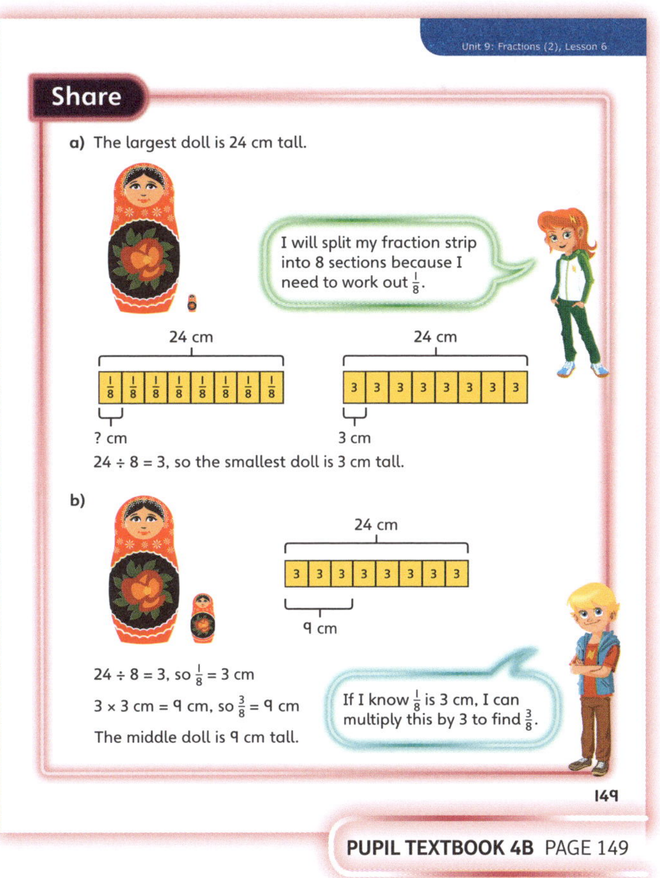

PUPIL TEXTBOOK 4B PAGE 149

Think together

 Whole class teacher led (I do, We do, You do)

ASK

- Question **2**: *What is the connection between part a) and part b)? Does your answer to part a) help you to answer part b)?*
- Question **3**: *What is the same and what is different between the first two calculations? How will your working out differ? What should you work out first for the third calculation?*

IN FOCUS Question **3** gives children the opportunity to begin calculating different fractions of the same amount. Pay attention to their strategy. In particular, are children certain of how many parts their fraction strip needs to be split into? And are they calculating the unit fraction before taking it any further? For example, to calculate $\frac{3}{7}$ of a quantity, they should split their strip into sevenths and first calculate $\frac{1}{7}$.

STRENGTHEN To support understanding, ensure children have access to base 10 equipment so they can physically make the whole and split it up themselves. Ideally, it should be possible to arrange these cubes into a fraction strip. Perhaps encourage children to draw a fraction strip for their question once they have split up their cubes, and distribute their cubes into it. This will really help children to understand why they are performing certain calculations.

DEEPEN In question **3**, children complete calculations that include units such as measurement and money. They should realise that this does not make the question any more difficult. Question **4** also draws out a misconception that $\frac{3}{8}$ is bigger than $\frac{3}{4}$. Encourage children to think back to their prior learning to make correct predictions about the answer. Challenge children to complete more fraction calculations.

ASSESSMENT CHECKPOINT Question **2** will assess whether children have understood the main concept and whether they can set up and label their own fraction strip to represent a problem. Children should be encouraged to do this even if they think they do not need the diagram, as drawing the diagram will help them to consolidate their thinking.

ANSWERS

Question **1**: The engine is 12 cm long.

Question **2** a): The rubber is 3 cm long.

Question **2** b): The pencil is 21 cm long.

Question **3**: $\frac{1}{3}$ of 15 cm = 5 cm

$\frac{1}{5}$ of 15 cm = 3 cm

$\frac{3}{7}$ of £42 = £18

Question **4**: Aki has £18. Ebo has £9. So, Aki has more money than Ebo.
Dexter is incorrect because 8 and 4 are the denominators so they cannot be compared alone. Dexter needs to work out the unit fractions ($\frac{1}{4}$ and $\frac{1}{8}$) and then multiply them both by 3 to see how much Aki and Ebo have respectively.

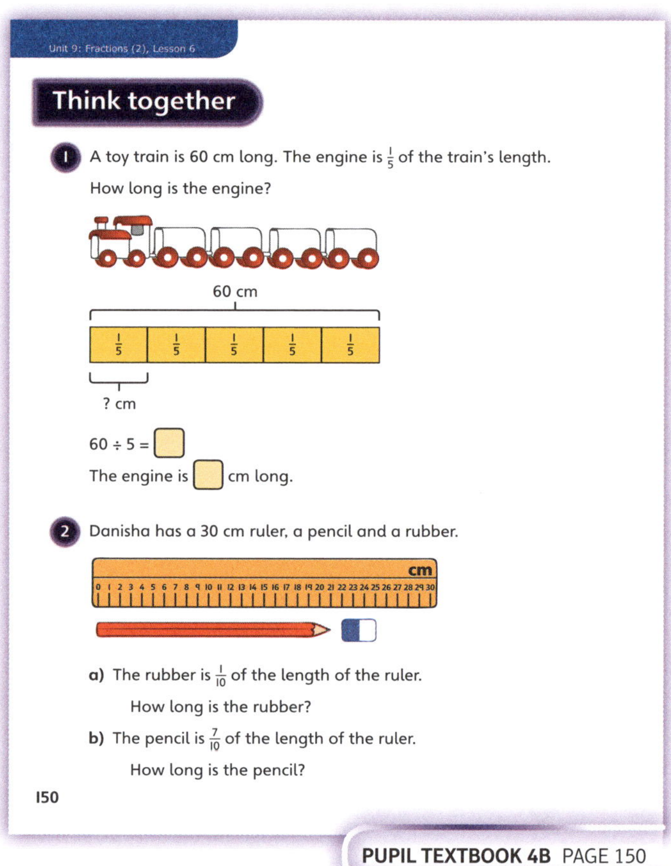

PUPIL TEXTBOOK 4B PAGE 150

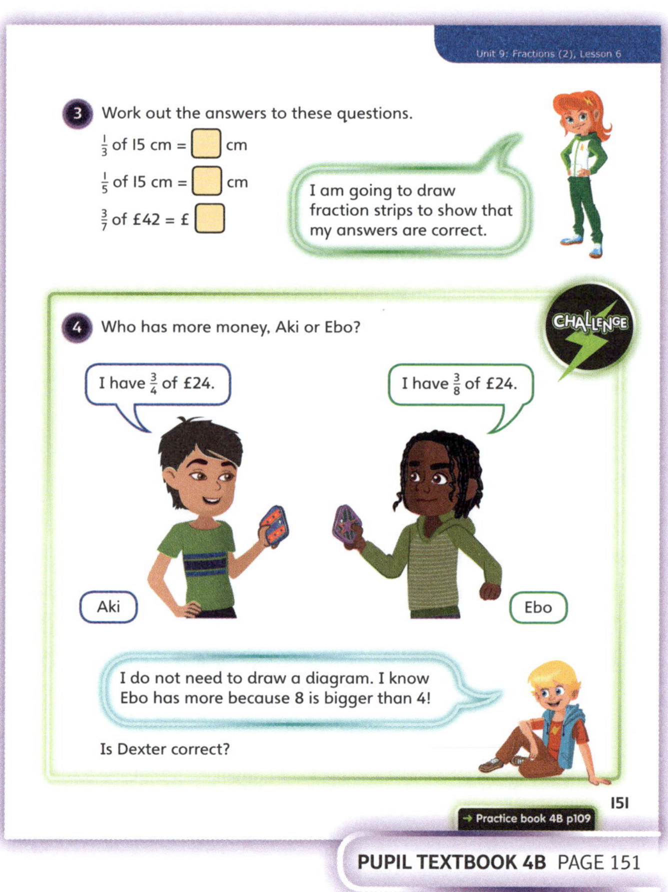

PUPIL TEXTBOOK 4B PAGE 151

Practice

WAYS OF WORKING Independent work

IN FOCUS Question **3** is purposefully completely abstract and designed in such a way that children cannot just guess the answer: they need to show full working out. This question highlights how two things that look completely different can be equal.

STRENGTHEN Provide concrete manipulatives throughout the exercise and encourage children to draw fraction strips to represent their answers; this will consolidate their understanding and prevent them finding short cuts or following processes they may have discovered.

DEEPEN Challenge children to create their own matching game, like that in question **4**. They should then swap with a partner to answer each other's questions. Question **6** challenges children to complete a multi-step word problem.

THINK DIFFERENTLY In question **5**, children are not just calculating a fraction of a quantity; they have to complete the question in different orders, and really think about what is being asked. This gives children the opportunity to deepen their understanding by applying what they have learnt in other contexts. This means they cannot rely on a rehearsed process rather than actual understanding. Encourage children to label the information they have been given on a fraction strip and decide on the correct steps they need to take to get to their answer.

ASSESSMENT CHECKPOINT If children can answer questions **1** to **4** correctly, this is confirmation that they have understood the lesson and can find a fraction of a quantity. Children who need help by the time they get to question **4** may need some more time to consolidate their understanding, for example by using concrete objects, or by talking about mental strategies they could use.

ANSWERS Answers for the **Practice** part of the lesson appear in the separate **Practice and Reflect answer guide**.

Reflect

WAYS OF WORKING Pair work

IN FOCUS This simple activity will check that children thoroughly understand the lesson and that they have not followed a process. If they cannot see what a fraction strip is representing and write a question to go with it, then their understanding is unlikely to be deep enough to move on.

ASSESSMENT CHECKPOINT Ideally, children should be able to come up with a variety of different questions that this fraction strip would support. They should be able to put it in different contexts, and also answer each question that they write.

ANSWERS Answers for the **Reflect** part of the lesson appear in the separate **Practice and Reflect answer guide**.

After the lesson 

- Can children calculate a unit fraction of an amount?
- Can children find a non-unit fraction of an amount?
- Can children compare different fractions of amounts?
- Can children calculate the whole when given a fraction of an amount?

183

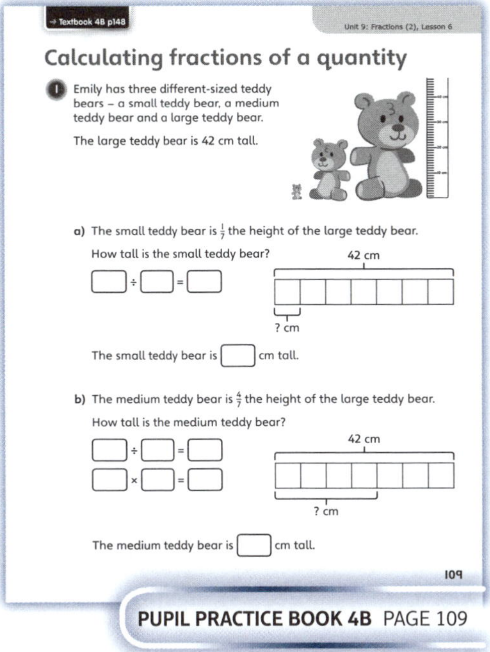

PUPIL PRACTICE BOOK 4B PAGE 109

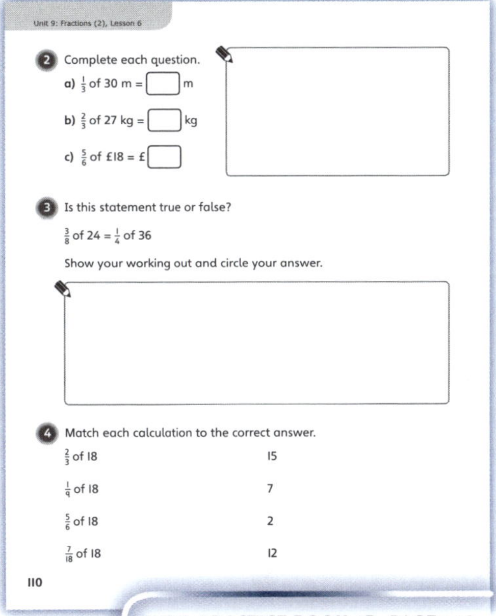

PUPIL PRACTICE BOOK 4B PAGE 110

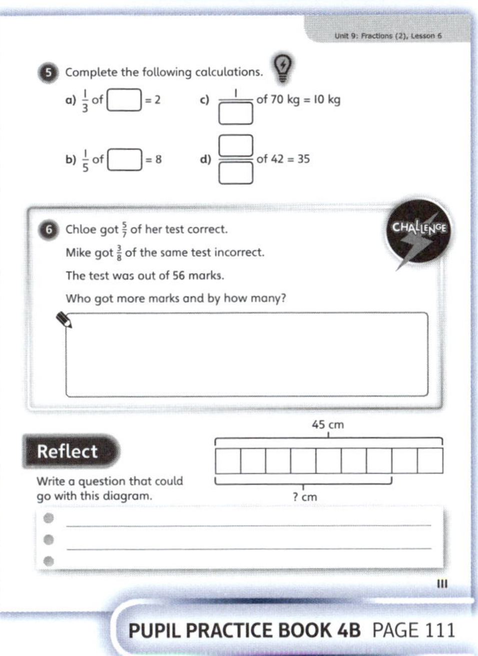

PUPIL PRACTICE BOOK 4B PAGE 111

Problem solving – fraction of a quantity ❶

Learning focus

In this lesson, children will calculate the whole when a fraction of an amount is given or when a part is given as a quantity.

Small steps

→ Previous step: Calculating fractions of a quantity
→ **This step: Problem solving – fraction of a quantity (1)**
→ Next step: Problem solving – fraction of a quantity (2)

NATIONAL CURRICULUM LINKS

Year 4 Number – Fractions (Including Decimals)

Solve problems involving increasingly harder fractions to calculate quantities, and fractions to divide quantities, including non-unit fractions where the answer is a whole number.

ASSESSING MASTERY

Children can recognise that the quantity given is a fraction of an amount and can use this to work out the whole.

COMMON MISCONCEPTIONS

Children may find the fraction of the part given and interpret this as the whole. Ask:
• *Have you found a fraction of a whole, or a whole?*

Children may confuse which operations to use when finding fractions of amounts. Ask:
• *Do you need to add, subtract, multiply or divide?*

STRENGTHENING UNDERSTANDING

If children need support to understand questions that involve the fraction of a quantity, provide real world objects so they can represent situations. For example, supply jugs and water, paper plates that can be cut up into fractions, or packets of sweets or bags of counters that can be divided up.

GOING DEEPER

Challenge children by asking them more questions which develop their understanding of parts of a whole as fractions. For example: *Harry eats $\frac{1}{2}$ a bag of sweets. Harry gives Jamie half of the sweets that are left. Harry has 7 sweets left. How many sweets did Harry start with?* Children could use fraction strips to represent this.

KEY LANGUAGE

In lesson: problem solving, fraction, proper fraction, improper fraction, mixed number, whole, equal parts, numerator, denominator, add, subtract, partition

Other language to be used by the teacher: sum, difference, total, multiply (×), divide (÷), represent, quantity, gram (g), millilitre (ml), kilometre (km), money, pound (£)

STRUCTURES AND REPRESENTATIONS

fraction strips

RESOURCES

Optional: base 10 equipment, fraction number lines

 In the eTextbook of this lesson, you will find interactive links to a selection of teaching tools.

Before you teach 

• Do children know that the denominator has to be the same in order to add and to subtract fractions?
• Can children use efficient methods to add and subtract fractions?

Discover

 Pair work

ASK

- Question ❶ a): *How much jam is left? What fraction is left? How can you use this information to work out how much jam was in the jar when it was full?*
- Question ❶ b): *How much cheese does Reena eat? What fraction of cheese is left? What fraction does Reena eat? How are these different? How can you use this information to work out how much cheese is left?*

IN FOCUS   Question ❶ a) encourages children to recognise that the quantity given is a unit fraction of an amount and not the whole amount.

Question ❶ b) encourages children to recognise that the quantity given is a non-unit fraction of an amount, and so builds on skills used in question ❶ a). Children also discover that the quantity given and the fraction given are not describing the same part.

PRACTICAL TIPS   Children could be given paper plates to divide up, like the block of cheese, into fractions. They could be encouraged to label each fraction with the number of grams it represents.

ANSWERS

Question ❶ a): There are 600 g of jam in the jar when it is full.

Question ❶ b): There are 90 g of cheese left.

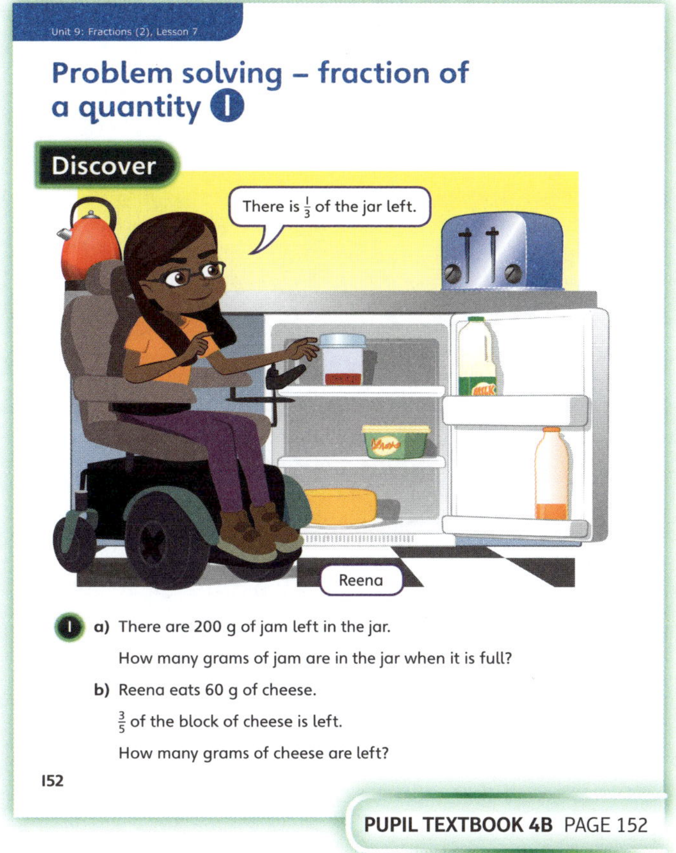

PUPIL TEXTBOOK 4B PAGE 152

Share

  Whole class teacher led

ASK

- Question ❶ a): *What fraction of jam is left? How many lots of 200 g will you need to make the whole? How can you show this as a fraction strip and a number sentence?*
- Question ❶ b): *How much does Reena eat? What is this as a fraction? If you know $\frac{2}{5}$ is worth 60 g, how can you use this to work out what fraction is left?*

IN FOCUS   In question ❶ a), children first establish how much jam is left and identify what fraction this is. Encourage children to use the denominator to make links with multiplication to find the whole. Children can use a fraction strip to represent the problem. Discuss with children the same process for ❶ b). Children establish that the fraction left is different from the amount Reena has eaten. Encourage children to use the numerator to make links with division to find out how many grams one of the parts will be worth. Discuss how finding this out helps children to work out what fraction is left. Again, children may use fraction strips to show this.

By the end of the **Share** section children should be able to identify what a fraction of a quantity represents and use their understanding of multiplication and division, along with an understanding of what the numerator and denominator represent, to work out the whole or the unknown part.

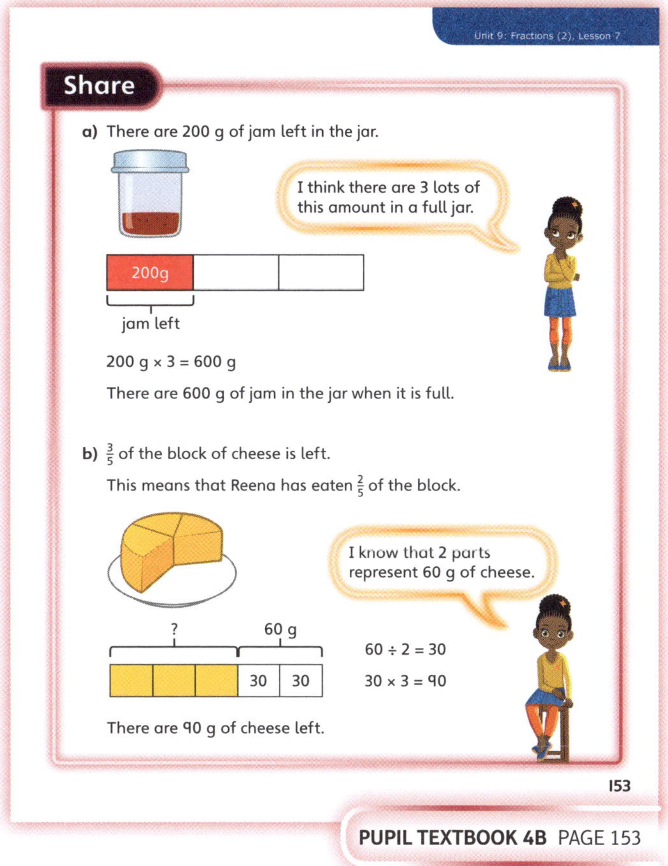

PUPIL TEXTBOOK 4B PAGE 153

Think together

WAYS OF WORKING Whole class teacher led (I do, We do, You do)

ASK

- Question ❶: *What fraction is left? How much is left? What will you multiply the amount left by? What is the total?*
- Question ❷: *What fraction is left? How much is left? How many parts will make the whole? What will you multiply together to find the total?*
- Question ❸: *What information do you know? How much does each person have left? What fraction do they have left? How can you use this to find out how many chocolates they had at the start?*

IN FOCUS In question ❶, children use a unit fraction of an amount to calculate the whole. Encourage children to explain what the value represents and to use the fraction strip to help them with this.

Question ❷ encourages children to see that the amount given and the fraction of an amount given are different but that both are needed to work out what one unit part is worth. Children could explore different ways of finding the whole. For example: $7 \times 7 = 49$, or $35 + 14\ (\frac{2}{7}) = 49$

STRENGTHEN To support children's understanding, real objects such as sweets or counters could be used by children to act out the questions posed.

DEEPEN Challenge children to use the fraction strips they have created to write other questions based on the same models. For example, in question ❷, if Erik ate double the amount of sweets, how many would he have left?

ASSESSMENT CHECKPOINT Question ❸ requires children to apply skills developed in questions ❶ and ❷. If they can complete this question successfully, they will have demonstrated that they can solve problems involving a fraction of a quantity. Can children explain which operation to use for each problem and at what point? Can they explain what the amount given represents and how to use this information to answer the question?

ANSWERS

Question ❶: $5 \times 100 = 500$ ml
A full bottle has 500 ml of juice.

Question ❷: $35 \div 5 = 7$
$7 \times 7 = 49$
A full packet of sweets has 49 sweets.

Question ❸: Lexi had more chocolates to start with. She had 49 chocolates; Andy had 40 chocolates.

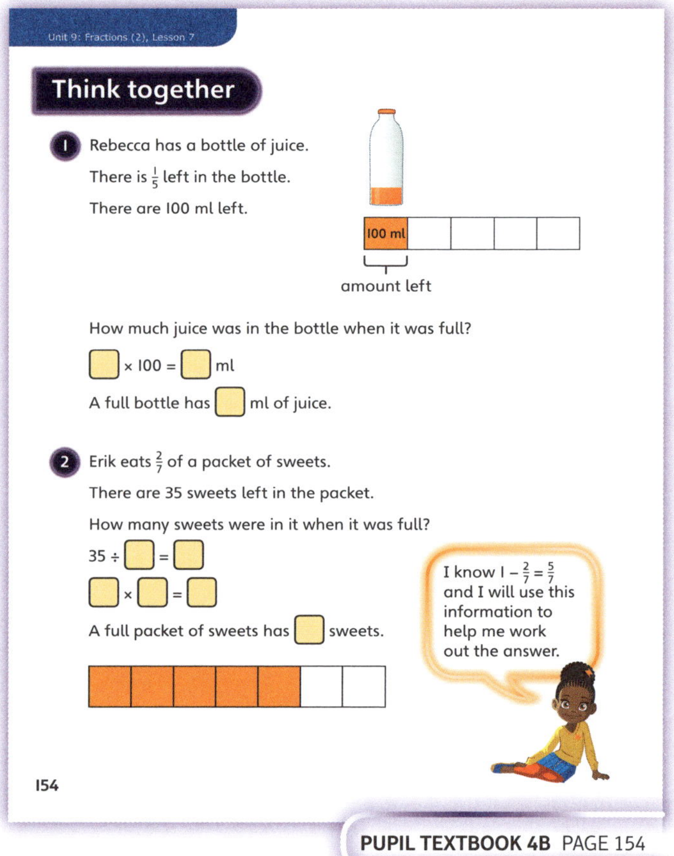

PUPIL TEXTBOOK 4B PAGE 154

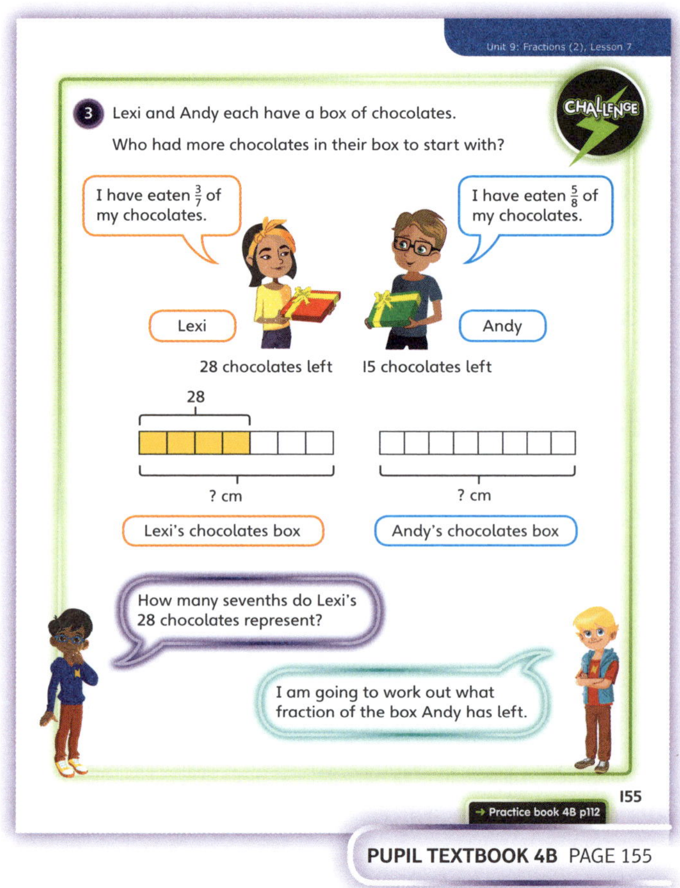

PUPIL TEXTBOOK 4B PAGE 155

Practice

WAYS OF WORKING Independent work

IN FOCUS Questions **1**, **2** and **3** focus on children finding the whole, when they are given a part. Check that children do not find the fraction of the given amount. Question **4** provides a given part and children have to find the other part, and not the whole. Some children may want to find the whole first and then find the missing part; however, encourage them to find the parts and then multiply to find the missing parts.

STRENGTHEN Throughout this practice, children are given visual representations to help them to understand the problems. For those children making common mistakes, work through the problems step by step. Ask: *Are you given the whole? How do you know you are only given a part?*

DEEPEN Question **6** provides children with a word problem with difficult wording. Encourage children to approach the problem step by step. They need to realise that $\frac{5}{9} + \frac{4}{9}$ makes the whole and that $\frac{5}{9}$ is equal to 40. Ask them to draw a fraction strip as this will support their understanding.

THINK DIFFERENTLY In question **5** children are given word-style problems but in equation form. This is identical to question **2**, where they are given a number and told a fraction of it.

ASSESSMENT CHECKPOINT By the end of the practice, children should be able to find the whole or another part when they are given a fraction of a number.

ANSWERS Answers for the **Practice** part of the lesson appear in the separate **Practice and Reflect answer guide**.

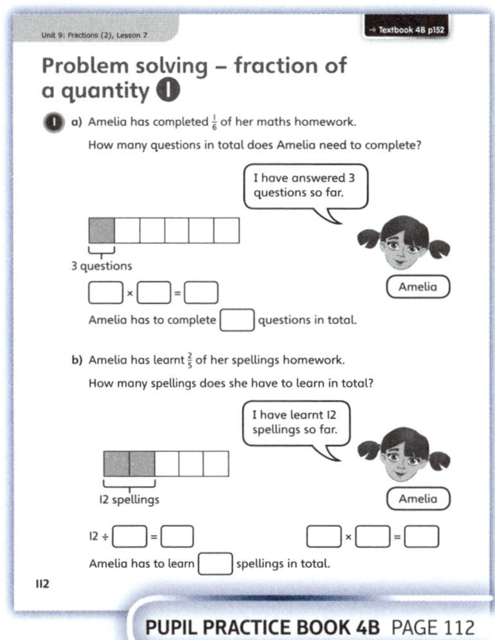

PUPIL PRACTICE BOOK 4B PAGE 112

PUPIL PRACTICE BOOK 4B PAGE 113

Reflect

WAYS OF WORKING Pair work

IN FOCUS This **Reflect** section checks that children understand the main objective of the lesson. Children show their understanding by drawing a fraction strip to explain how to work out the whole, given that $\frac{3}{5}$ is equal to 60. The numbers have been chosen to identify children who make the common mistake of just finding $\frac{3}{5}$ of £60. Discuss why this is incorrect.

ASSESSMENT CHECKPOINT Children can explain how to find the whole when they are given a fraction of an amount.

ANSWERS Answers for the **Reflect** part of the lesson appear in the separate **Practice and Reflect answer guide**.

After the lesson ⏸

- Can children draw a fraction strip to represent problems where they are given a fraction of an amount?
- Can children find the value of one part and use this to find the whole or other part?
- Do children know the difference between finding a fraction of an amount and find the whole when they are given a fraction of an amount?

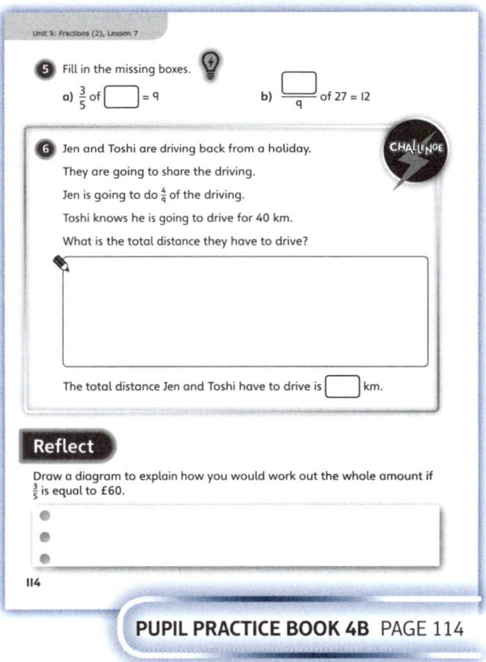

PUPIL PRACTICE BOOK 4B PAGE 114

Problem solving – fraction of a quantity ②

Learning focus

In this lesson, children will solve multi-step problem-solving questions involving finding a fraction of a quantity and finding the whole.

Small steps

→ Previous step: Problem solving – fraction of a quantity (1)
→ **This step: Problem solving – fraction of a quantity (2)**
→ Next step: Tenths (1)

NATIONAL CURRICULUM LINKS

Year 4 Number – Fractions (Including Decimals)

Solve problems involving increasingly harder fractions to calculate quantities, and fractions to divide quantities, including non-unit fractions where the answer is a whole number.

ASSESSING MASTERY

Children use their knowledge from previous lessons and apply it to solve multi-step problems involving finding a fraction of a quantity and finding the whole.

COMMON MISCONCEPTIONS

Children may find the fraction of the part given and interpret this as the whole. For example, when told that $\frac{2}{3}$ is equal to 18 and they have to find the whole, they often find $\frac{2}{3}$ of 18. To overcome this problem, children may find it useful to represent it visually, for example with a fraction strip. Ask:
- *What number represents the fraction? How would you represent this question on a fraction strip?*

STRENGTHENING UNDERSTANDING

Children who need more support could represent problems visually, with a fraction strip. Children can use the fraction strip to work out whether they have to find a fraction of an amount or whether they have to find the whole. For some questions, children might find that using concrete manipulatives helps their understanding. They can make an array of cubes: for example, if they know that there are 24 cubes and that $\frac{2}{3}$ of them are red, children may find it useful to show this as a 3×8 array. This will allow children to see why there are 16 red cubes – they can see the $\frac{2}{3}$.

GOING DEEPER

To deepen children's understanding, ask them to consider finding a fraction of a fraction and finding the whole from this. For example, children could be presented with problems where they know that $\frac{1}{3}$ of the cubes in a box are red, $\frac{1}{5}$ of the remaining cubes are blue and the rest of the cubes are yellow. If there are 20 yellow cubes, they can then find out how many cubes there are in total.

KEY LANGUAGE

In lesson: problem solving, fraction, proper fraction, improper fraction, mixed number, whole, equal parts, numerator, denominator, add, subtract, partition, quantity, represent, unit fraction, remaining

Other language to be used by the teacher: more, most, less, least, equal, multiply, divide, sum, difference, total, non-unit fraction

STRUCTURES AND REPRESENTATIONS

fraction strips

RESOURCES

Optional: base 10 equipment, cubes, counters, place value counters

 In the eTextbook of this lesson, you will find interactive links to a selection of teaching tools.

Before you teach 

- Do children know how to find a fraction of an amount?
- Do they know how to find the whole when given information about a part?
- Do children know how to represent such problems using fraction strips?

Discover

 Pair work

- Question ① a): *What fraction of Danny's tower is made of red cubes? How many red cubes is this? What is $\frac{1}{5}$? How can you use this to find $\frac{2}{5}$?*
- Question ① b): *What do you need to work out first? How can you represent this on a fraction strip? How will this help you work out how many cubes Lee has? How can you now work out how many more cubes Danny has?*

 The problem presented in each question part is a 2-step problem, which allows children to show the distinct steps they would take to solve a multi-step problem. Question ① a) asks children to find a simple fraction of an amount. Children may represent this using a fraction strip or they may know that they need to find a fifth (by dividing by 5) and then multiply by 2. Stop children from using phrases such as 'divide by the denominator' and 'multiply by the numerator'. Question ① b) asks children to use information from ① a) to work out how many cubes Lee has. This time, children should see that they are given a part and they have to work out the whole. Children then have to use this information to work out how many more cubes Danny has.

 Children could use cubes to represent the question visually and recreate the **Discover** context.

Question ① a): There are 12 red cubes in Danny's tower.

Question ① b): Danny's tower has more cubes (30). (Lee's tower has 16.) Danny's tower has 14 more cubes than Lee's tower.

Share

 Whole class teacher led

- Question ① a): *How does the diagram show you that you have to work out $\frac{2}{5}$ of 30? Can you explain each part of the fraction strip? What are the steps involved in finding a fraction of an amount?*
- Question ① b): *Why don't you find $\frac{2}{3}$ of 12? Are you finding the whole? How do you know?*

 For question ① a), explain to children that they need to find $\frac{2}{5}$ of 30. They do this first by finding $\frac{1}{5}$ and then multiplying by 2. Use the fraction strip to explain the steps, building up the model step by step. Discuss with children how they can do this without the fraction strip. Astrid's comment should show children that they have to find the whole in this question. They will then subtract in order to find the difference. Some children may forget this final step.

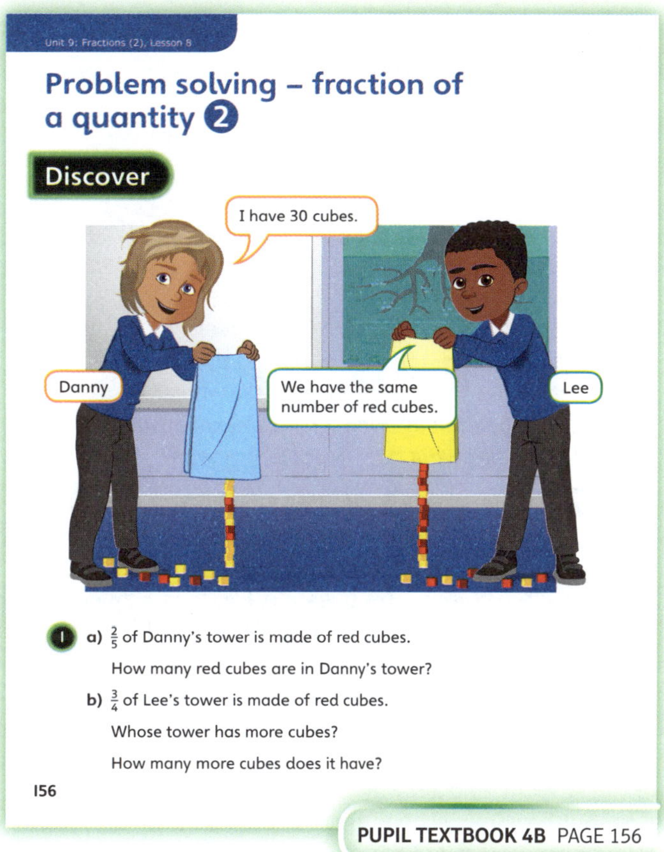

PUPIL TEXTBOOK 4B PAGE 156

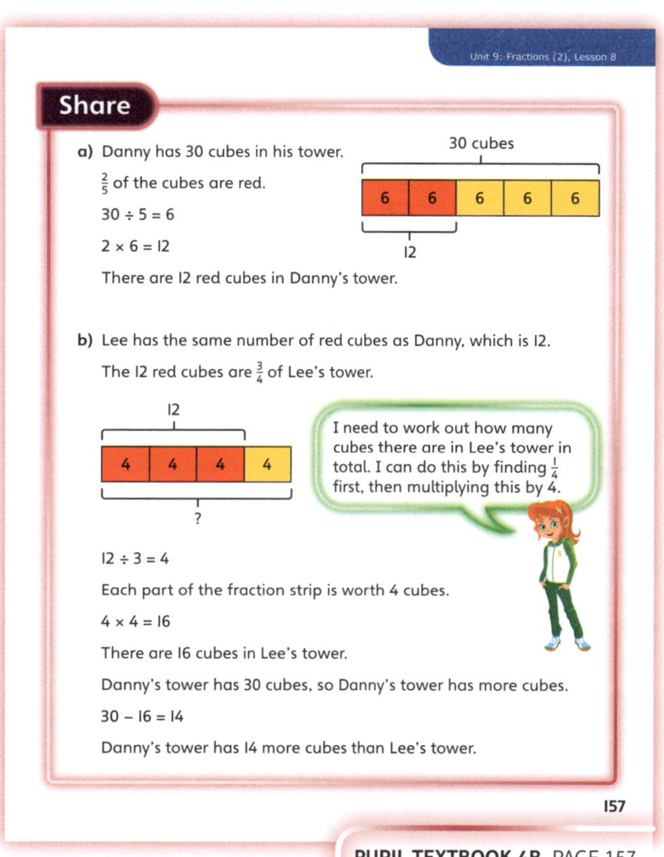

PUPIL TEXTBOOK 4B PAGE 157

Think together

WAYS OF WORKING Whole class teacher led (I do, We do, You do)

ASK

- Question ❶: *What information has been given? What should you do first? Is there any way you can do this as one step?*
- Question ❷: *What information are you given? Which information is useful? Is 18 the total number of cubes or just part of the tower? How can you find the whole?*
- Question ❸: *What information do you know? Are you given a part or whole? What does 'remaining' mean?*

IN FOCUS In question ❶, discuss with children that they need to find out how many more red cubes Lee's tower has than Danny's. Children may think that they have to find the whole at some point in this question. Explain to children that they have been told the whole already. Children should use fraction strips to help them understand the steps they need to take. In question ❷, children are given part of the tower in both parts and they have to work out the whole. Some children may jump straight in and just find a fraction of the amount. Explain why this is incorrect: they have to work out how many cubes are in the tower when they are told only part of the tower.

STRENGTHEN Use fraction strips to support understanding. This should help children see whether or not they are working out a part or the original whole. Discuss with children the steps they need to take and break down the question into its parts. You may find it useful for children to use cubes to help them. Some children may be overwhelmed by the amount of information in each question. If this is the case, ask children to highlight keywords, read the problem aloud and complete one step at a time.

DEEPEN For question ❸, instead of telling children the total, tell them that there are 63 white roses and see if they can work out how many roses there are in total.

ASSESSMENT CHECKPOINT Question ❸ will assess whether children can work out 2-step or multi-step problems involving finding a fraction of an amount, including problems where children also have to find the whole.

ANSWERS

Question ❶: $24 \div 6 = 4$
Danny's tower has 4 red cubes.
$24 \div 8 = 3$
$3 \times 5 = 15$
Lee's tower has 15 red cubes.
$15 - 4 = 11$
Lee's tower has 11 more red cubes than Danny's tower.

Question ❷: Danny's tower has more cubes. It has 28 more cubes. (Danny's tower has 48 cubes. Lee's tower has 20 cubes.)

Question ❸: There are 20 red roses, 18 yellow roses and 42 white roses.

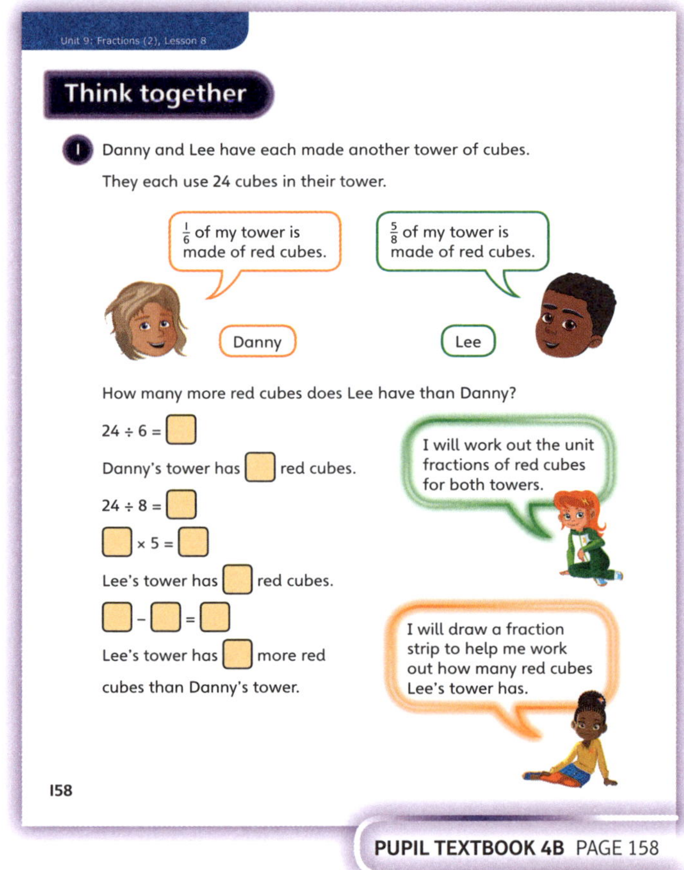

PUPIL TEXTBOOK 4B PAGE 158

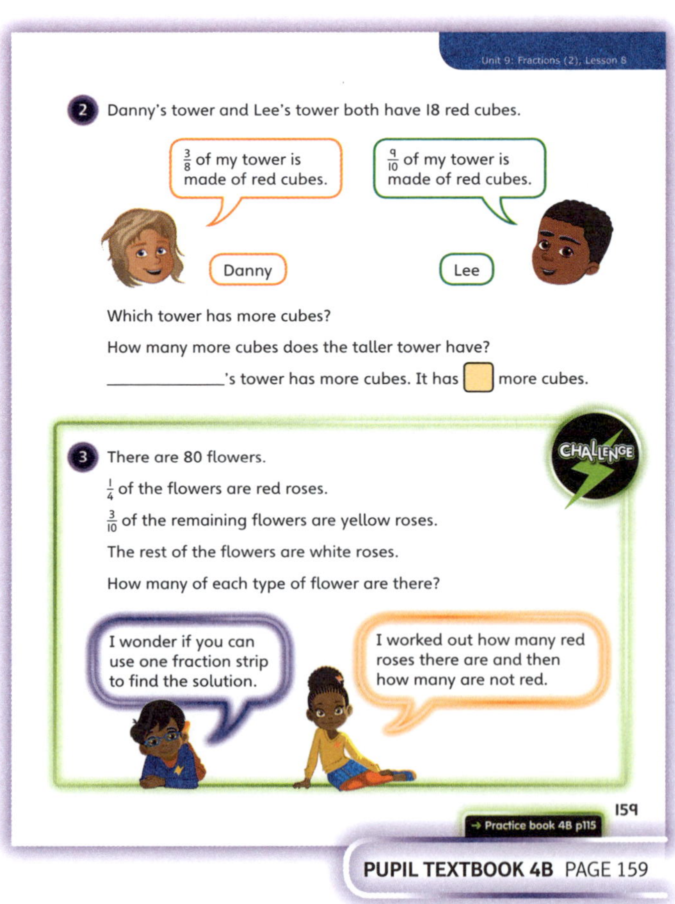

PUPIL TEXTBOOK 4B PAGE 159

Practice

WAYS OF WORKING Independent work

IN FOCUS Question ② presents some fraction strips which are context free. The objective is for the children to work out the missing part when given information about another part of the fraction strip. This will help children when it comes to drawing their own models later.

In Question ④, children find fractions of an amount and then work out what number a child has made from these amounts.

There is a mix of problems where children have to find the whole and some where they have to find a part. Discuss with children strategies for extracting information from questions and how to draw a related fraction strip. You might want to discuss which questions they found easy and which ones they found more challenging.

STRENGTHEN Ask children to draw their own fraction strips for questions that do not provide them. This should help children to understand whether or not they are working out a part or the original whole. Discuss with children the steps they need to take and break down each question.

DEEPEN For question ②, ask children to write a question for each of the fraction strips. Get them to explain why the situation fits the diagram.

ASSESSMENT CHECKPOINT By the end of the practice children should be able to solve difficult multi-step problems that involve finding fractions of an amount. Question ③ will be a good reflection of children's understanding.

ANSWERS Answers for the **Practice** part of the lesson appear in the separate **Practice and Reflect answer guide**.

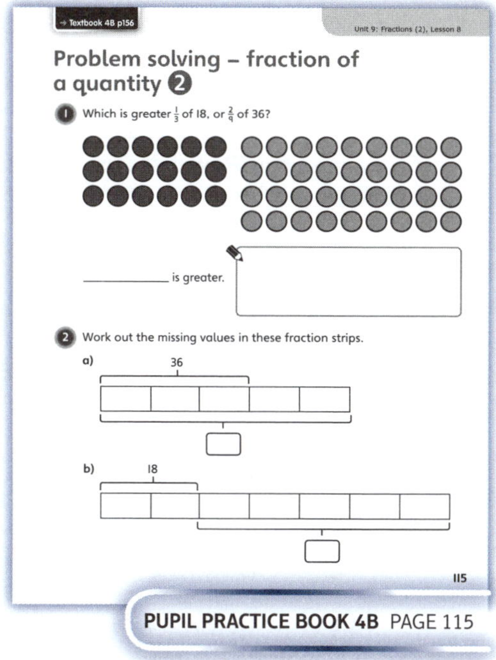

PUPIL PRACTICE BOOK 4B PAGE 115

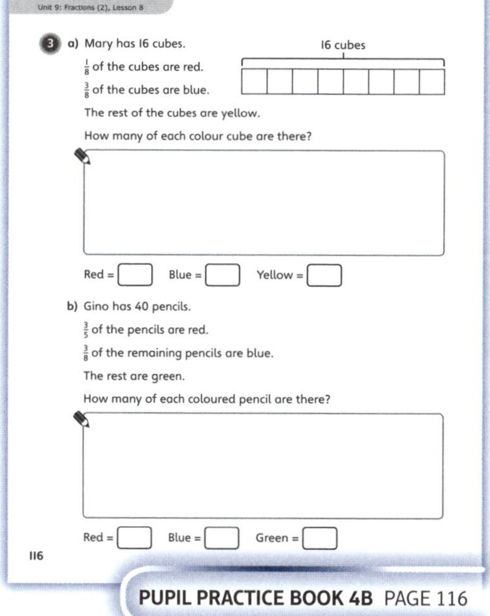

PUPIL PRACTICE BOOK 4B PAGE 116

Reflect

WAYS OF WORKING Pair work

IN FOCUS This section checks whether children can find the whole if they are given a fraction of the whole. The numbers have been chosen to highlight a common mistake that children often make. Discuss Eva's answer with children and get them to show with a fraction strip why the answer is actually 27. If children think Eva is correct, ask them to consider if the answer can actually be smaller or whether it should be bigger.

ASSESSMENT CHECKPOINT Children can explain the mistake that Eva has made and are able to explain why the answer should be 27 and not 12.

ANSWERS Answers for the **Reflect** part of the lesson appear in the separate **Practice and Reflect answer guide**.

After the lesson ⏸

- Can children find a fraction of an amount?
- Can they find the whole when given a fraction of an amount?
- Can children solve simple 2-step problems involving the above?

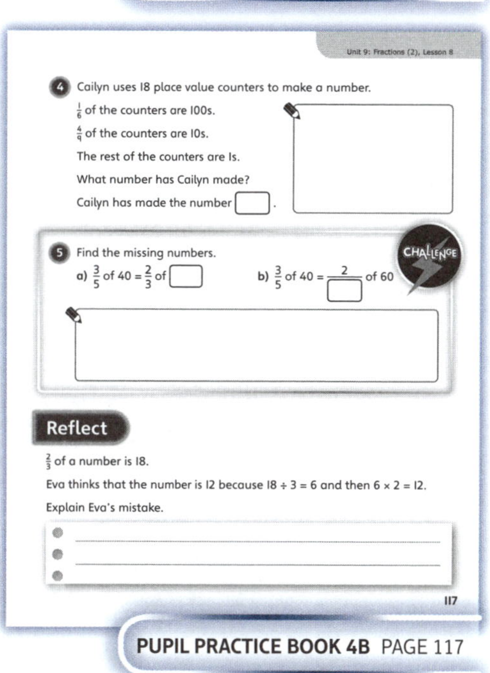

PUPIL PRACTICE BOOK 4B PAGE 117

End of unit check

Don't forget the *Power Maths* unit assessment grid on p26.

WAYS OF WORKING Independent work

IN FOCUS This **End of unit check** will allow you to focus on children's understanding of adding and subtracting fractions where the answers are greater than 1. The questions within the **End of unit check** cover the full range of questions in the unit and are designed to draw out particular misconceptions or misunderstandings.

- Questions **1** and **2** assess children's ability to add and subtract fractions where the denominators are the same, using both mixed numbers and improper fractions.
- Questions **3** and **4** assess children's ability to find a unit or non-unit fraction of a quantity when given the whole and fraction of the whole.
- Question **5** assesses children's ability to find a whole given a part and fraction of a quantity.
- Question **6** is a SATs-style question and assesses children's ability to find a fraction of an amount given another fraction and the whole.

ANSWERS AND COMMENTARY Children who have mastered the concepts in this unit will be able to add and subtract fractions with the same denominator, writing their answers as either mixed numbers or improper fractions. They will be able to subtract a fraction from a whole number. Children will be able to find a fraction of an amount and also find the whole when given a fraction of an amount.

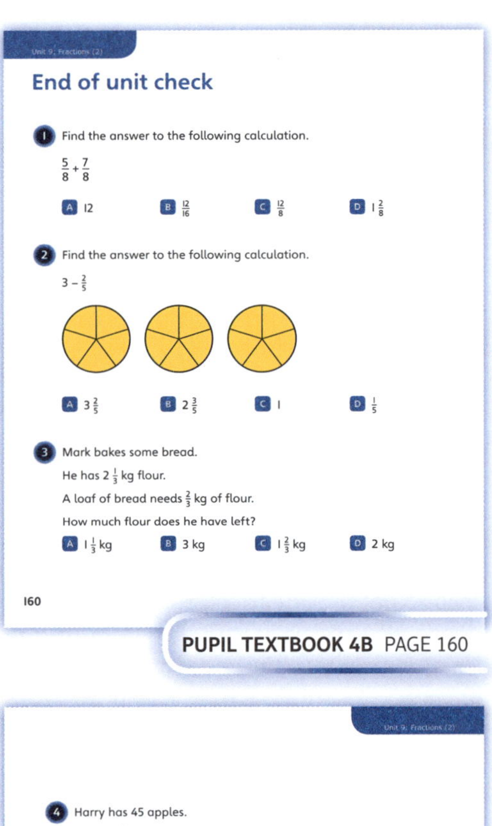

PUPIL TEXTBOOK 4B PAGE 160

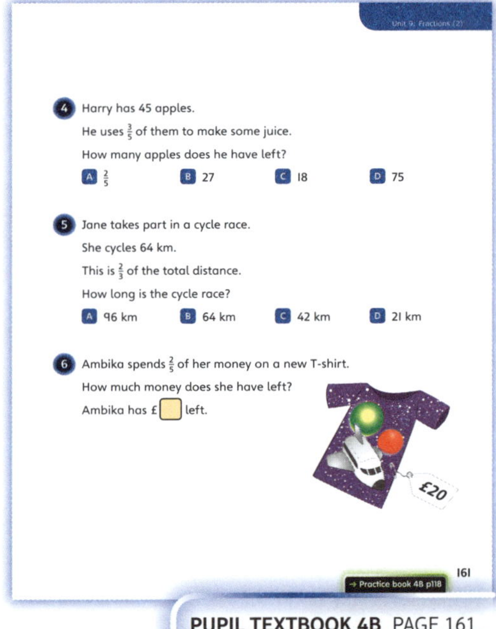

PUPIL TEXTBOOK 4B PAGE 161

Q	A	WRONG ANSWERS AND MISCONCEPTIONS	STRENGTHENING UNDERSTANDING
1	C	A suggests ignoring the denominator. B suggests adding the denominators. D suggests thinking $\frac{12}{8}$ is equal to $1\frac{2}{8}$.	Use a fraction strip above a number line to help children with adding and subtracting fractions.
2	B	A suggests adding. C suggests subtracting a whole. D suggests finding $3 - 2 = 1$ then adding the denominator $\frac{1}{5}$.	For questions 5 and 6 use clearly labelled fraction strips to explain to children the difference between finding a fraction of an amount and finding the whole when given the fraction of an amount. Ask children to look for keywords in the question that will help them with drawing their fraction strips.
3	C	A suggests subtracting a whole. B suggests adding the two fractions.	
4	C	A is the fraction that is left. B is $\frac{3}{5}$ of 45. D suggests dividing by 3 and then multiplying by 5.	
5	A	C suggests trying to find $\frac{2}{3}$ of 64.	
6	£30	Children might find $\frac{2}{5}$ of 20 and get £8.	

My journal

WAYS OF WORKING Independent thinking

ANSWERS AND COMMENTARY In this journal, children explain why certain fractions are equal to others, which highlights particular concepts they have covered within this unit.

First, children explain why an improper fraction and mixed number are equal. They do this not by formally converting, but by showing on a number line why these fractions are equivalent. Children should mark on the number line $1\frac{5}{6}$ and $\frac{11}{6}$ showing that they are the same. They should then explain this in words.

Part b) asks children to explain why the fraction addition is correct; do not accept answers where children just say you add the numerators when the denominator is the same. This is true, but children should try to explain why this is the case.

In part c) again look for explanations. Some children will convert 2 into $1\frac{6}{6}$ and then subtract.

Power check

WAYS OF WORKING Independent thinking

ASK

- *Can you add/subtract the following fraction …?*
- *Tell me two fractions that add to make $\frac{7}{10}$. What do you notice about the fractions?*
- *Find $\frac{2}{3}$ of 24.*

Power puzzle

WAYS OF WORKING Independent working

IN FOCUS The **Power puzzle** will show whether children can successfully follow the story and find a fraction of an amount. Look for accuracy rather than speed. The point is not just to give children lots of practice, but for them to notice when they have made a mistake. For example, children may come across an answer that cannot be true (for example, a fraction of a grape). If this is the case, encourage children to go back and identify where they could have made a mistake.

ANSWERS AND COMMENTARY

Emma gets 8 grapes and Holly has 40 left.
Holly eats 1 grape and now has 39 left.
Andy gets 13 grapes and Holly has 26 left.
Holly eats 2 grapes and she has 24 left.
Reena gets 9 grapes and Holly has 15 left.
Holly eats 3 grapes and now has 12 left.
Holly gives Lee 9 grapes so she has 3 left.
Holly eats 9 grapes in total.

Andy gets the most grapes.

After the unit ⏸

- Can children add or subtract fractions with the same denominator?
- Can they subtract a fraction from a whole?
- Can children find a fraction of an amount?

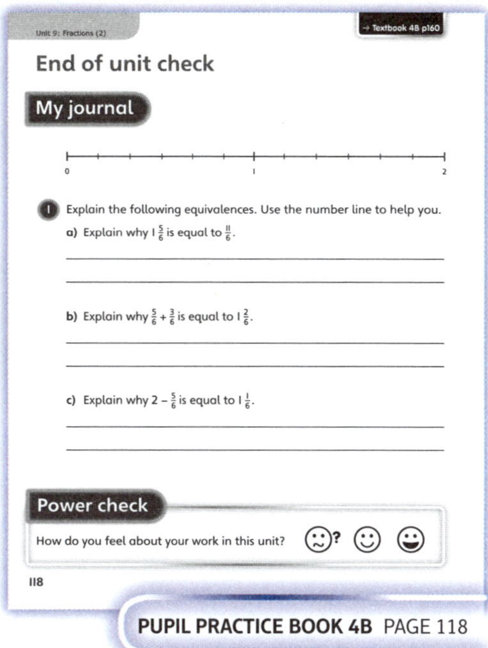

PUPIL PRACTICE BOOK 4B PAGE 118

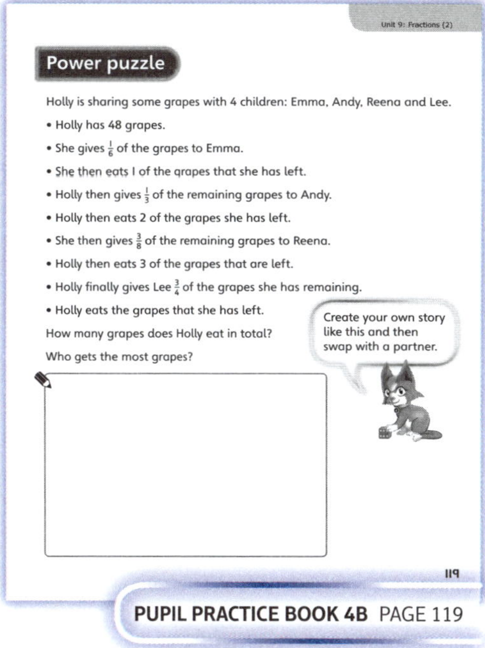

PUPIL PRACTICE BOOK 4B PAGE 119

Strengthen and **Deepen** activities for this unit can be found in the *Power Maths* online subscription.

Unit 10
Decimals ❶

Don't forget to watch the Unit 10 video!

WHY THIS UNIT IS IMPORTANT

This unit is important as it is the first time children have encountered decimals and therefore the decimal point and the tenth and hundredth columns. It sets the foundations for key concepts and future units, where children will be asked to order and round decimals as well as work with decimals in money.

WHERE THIS UNIT FITS

→ Unit 9: Fractions (2)
→ **Unit 10: Decimals (1)**
→ Unit 11: Decimals (2)

This is the first time children have covered decimals, but it builds directly on content covered within previous fraction units. The unit introduces children to writing fractional amounts in decimal notation and, in doing so, introduces the decimal point and the tenth and hundredth columns. As key learning points, tenths and hundredths are covered in detail; dividing by 10 and 100 to result in answers containing decimal numbers is also a major focus. In the next unit, children will explore decimals in greater depth, and learn about their relationship with fractions.

Before they start this unit, it is expected that children:
- know how to describe fractional amounts using the language of tenths and hundredths
- understand the place value system and can therefore extend this understanding
- understand the concept of regrouping a quantity in different ways, using place value knowledge.

ASSESSING MASTERY

Children who master this unit will be able to identify the value of any digit within a number up to two decimal places. They will be able to count forwards and backwards in tenths and hundredths and write each step accurately. Children will also be able to divide 1- and 2-digit numbers by 10 and 100, writing the solutions as decimal numbers.

COMMON MISCONCEPTIONS	STRENGTHENING UNDERSTANDING	GOING DEEPER
Children may find it difficult to count through the boundaries of the new place value columns and are likely to count 0·8, 0·9, 0·10 etc.	Children should have the opportunity to count together going up or down from a given value, using a number line to help as required.	Children should be encouraged to spot patterns between the calculations they complete and make links to areas of maths previously covered.
Children may learn procedural short-cuts to divide a number by 10 or 100, rather than fully understanding the concept. This may lead to errors in future work.	Give children the opportunity to use place value counters, ten frames and place value grids in order to understand the key concepts presented to them.	Give children the opportunity to create their own problems or equations that match a context, or produce answers within given criteria.

Unit 10: Decimals ❶

Use these pages to introduce decimals to children. You can use the characters to explore key language and visual representations needed for the unit.

STRUCTURES AND REPRESENTATIONS

Ten frame: This model helps children to understand how a quantity can be split into 10 equal parts and how 10 of these parts make 1. This resource will be invaluable to stop children counting in tenths incorrectly; for example: 0·9, 0·10, 0·11, and so on. Using place value counters on a ten frame that has tenths recorded as $\frac{1}{10}$ and 0·1 will allow children to make links with to fractions, understanding this concept in greater depth.

Number line: This model helps children to see the position of decimal numbers and their fraction equivalents within given integers and helps them to count on and back in decimal amounts. It Is also an important representation to allow children to make links with measure.

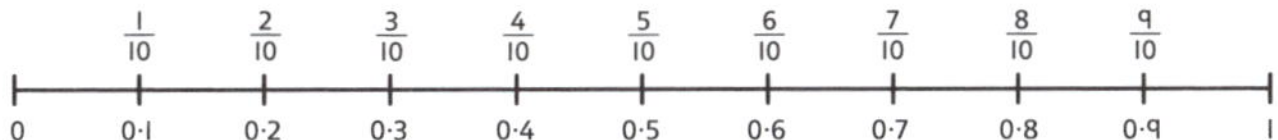

Place value grid: This is an important model to show how the place value columns relate to each other. Use it to introduce the tenths and hundredths columns and to visually show the value of each digit within a decimal number, as well as how numbers can be regrouped in different ways to show the same amount.

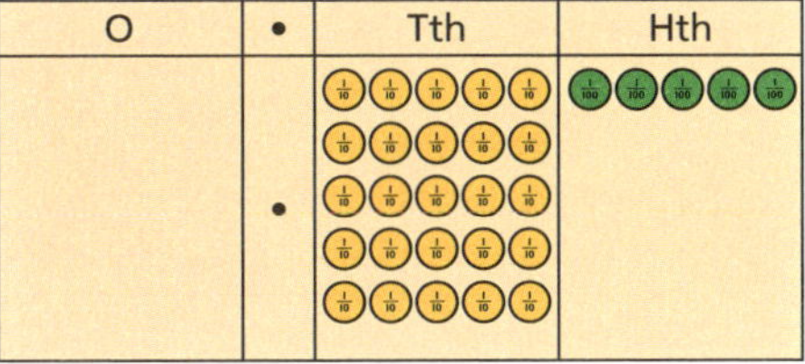

Part-whole model: This model shows how an amount can be split into different parts, which is useful to see when a part can or cannot be divided by a required amount.

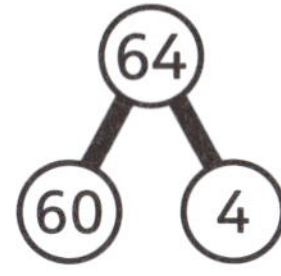

KEY LANGUAGE

There is some key language that children will need to know as a part of the learning in this unit.

→ decimal point, whole, tenths, hundredths, integer, tenths column, hundredths column

→ one more, one less, greater than, less than, increase, decrease

→ divide, regroup, equivalent, partition

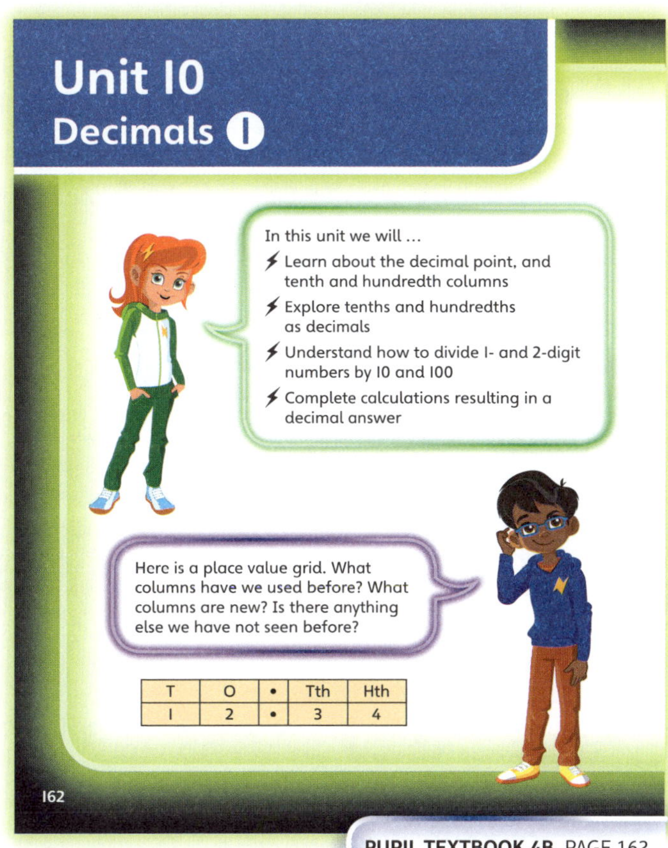

PUPIL TEXTBOOK 4B PAGE 162

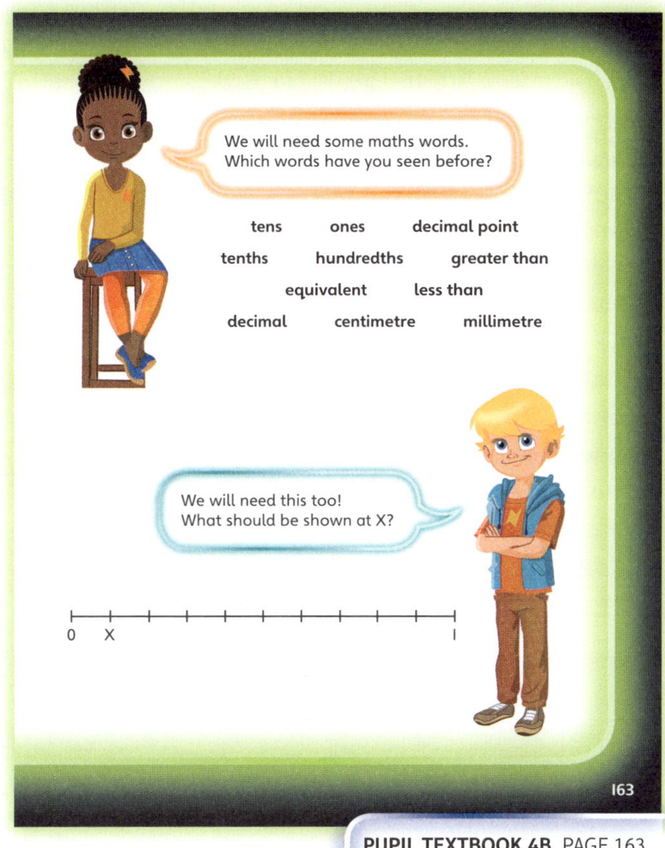

PUPIL TEXTBOOK 4B PAGE 163

Tenths ❶

Learning focus

In this lesson, children will be introduced to the decimal point and how it can be used to write tenths as decimals. Children also count in tenths and record these as decimals.

Small steps

→ Previous step: Problem solving – fraction of a quantity (2)
→ **This step: Tenths (1)**
→ Next step: Tenths (2)

NATIONAL CURRICULUM LINKS

Year 4 Number – Fractions (Including Decimals)

Recognise and write decimal equivalents of any number of tenths or hundredths.

ASSESSING MASTERY

Children can explain the value of each digit within a decimal number and can write any number of tenths, within 1, in their decimal form. Children can count and record sequences of numbers, increasing or decreasing in tenths, in their decimal form, as well as completing missing information on a number line.

COMMON MISCONCEPTIONS

Children may write 10 tenths as 0·10 rather than understanding what it represents. Ask:
• *What are 10 tenths the same as?*

Children may believe that 0·9 is greater than 1 because the digit 9 is greater than the digit 1. Ask:
• *What place value column is the digit 9 in? Is the value greater or smaller than the ones column?*

STRENGTHENING UNDERSTANDING

Children should be given the opportunity to use place value counters, ten frames and multi-linking cubes to help them to understand that 10 lots of 0·1 make a whole and that 0·1 is equivalent to $\frac{1}{10}$. Adding these counters one at a time to the ten frame will also help with counting in tenths.

GOING DEEPER

Ask children to count forwards and backwards in tenths from different starting numbers, in pairs. Listen for conflicting answers.

KEY LANGUAGE

In lesson: decimal, decimal point, tenths, whole, fraction, equivalent, denominator, numerator

Other language to be used by the teacher: decimal notation

STRUCTURES AND REPRESENTATIONS

ten frame, number line, place value grid

RESOURCES

Mandatory: place value counters

 In the eTextbook of this lesson, you will find interactive links to a selection of teaching tools.

Before you teach ❚❚

• Do children understand what a tenth is?
• Do they need visual resources to provide support counting in tenths?

Discover

WAYS OF WORKING Pair work

ASK

- Question **1** a): *How are the ten frames different? How many spaces in total does each ten frame have?*
- Question **1** b): *What do the numbers 5 and 10 represent in the fraction $\frac{5}{10}$? What do you know about numbers with a value less than 1? Is $\frac{5}{10}$ less than 1?*

IN FOCUS Question **1** a) begins with representations that children should be familiar with, where all denominators are 10; therefore the focus is on tenths. Children may initially say that $\frac{5}{10}$ can be written as a half and that this may be represented in different ways. They may not, at this stage, be able to write this fraction as a decimal. It is useful at this point to highlight that 0·5 is half-way along the number line between 0 and 1.

PRACTICAL TIPS Use place value counters that have $\frac{1}{10}$ written on one side and 0·1 written on the other, to help children make links between fractions and decimals.

Use place value counters with ten frames to help children understand that ten tenths make a whole.

ANSWERS

Question **1** a): The ten frame that represents $\frac{5}{10}$ is the one with counters on 5 of the 10 parts.

Question **1** b): $\frac{5}{10}$ can be represented as 0·5.

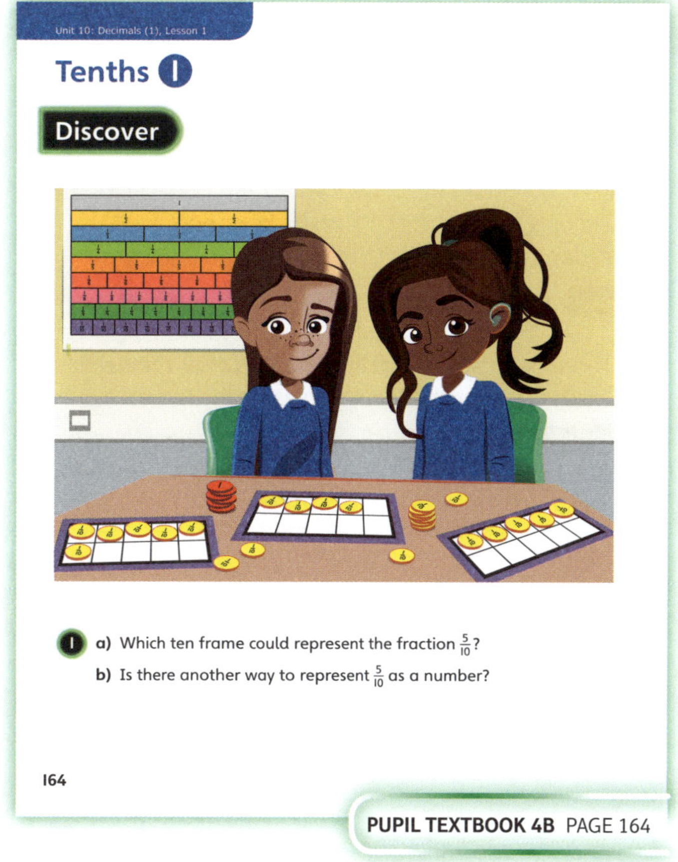

PUPIL TEXTBOOK 4B PAGE 164

Share

WAYS OF WORKING Whole class teacher led

ASK

- Question **1** a): *What is the name given to the place value column of the smallest value?*
- Question **1** b): *What separates the ones column from the tenths column?*

IN FOCUS Question **1** b) introduces children to the decimal point for the first time. Discuss Sparks's comment: the value of each digit within 0·5 and the new place value column, tenths. Make as many links as possible between tenths written in their fractional and decimal forms.

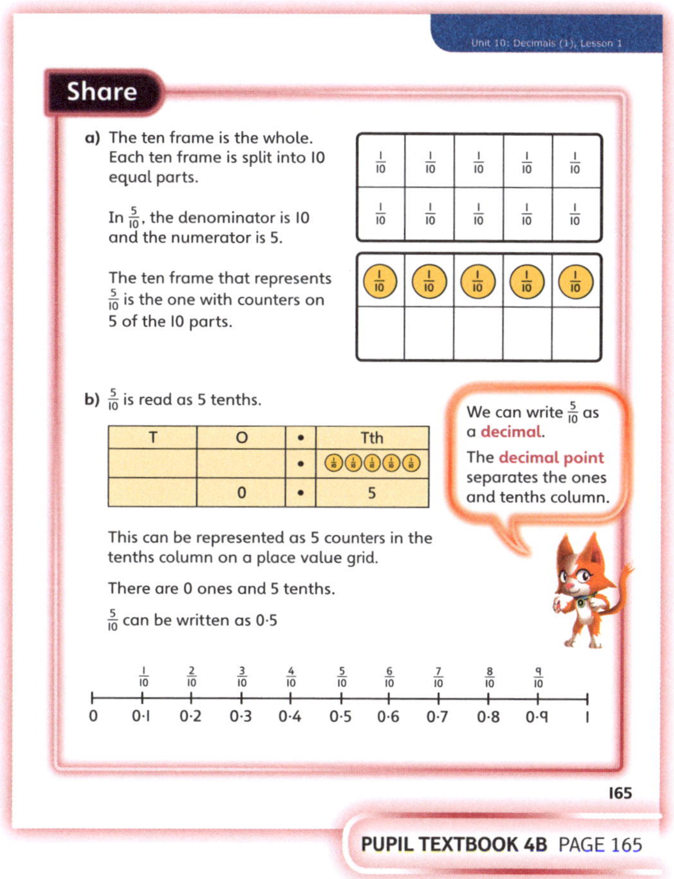

PUPIL TEXTBOOK 4B PAGE 165

Think together

 Whole class teacher led (I do, We do, You do)

- Question **1**: *What does 10 tenths represent?*
- Question **3**: *What comes after 0·9?*

 Question **3** tackles the misconception that children are most likely to have: they may continue to count in 1s through 10, writing all of the digits after the decimal point, rather than considering the value of 10 tenths.

 Children who find it difficult to understand why 10 tenths is written as 1·0, rather than 0·10, should use place value counters on a ten frame. This will show them that when 10 counters have been used, the whole frame has been covered.

 Children could extend the task presented in question **3** and begin at different starting numbers to investigate when they would say the same number at the same time. For example, one child starts at 0 and counts upwards while the other starts at 0·9 and counts backwards. Will they say the same number at the same time?

 Children should be able to explain the value of each digit within any decimal number. They should be able to identify the decimal point and explain that its role is to separate the ones and tenths columns.

Question **1** a): $\frac{3}{10}$ = 0·3

Question **1** b): $\frac{6}{10}$ = 0·6

Question **2**: The shaded counters show $\frac{7}{10}$. This can be written as 0·7.
The unshaded counters show $\frac{3}{10}$. This can be written as 0·3.

Question **3** a): Max has said 0·10 and he should have said 1·0 as 10 tenths represents 1 whole.

Question **3** b): Both children will say 0·5 at the same time.

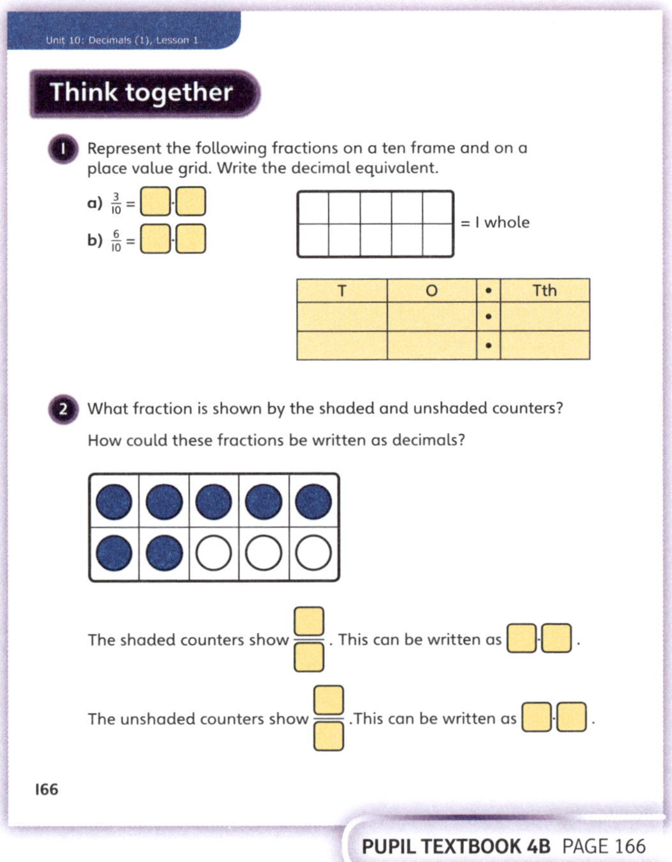

PUPIL TEXTBOOK 4B PAGE 166

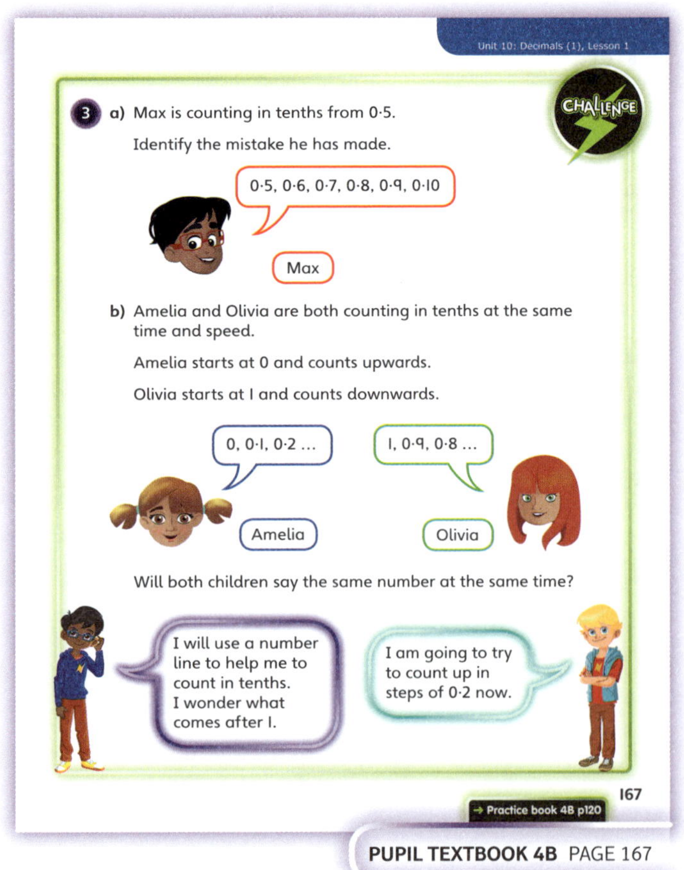

PUPIL TEXTBOOK 4B PAGE 167

Practice

WAYS OF WORKING Independent thinking

IN FOCUS Question **1** presents children with different representations that show tenths and requires them to record these as both fractions and decimals.

Question **2** presents decimals and asks children to show how they would be represented pictorially.

During this practice, children should be encouraged to see decimals as building on their understanding of fractions, rather than as totally new learning.

STRENGTHEN Use place value counters on a ten frame to assist children who find it difficult to write numbers as decimals. They may also benefit from a number line that has fractions and decimals written on it, to strengthen their understanding of the equivalent forms.

DEEPEN Children could play games in pairs, similar to question **6**, where one child thinks of a decimal number and the other child has to ask questions to find out what the number is. These questions should contain mathematical language; for example, 'odd' or 'even', rather than just guessing the decimal number.

ASSESSMENT CHECKPOINT Children should be confident writing any decimal number from 0 to 1 and making any decimal number in different ways with different pictorial and concrete resources. They should be able to give the value of each digit within a decimal number and relate these to the place value headings of ones and tenths.

ANSWERS Answers for the **Practice** part of the lesson appear in the separate **Practice and Reflect answer guide**.

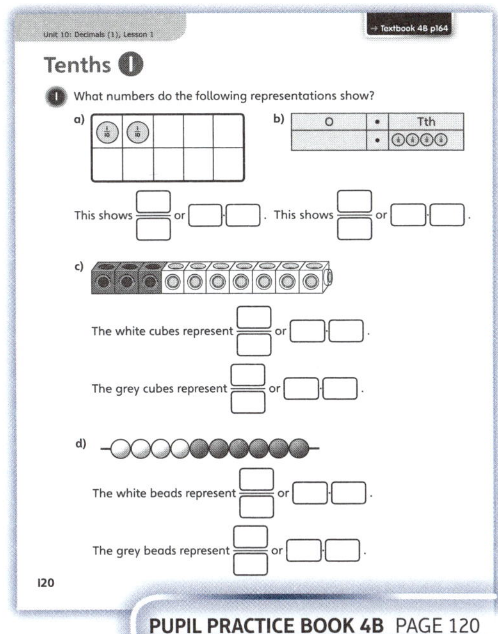

PUPIL PRACTICE BOOK 4B PAGE 120

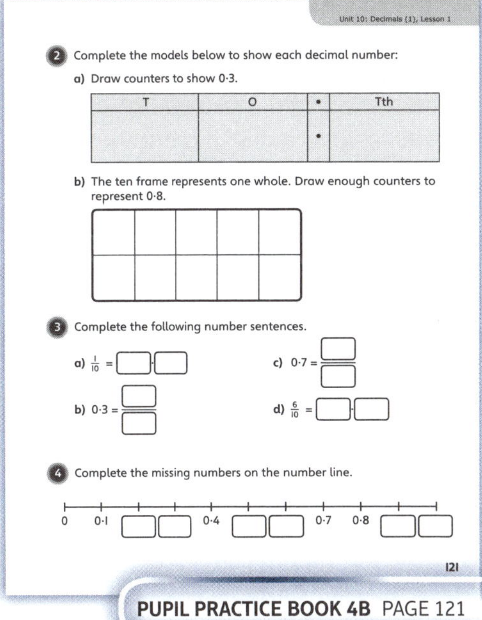

PUPIL PRACTICE BOOK 4B PAGE 121

Reflect

WAYS OF WORKING Independent thinking

IN FOCUS Children should be confident with different representations, either pictorial or concrete, to represent a given decimal. Primarily, the focus should be on writing the fractional equivalent, but children should be encouraged to use as many other representations as possible, to make links with other areas of their learning.

ASSESSMENT CHECKPOINT The variety of representations that children are able to use to represent 0·6 will be an indicator of the depth of their understanding of fractions, decimals and the links between them.

ANSWERS Answers for the **Reflect** part of the lesson appear in the separate **Practice and Reflect answer guide**.

After the lesson 

- Do children still have any misconceptions when counting in decimals beyond 1?
- Are all children secure in their understanding of decimals? If not, what additional support is needed to consolidate basic understanding?

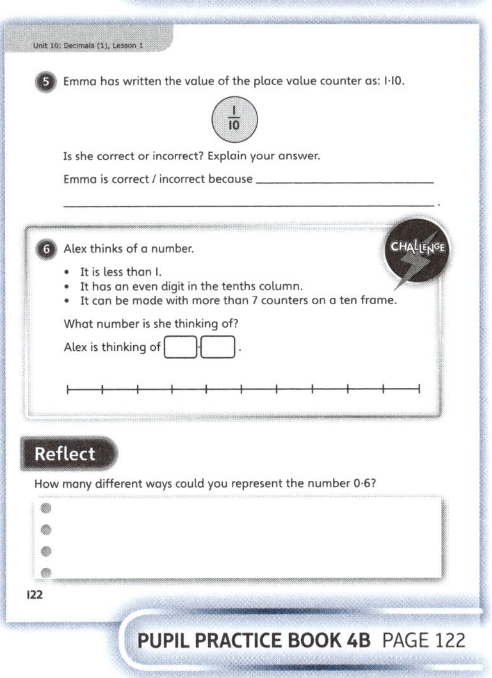

PUPIL PRACTICE BOOK 4B PAGE 122

199

Tenths ②

Learning focus

In this lesson, children will build on their understanding of tenths and extend this to numbers greater than 1. They will explore the place value of numbers, using a place value grid, with one decimal place.

Small steps

→ Previous step: Tenths (1)
→ **This step: Tenths (2)**
→ Next step: Tenths (3)

NATIONAL CURRICULUM LINKS

Year 4 Number – Fractions (Including Decimals)

Recognise and write decimal equivalents of any number of tenths or hundredths.

ASSESSING MASTERY

Children can explain the value of all digits within numbers that contain a decimal. Children can count forwards and backwards in tenths from any number, including counting through 1.

COMMON MISCONCEPTIONS

Children may continue to count through the ones boundary incorrectly, for example, 1·9, 1·10. Ask:
• *What does 10 tenths represent? What effect does this have on the number of ones in the number?*

When required to position a decimal number greater than 1 on a number line, children may work in an inefficient way, always beginning at 0. Ask:
• *How does each digit within a number help you to accurately mark the location of the number?*

STRENGTHENING UNDERSTANDING

If children find it difficult to understand the value of each digit within the numbers, use ten frames to make each number with 0·1 counters. Once they see how numbers greater than 1 can be made in this way, progress to a place value grid. This will allow children to see 10 ones as a whole ten frame, and the number of tenths as the fractional part of another ten frame.

GOING DEEPER

To extend learning, children could be given digit cards and criteria (similar to question ❸ in the **Think together** section), where they must explore how many different numbers can be created. They could work in pairs to set each other criteria and then justify how they know the number they have made is correct.

KEY LANGUAGE

In lesson: tenths, value

Other language to be used by the teacher: one more, one less, greater than, less than, decimal point

STRUCTURES AND REPRESENTATIONS

ten frame, number line, place value grid

RESOURCES

Mandatory: place value counters

Optional: digit cards

 In the eTextbook of this lesson, you will find interactive links to a selection of teaching tools.

Before you teach 

• Are children confident using place value grids and ten frames?
• Are all children secure in their understanding of writing any number of tenths, less than 1, as a decimal?

Discover

 Pair work

ASK

- Question **1** a): *What place value columns does a decimal point separate? What is the name given to the column to the left of a decimal point? What is the name given to the column to the right of the decimal point?*
- Question **1** b): *What does each mark on the number line represent?*

IN FOCUS Question **1** a) introduces children to numbers that include a decimal point and are greater than 1. The focus should be on the value of each digit within these numbers; children need to understand this in order to understand the value of the number as a whole.

PRACTICAL TIPS Use ten frames and place value grids to represent numbers encountered in the lesson.

Ask children to say what is the same and what is different about the representations, to show they understand the concept of how many tenths there are within larger numbers.

ANSWERS

Question **1** a): 0·3 is 0 ones and 3 tenths.
2·3 is 2 ones and 3 tenths.
3·1 is 3 ones and 1 tenth.
1·3 is 1 one and 3 tenths.

Question **1** b):

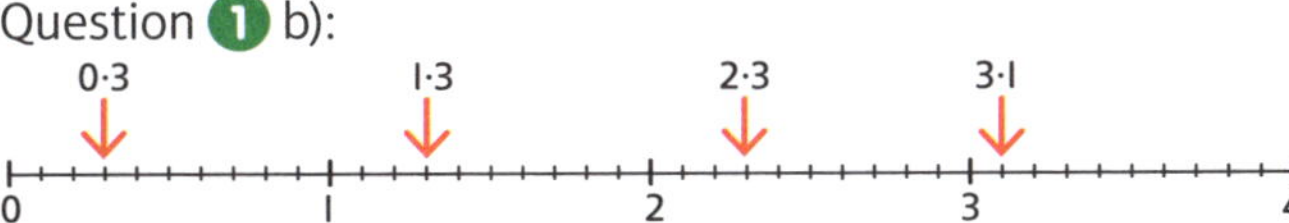

Share

 Whole class teacher led

ASK

- Question **1** a): *Why is there a zero before the decimal point in one number, but not in the others?*
- Question **1** b): *Which numbers are greater than 1? How do you know?*

IN FOCUS Question **1** a) focuses on a set of numbers with a decimal point. Children must understand what each digit within a decimal number represents and begin to recognise the significance and relative size of each digit.

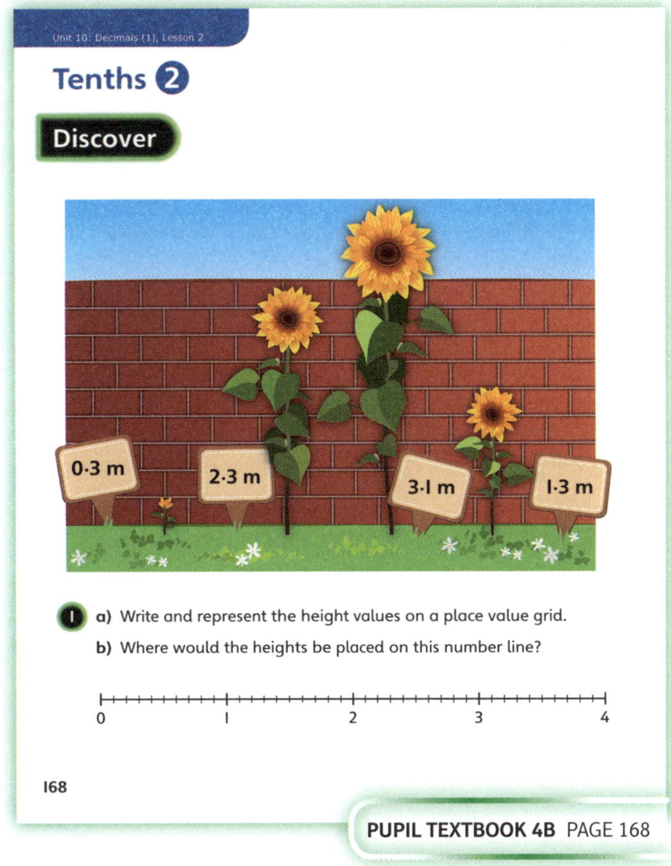

PUPIL TEXTBOOK 4B PAGE 168

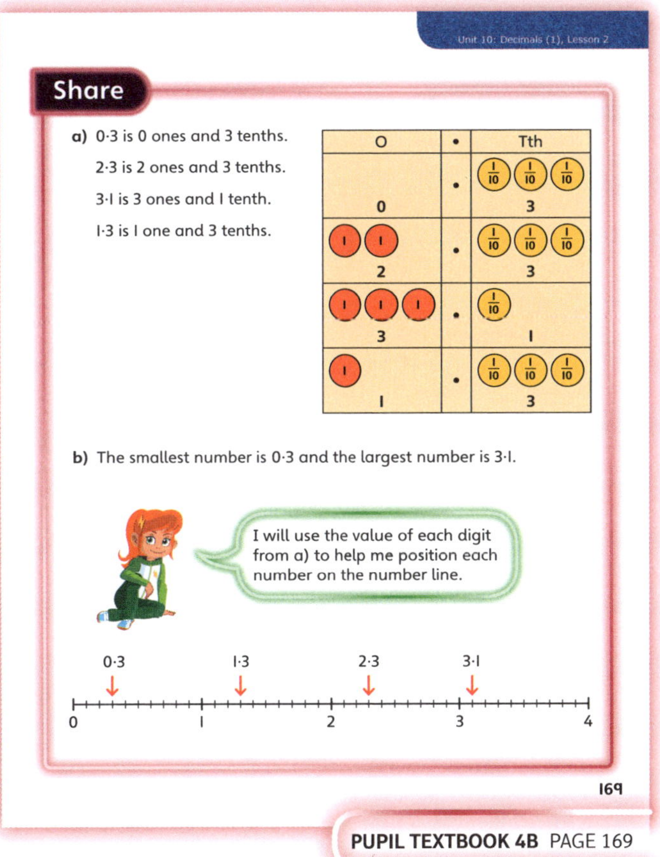

PUPIL TEXTBOOK 4B PAGE 169

Think together

WAYS OF WORKING Whole class teacher led (I do, We do, You do)

ASK

- Question ❶ b): *How can you use place value counters to support your understanding?*
- Question ❸: *What does each digit within a number represent?*

IN FOCUS Question ❸ allows children to create their own numbers that include a decimal point to meet certain criteria. This question allows children to consider the different values of the digits within a decimal number in a more open way. Ensure that children give numbers with 10s, 1s and 1 decimal place. The same digit card should not be used twice in the same question.

STRENGTHEN Continue to use place value counters on a place value grid to help children to understand the value of each digit within a number. Making numbers in this way will strengthen children's understanding of the key concept.

DEEPEN Provide children with criteria, similar to those in question ❸, that decimal numbers must meet. They could be given restrictions; for example, they are only able to use certain digits.

ASSESSMENT CHECKPOINT Challenge children to explain the value of any digit in an answer they have written. In question ❸, in particular, they should be able to explain the choices they have made and how the positioning of the different digits meets the criteria they have been set.

ANSWERS

Question ❶ a): 1·3 is shown.

Question ❶ b): Children should draw or use 3 ones and 4 tenths counters accurately to make 3·4.

Question ❷: Danny is incorrect: 1 tenth more than 0·9 is 1, not 0·10; all of the numbers following this number are also wrong.

Question ❸ a): 95·3

Question ❸ b): 0·1

Question ❸ c): 50·1

Question ❸ d): 31·5, 31·9

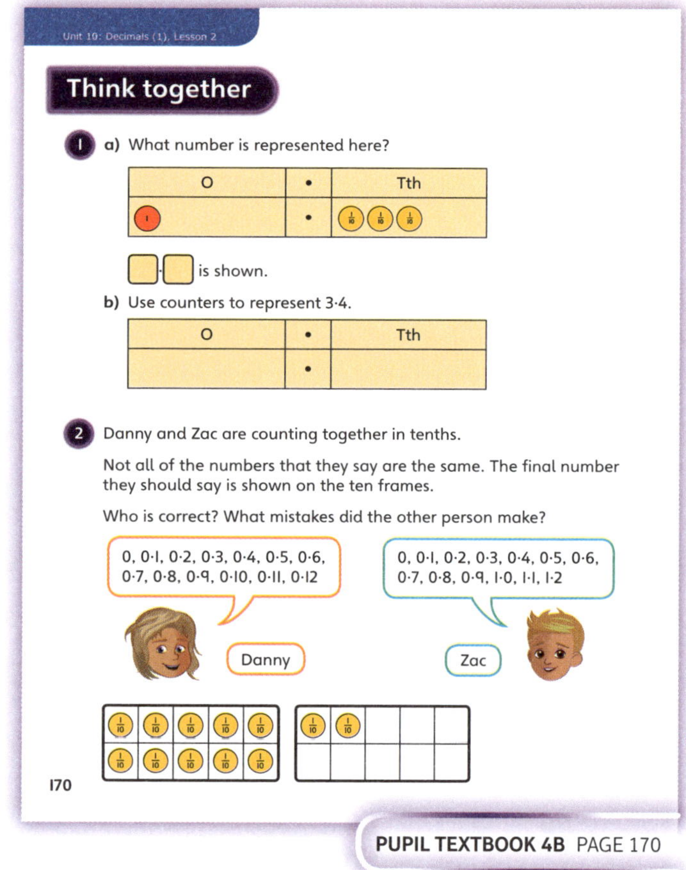

PUPIL TEXTBOOK 4B PAGE 170

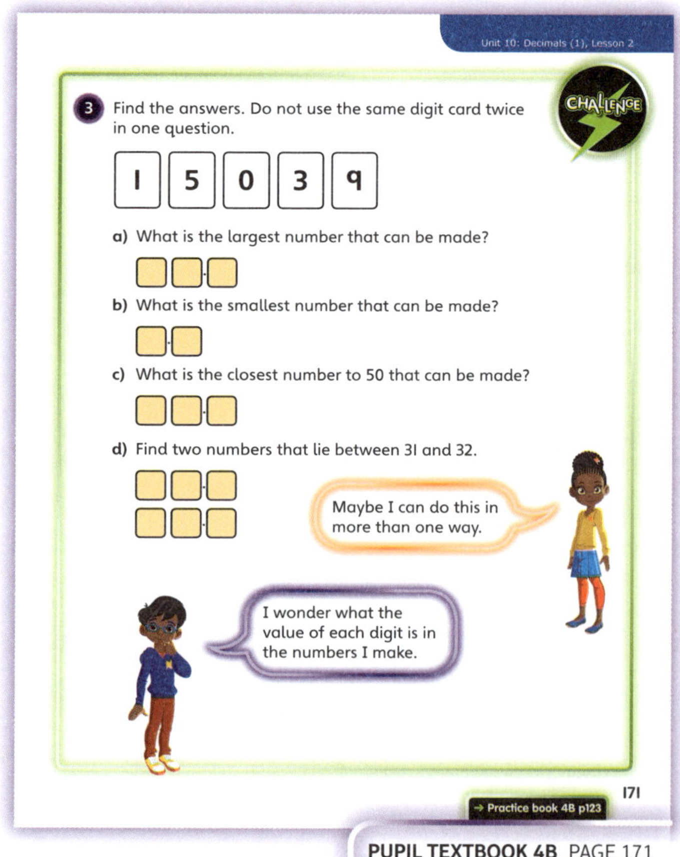

PUPIL TEXTBOOK 4B PAGE 171

Practice

WAYS OF WORKING Independent thinking

IN FOCUS Question ❶ allows children to focus on the value of each digit within decimal numbers.

Question ❷ re-emphasises the link between fractions and decimals and how they can be represented in different ways. This is important because it provides initial scaffolding to guide thinking, which will later be removed.

STRENGTHEN Use place value counters to strengthen understanding of the value of each digit within numbers.

Provide children with decimal numbers on a number line to strengthen their understanding of this concept.

DEEPEN Children should be given the opportunity to describe decimal numbers that they are presented with, or have made, in ways similar to those in question ❸. This should include what each digit within the number represents and what the number is greater and smaller than.

THINK DIFFERENTLY This question highlights the equivalence of 10 tenths and 1 whole, an important concept where misconceptions can often arise. As suggested, a ten frame should be used to highlight this key principle.

ASSESSMENT CHECKPOINT Children should be able to explain the value of each digit within any decimal number they write or read. Asking children to explain their choices for more challenging questions, for example question ❺, will allow more accurate assessments to be made.

ANSWERS Answers for the **Practice** part of the lesson appear in the separate **Practice and Reflect answer guide**.

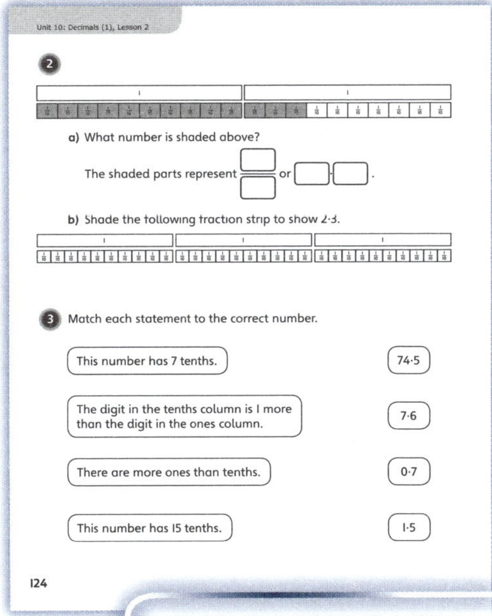

PUPIL PRACTICE BOOK 4B PAGE 123

PUPIL PRACTICE BOOK 4B PAGE 124

Reflect

WAYS OF WORKING Independent thinking

IN FOCUS Building on the **Reflect** section in the previous lesson, children should be encouraged to represent the number 12·3 in different ways, demonstrating the links they can make with previous learning and with other areas of maths.

ASSESSMENT CHECKPOINT The number of different ways in which children can confidently represent the same number will demonstrate the depth of their understanding of decimal numbers.

ANSWERS Answers for the **Reflect** part of the lesson appear in the separate **Practice and Reflect answer guide**.

After the lesson 

- Are children able to link the learning in this lesson with the previous lesson?
- Are children confident identifying the place value of each digit in a decimal number?

PUPIL PRACTICE BOOK 4B PAGE 125

Tenths ③

Learning focus

In this lesson, children will represent tenths on a number line extending beyond 1.

Small steps

→ Previous step: Tenths (2)
→ **This step: Tenths (3)**
→ Next step: Dividing by 10 (1)

NATIONAL CURRICULUM LINKS

Year 4 Number – Fractions (Including Decimals)
- Recognise and write decimal equivalents of any number of tenths or hundredths.
- Solve simple measure and money problems involving fractions and decimals to two decimal places.

ASSESSING MASTERY

Children can count forwards and backwards in tenths from any whole or decimal number. They can position decimal numbers accurately on a number line as a result of their understanding of how many tens, ones and tenths the number has.

COMMON MISCONCEPTIONS

Children may count backwards from the wrong whole number and, for example, record 3·7 cm rather than 2·7 cm. Ask:
- *What is the whole number of centimetres? How many more tenths than this amount is the length of the object?*

Children may continue to count across 1 incorrectly. Ask:
- *What does 10 tenths represent? What effect does this have on the number of ones in the number?*

STRENGTHENING UNDERSTANDING

Give children the opportunity to use rulers around the classroom, measuring and recording lengths of objects in centimetres to one decimal place. For extra support, create rulers for children where each millimetre interval is marked on the ruler as a decimal number. Encourage children to use rulers as number lines for an additional resource.

GOING DEEPER

Children could work with a partner and begin at different whole or decimal numbers and then count forwards or backwards in tenths, in both fractional and decimal notations, to see which numbers with equal value they will say at the same time. If children are encouraged to explore when this does or does not happen, their ability to count forwards and backwards in decimals will improve.

KEY LANGUAGE

In lesson: tenths, ones, fraction, decimal, equivalent, centimetre (cm), millimetre (mm)

Other language to be used by the teacher: decimal place, integer, increase, decrease

STRUCTURES AND REPRESENTATIONS

number line

RESOURCES

Mandatory: place value counters

Optional: ruler with each millimetre labelled

 In the eTextbook of this lesson, you will find interactive links to a selection of teaching tools.

Before you teach 

- Are all children secure in their knowledge of how to measure the length of an object?

Discover

 Pair work

ASK

- Question **1** a): *How many parts is each centimetre split into? Do you know what we call these parts? What whole centimetre values is the beetle between? How does knowing the whole centimetre values help you to identify the number of wholes in the length of the beetle?*
- Question **1** b): *What could you use to work out the length of the caterpillar?*

IN FOCUS Question **1** a) uses the context of length as a real life situation in which decimals are used. Remind children that when they measure the length of an object, they are counting along the ruler in the number of whole units and then counting on in the smaller decimal parts.

PRACTICAL TIPS Allow children to use rulers to measure the length of items around the classroom to the nearest decimal. This will provide a fun, hands-on way for them to practise the concept of decimals. Use the results collected by children to practise ordering decimals on a number line.

ANSWERS

Question **1** a): The beetle is $3\frac{4}{10}$ cm long. This can also be written as 3·4 cm.

Question **1** b): The caterpillar is 4·1 cm long.

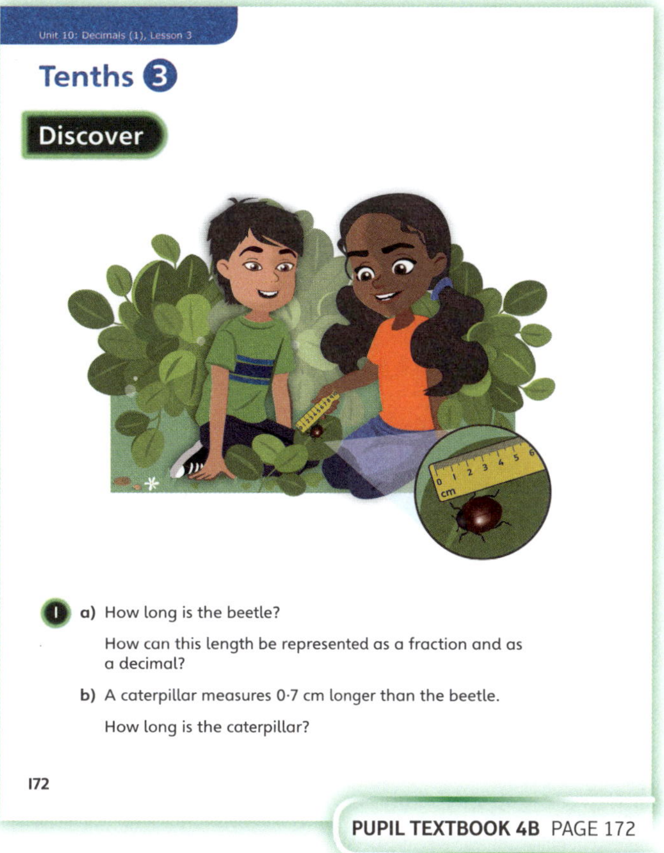

PUPIL TEXTBOOK 4B PAGE 172

Share

 Whole class teacher led

ASK

- Question **1** a): *Can you count forwards and backwards to find the length of the beetle? A ruler and a number line have been used. What is the same and what is different about these representations?*

IN FOCUS Question **1** a) shows a ruler and a number line. Children should understand that both representations can help them to think about the length of the beetle. Children should focus on the number line, which displays the fractional and decimal notation of the same lengths, and ensure they understand their equivalence.

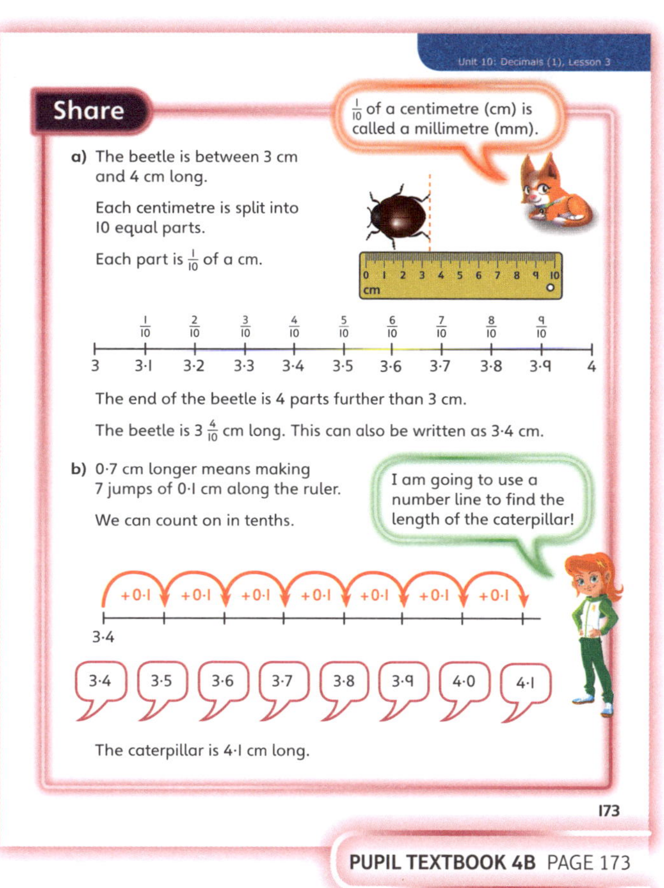

PUPIL TEXTBOOK 4B PAGE 173

Think together

WAYS OF WORKING Whole class teacher led (I do, We do, You do)

ASK

- Question **2**: *How many parts is each centimetre split into?*
- Question **3**: *Can the same number be represented as a decimal and as a fraction? How could a ruler be used in place of a number line?*

IN FOCUS In question **1**, children are not provided with the integer values on the number line. They should be encouraged to identify where these occur and might find it useful to write these numbers on a number line in their books to the same scale. This will help to emphasise that 10 tenths make each 1 and that it is possible to count on from each integer in tenths to identify the length of an object.

STRENGTHEN In question **1** a), children who find it difficult to interpret the lengths of items, or positions of lengths on a number line, should be given a ruler so they can see values more clearly. The more they practise measuring real life objects and writing their lengths to one decimal place, the stronger their understanding will become.

DEEPEN Children should be given the opportunity to count forwards and backwards from different numbers in tenths. To further deepen their understanding of decimals, they could be encouraged to count in a different number of tenths with the aid of a ruler or a number line. This will deepen the links that they make between decimals and fractions and support their understanding of how to count in different quantities of tenths.

ASSESSMENT CHECKPOINT Children should be able to identify the whole number of centimetres that they have written for each length and link this to the place value of each number. They be able to count forwards and backwards from any number in tenths and understand the fractional and decimal notation of each value.

ANSWERS

Question **1** a): Tom would be 35 partitions along the number line.

Question **1** b): Tom has travelled 3·5 km.

Question **2**: Both children are correct: 9·3 is equivalent to $9\frac{3}{10}$.

Question **3** a): Ambika and Emma will reach 5·5 at the same time. Ambika will say 5·5 and Emma will say $5\frac{5}{10}$.

Question **3** b): Emma will say $7\frac{2}{10}$.

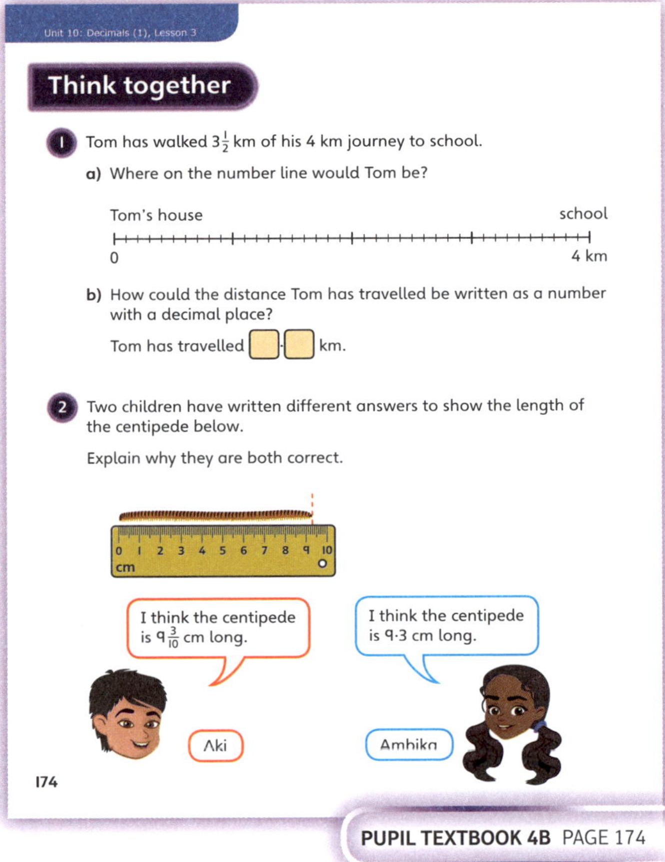

PUPIL TEXTBOOK 4B PAGE 174

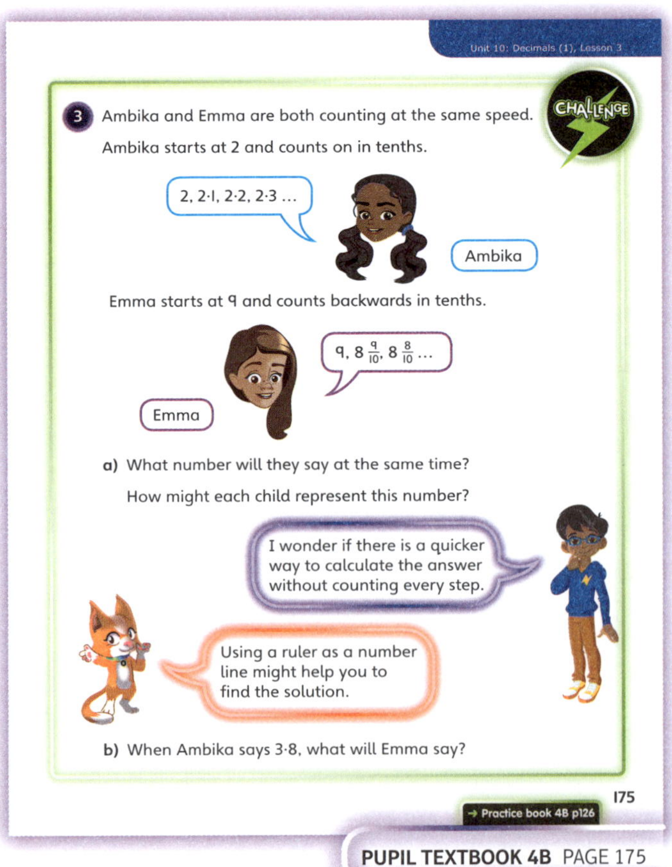

PUPIL TEXTBOOK 4B PAGE 175

Practice

WAYS OF WORKING Independent thinking

IN FOCUS Questions ➊ and ➋ give different contexts for measuring. Question ➌ requires children to count back in tenths, and question ➍ presents a number line for children to add decimal equivalents to the fractions shown. Question ➎ presents children with missing information within a number line; they must choose a suitable starting point in order to complete the required information. This allows them to draw on what they have learnt and to count both forwards and backwards in tenths to complete the number line.

STRENGTHEN Use place value counters with children who find it difficult to count forwards and backwards from a decimal number. Ask them to make the starting number with a place value counter and then to add or remove counters as necessary to find the subsequent numbers. Within this process, children will need to regroup 10 tenths to be 1 whole, or 1 whole to be 10 tenths.

DEEPEN Give children the opportunity to measure different items in the classroom and write their lengths to one decimal place. They could compare the lengths they have measured, using the symbols < or >, or place them in ascending or descending order on a number line.

ASSESSMENT CHECKPOINT Children should be able to show where any given length would occur on a ruler or number line by first identifying the number of whole centimetres, or tens and ones in a number, and then using the value of the digit in the tenths column accordingly.

ANSWERS Answers for the **Practice** part of the lesson appear in the separate **Practice and Reflect answer guide**.

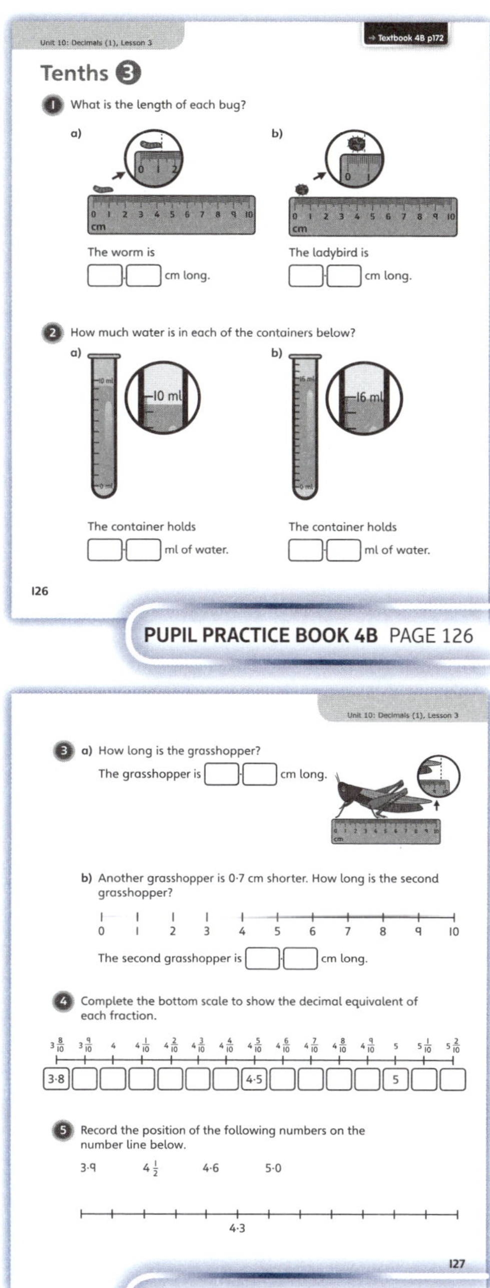

PUPIL PRACTICE BOOK 4B PAGE 126

PUPIL PRACTICE BOOK 4B PAGE 127

Reflect

WAYS OF WORKING Independent thinking

IN FOCUS Children are required to show their understanding of what is the same and what is different about the fractional and decimal notations of the same number. They should be encouraged to consider different visual and concrete representations they could use to explain their thinking.

ASSESSMENT CHECKPOINT The strength of children's written response will allow you to assess their confidence when using the fractional and decimal notations of the same number. The depth of their understanding can also be assessed based on their use of different resources to explain and represent their thinking.

ANSWERS Answers for the **Reflect** part of the lesson appear in the separate **Practice and Reflect answer guide**.

After the lesson 

- Are all children secure in counting across 1 when counting in tenths?
- Are children able to record lengths accurately to one decimal place?

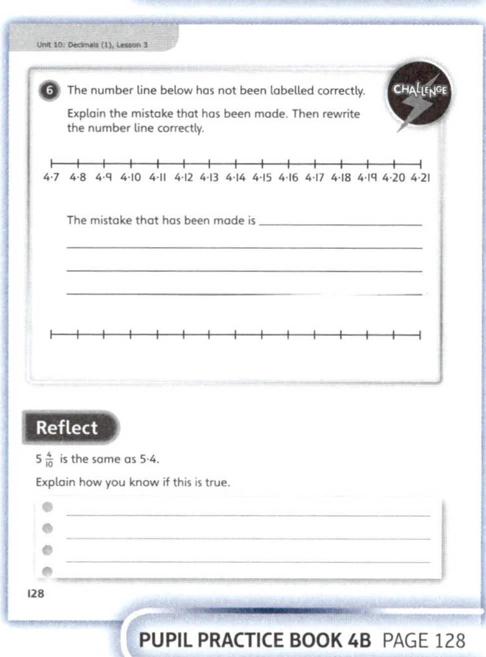

PUPIL PRACTICE BOOK 4B PAGE 128

Dividing by 10 ①

Learning focus

In this lesson, children will understand what happens when you divide a 1-digit number by 10, making connections with tenths during this process.

Small steps

→ Previous step: Tenths (3)
→ **This step: Dividing by 10 (1)**
→ Next step: Dividing by 10 (2)

NATIONAL CURRICULUM LINKS

Year 4 Number – Fractions (Including Decimals)

Find the effect of dividing a 1- or 2-digit number by 10 and 100, identifying the value of the digits in the answer as ones, tenths and hundredths.

ASSESSING MASTERY

Children can regroup a 1-digit number into the correct number of tenths and therefore divide a number by 10. Children can explain the steps of this calculation as they use resources and use this conceptual understanding to explain how they know their answers are correct.

COMMON MISCONCEPTIONS

Children may find it confusing that previously they were not able to record a digit greater than 9 in any column and now they are required to regroup the quantity of ones to be a multiple of 10 tenths. Ask:

- *How many tenths is 3 ones equivalent to? How does this help you to divide by 10?*

Children may begin to spot patterns within their answers, but not be able to explain why this occurs. Ask:

- *Can you use counters to check your answer and explain the concept of dividing by 10?*

Children may write their calculations incorrectly; for example, 10 ÷ 9 rather than 9 ÷ 10. Ask:

- *What is the value of the whole? How many parts must this be split into?*

STRENGTHENING UNDERSTANDING

Children should practise creating the number that they are required to divide by 10. Use place value counters and then exchange this number of ones for the required number of tenths so this is possible. Use a ten frame to represent the 'whole number' of tenths to help children to understand this process. To quicken the process, use a cut-out of 10 tenths joined together, to exchange a one counter for 10 tenths in this representation.

GOING DEEPER

Children could be encouraged to spot patterns within the calculations they complete and should be challenged to explain why these occur; for example, that 0·3 is $\frac{1}{10}$ of the size of 3, rather than just that the digit moves place or that the decimal point moves place.

KEY LANGUAGE

In lesson: divide, tenth, calculation

Other language to be used by the teacher: whole, parts, equivalent, regroup

STRUCTURES AND REPRESENTATIONS

place value grid, bar model

RESOURCES

Mandatory: place value counters

Optional: 10 tenth counters joined together, ten frames

 In the eTextbook of this lesson, you will find interactive links to a selection of teaching tools.

Before you teach 

- Are all children secure in their understanding of the links between place value columns?
- Do less confident children require support when exchanging any given number of ones for the correct number of tenths?

Discover

WAYS OF WORKING Pair work

ASK

- Question ❶ a): *How many parts must the whole be split into? Is it possible to split a 1-digit number into 10 parts?*
- Question ❶ b): *What representation could be used to show this calculation?*

IN FOCUS Question ❶ a) introduces dividing an amount by a quantity larger than the starting number. Discuss with children whether this is possible, building on their understanding of decimals from the previous lessons and providing the foundation for this lesson.

PRACTICAL TIPS Give children the opportunity to practise creating the number that they are required to divide by 10, using place value counters and then exchanging this number of ones for the required number of tenths.

Use a ten frame to represent the 'whole number' of tenths, as in the first lesson, to help children understand this process.

ANSWERS

Question ❶ a): Each new piece of rope would be $\frac{3}{10}$ of a metre, or 0·3 m long.

Question ❶ b): The calculation that shows what Holly has done is: 3 ÷ 10 = 0·3.
The 3 represents the length in metres of the original piece of rope. The 10 represents the number of pieces this rope is cut into. 0·3 represents the length of each new piece of rope.

Share

WAYS OF WORKING Whole class teacher led

ASK

- Question ❶ a): *What is the same and what is different about the representations of 3 ones and 30 tenths? Why has the starting number been regrouped in this way?*

IN FOCUS Question ❶ a) introduces the method that children will be encouraged to use when they are dividing a number by a number larger than itself. As it is not possible to divide 3 counters by 10, children must use their knowledge of equivalence, place value and tenths from previous lessons to regroup 3 ones into 30 tenths, which can then be divided by 10.

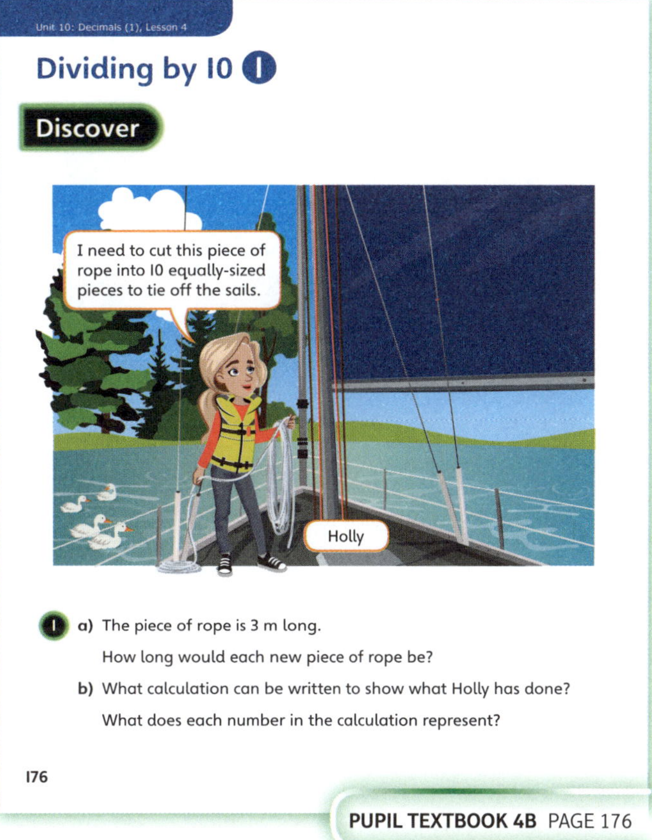

PUPIL TEXTBOOK 4B PAGE 176

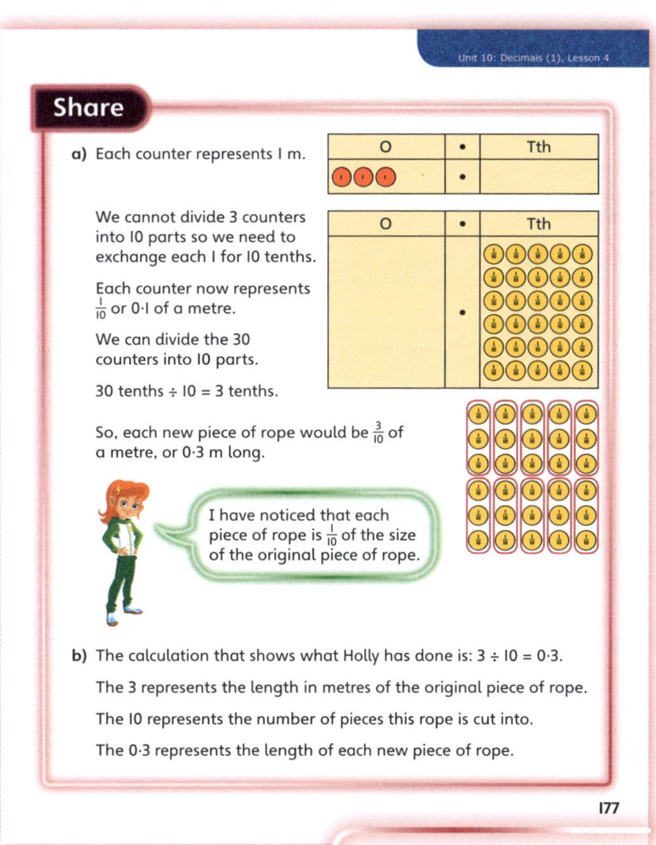

PUPIL TEXTBOOK 4B PAGE 177

Think together

ASK

- Question **1** : *How do the place value grids show the calculations?*
- Question **2** : *How could you visualise a place value grid and counters to speed up your calculations?*

IN FOCUS Question **2** presents children with a real life context that requires a measure to be divided by 10. The initial amount is a single-digit value and, as a result, each one must be exchanged for 10 tenths. Children should give their answer using the same unit as the question, rather than finding the answer as a result of converting to a different unit, in order to consolidate the learning objective.

STRENGTHEN Children who make mistakes when they are regrouping the number of ones into the required number of tenths could use a print-out of 10 tenths that are joined together. This will make the process quicker and allow them to see more easily how 10 of these counters are grouped together.

DEEPEN When children are ready to do so, they should be challenged to visualise the place value grid and the regrouping process of the ones counters into tenths, as suggested by Flo in question **2**. Ask children to explain what they see during this visualisation to remove the support of the counters and deepen their understanding of the concept.

ASSESSMENT CHECKPOINT The length of time for which children need to use place value counters to divide a single-digit number by 10, and the quality of their verbal explanations of the process will show how secure their understanding is.

ANSWERS

Question **1** : 40 tenths ÷ 10 = 4 tenths
There will be 0·4 kg of sweets in each bowl.

Question **2** : 2 ÷ 10 = 0·2
There will be 0·2 litres of water in each glass.

Question **3** a): Tom has written the calculation incorrectly. He should have written 9 ÷ 10, as 9 is the length of the whole fence and it is made out of 10 equally sized panels. He should have visualised these 9 ones as 90 tenths and used the fact that 90 tenths ÷ 10 = 9 tenths to complete the calculation as 9 ÷ 10 = 0·9.

Question **3** b): The counters would help Tom to visualise the 9 ones as 90 tenths.

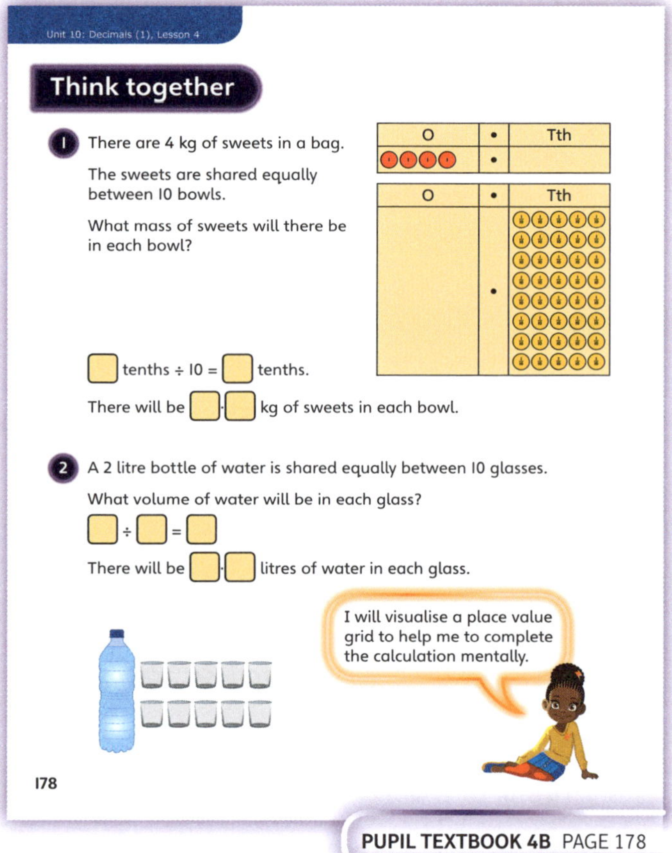

PUPIL TEXTBOOK 4B PAGE 178

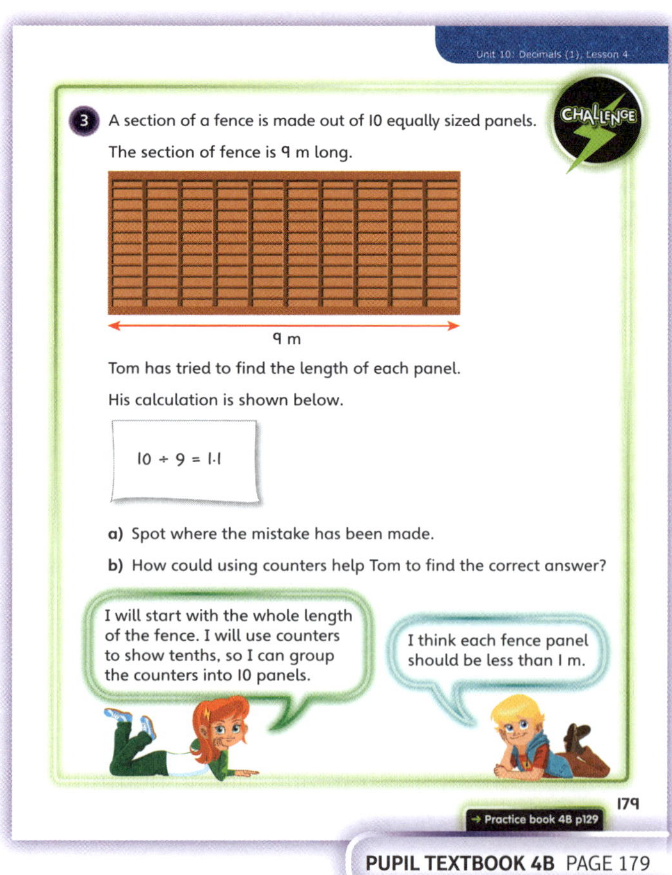

PUPIL TEXTBOOK 4B PAGE 179

Practice

WAYS OF WORKING Independent thinking

IN FOCUS In question **1**, children are initially given the calculation that they are required to complete. As they move through the question parts this scaffolding is removed, so it is wise to check that children are writing the calculations accurately. Children can use empty place value grids to draw counters or they can choose to visualise them and complete the calculations mentally.

STRENGTHEN Children who do not understand the concept should continue to use visual and concrete resources such as place value grids, value counters and bar models. If necessary, give children the calculation they are required to complete for each question so the emphasis is on understanding this concept, rather than interpreting the question.

DEEPEN Children could be asked to record all the calculations that are possible when dividing a single-digit number by 10; encourage them to work systematically to ensure that they record all possible calculations. Their understanding can be deepened further by asking them to record the solution first, followed by the division calculation in the form of _ = _ ÷ 10. This will familiarise them with all the different forms of the same calculation that they may come across.

THINK DIFFERENTLY This question will reveal a misconception and identify if children are reading calculations in the correct order. Remind them of the meaning of the placement of each number and sign.

ASSESSMENT CHECKPOINT Children should be able to explain any calculation they have completed. The strength of their explanations will indicate whether they have understood the concept or whether they have simply identified a pattern in how the digits move and used this to find the solutions.

ANSWERS Answers for the **Practice** part of the lesson appear in the separate **Practice and Reflect answer guide**.

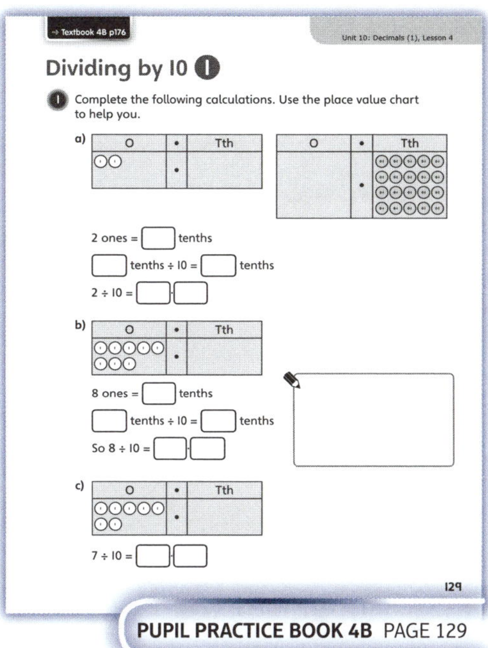

PUPIL PRACTICE BOOK 4B PAGE 129

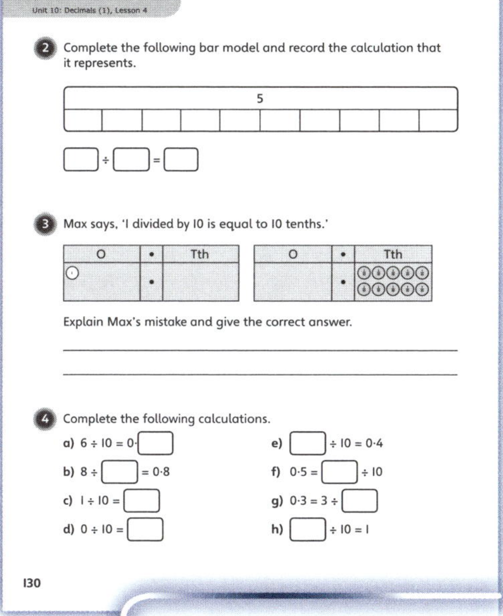

PUPIL PRACTICE BOOK 4B PAGE 130

Reflect

WAYS OF WORKING Independent thinking

IN FOCUS This question asks children to explain the process of dividing a single-digit number by 10. This explanation should match the process they have been exposed to, of regrouping the place value counters appropriately, rather than simply moving each digit. They should visualise using the counters as they write their explanation.

ASSESSMENT CHECKPOINT The quality of children's explanations and their ability to visualise the process will allow you to assess the strength of their understanding.

ANSWERS Answers for the **Reflect** part of the lesson appear in the separate **Practice and Reflect answer guide**.

After the lesson 

- Are all children secure in visualising and explaining the process of dividing a single-digit number by 10?
- Do any children still rely on counters to complete calculations?

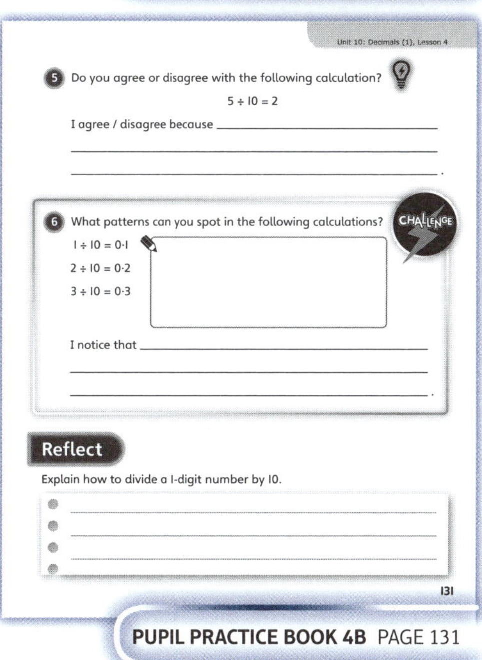

PUPIL PRACTICE BOOK 4B PAGE 131

Dividing by 10 ②

Learning focus

In this lesson, children will build on their previous understanding and extend the concept to divide 2-digit numbers by 10.

Small steps

→ Previous step: Dividing by 10 (1)
→ **This step: Dividing by 10 (2)**
→ Next step: Hundredths (1)

NATIONAL CURRICULUM LINKS

Year 4 Number – Fractions (Including Decimals)

Find the effect of dividing a 1- or 2-digit number by 10 and 100, identifying the value of the digits in the answer as ones, tenths and hundredths.

ASSESSING MASTERY

Children can divide any 2-digit number by 10 by understanding that each digit in the number is $\frac{1}{10}$ of its original size. They can explain the concept of dividing by 10 and spot patterns in the calculations they complete, helping them to solve subsequent calculations.

COMMON MISCONCEPTIONS

Children may believe that dividing any number by 10 will result in a solution with a decimal answer. Ask:
• *Can you visualise the calculation and consider what will happen to each digit when you divide by 10?*

Children may apply knowledge from the previous lesson incorrectly and only partially divide a 2-digit number by 10, or only divide 1 digit by 10. Ask:
• *Have you made the size of each digit in the number $\frac{1}{10}$ of its original size?*

STRENGTHENING UNDERSTANDING

To help children understand that each digit within the starting number becomes $\frac{1}{10}$ of its size, practise dividing 2-digit numbers by 10, using counters if required. Encourage children to spot patterns within their answers and give them cut-outs of 10 ones and tenths, presented in columns if needed.

GOING DEEPER

Give children a set of criteria, for example, $2 \cdot 6 < \_\cdot\_ < 3 \cdot 6$, and ask them to create as many division equations as they can that meet these criteria. This type of open question requires children to apply their understanding appropriately.

KEY LANGUAGE

In lesson: divide, tenth, partition

Other language to be used by the teacher: calculation, whole, parts, equivalent, place value

STRUCTURES AND REPRESENTATIONS

place value grid, bar model, part-whole model

RESOURCES

Mandatory: place value counters

Optional: grouped place value counters

 In the eTextbook of this lesson, you will find interactive links to a selection of teaching tools.

Before you teach ⏸

• Are all children secure in their knowledge of dividing a 1-digit number by 10?
• What can you use to provide scaffolding for those children who still need support?

Discover

 Pair work

ASK

- Question **1** a): *Can you apply what you learnt in the previous lesson to help you with these problems?*
- Question **1** b): *How many books are in the box?*

IN FOCUS Question **1** a) builds on the learning from the previous lesson as children are presented with a 2-digit number to be divided by 10. Encourage them to build on their understanding and apply similar strategies to the ones used in the previous lesson.

PRACTICAL TIPS Give children cubes or counters to represent the weight of the books so they can physically see the division happening.

ANSWERS

Question **1** a): Each book in Box A weighs 1·2 kg.

Question **1** b): Each book in Box B weighs 1·4 kg.
The digits in 14 and 1·4 are the same, but their position has changed.

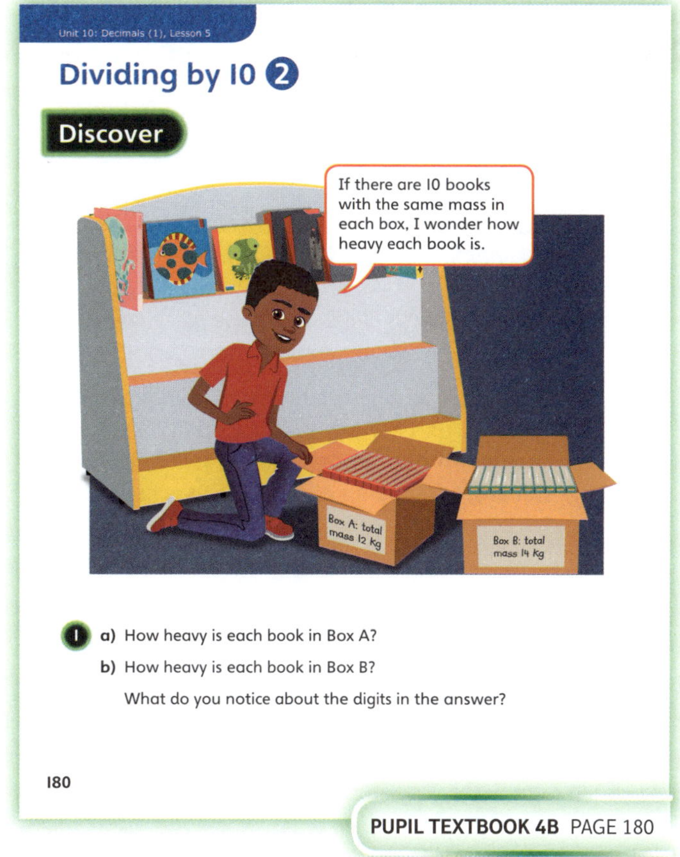

PUPIL TEXTBOOK 4B PAGE 180

Share

 Whole class teacher led

ASK

- Question **1** a): *Why has the number been partitioned in this way? Why have the counters been arranged in columns as they have?*

IN FOCUS Question **1** a) shows how each digit in the original 2-digit number can be remade as a result of regrouping in multiples of 10. Representing the regrouped quantity in columns of ten shows children that the new values are now easily divisible by 10.

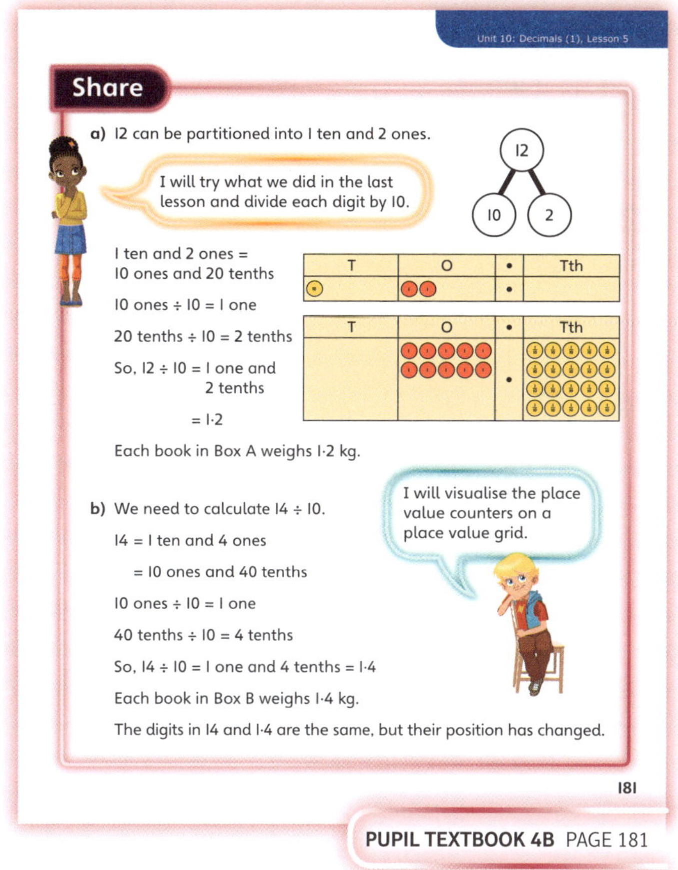

PUPIL TEXTBOOK 4B PAGE 181

Think together

WAYS OF WORKING Whole class teacher led (I do, We do, You do)

ASK

- Question ❶: *Can you spot any patterns between the starting and ending number when you divide a number by 10?*
- Question ❷: *Is it always necessary to use place value counters to find the solution to problems?*

IN FOCUS Question ❸ requires children to divide different 2-digit numbers by 10. They are encouraged to create and interpret pictorial representations and to identify patterns within the starting and ending numbers. This is important to ensure children understand the concept of dividing by 10 as well as recognising patterns in the starting and finishing numbers.

STRENGTHEN To encourage children to spot patterns, provide them with completed calculations and ask them to identify what is the same and what is different about the beginning and final numbers. Ask: *What does each digit in each number represent?* This will help children to understand that each digit within the starting number becomes $\frac{1}{10}$ of the size.

DEEPEN Challenge children to find as many different solutions as possible from given number cards, as in question ❸. They can then compare the values of these numbers and place them in increasing or decreasing order. This will encourage them to think more deeply about the value of each digit within calculations.

ASSESSMENT CHECKPOINT Children should be able to explain all of the calculations they have completed and how they know they are correct. They should also be able to identify why a calculation is incorrect, based on the value of the solution and their understanding of the method used.

ANSWERS

Question ❶: 20 ones ÷ 10 = 2 ones.
30 tenths ÷ 10 = 3 tenths.
23 ÷ 10 = 2 ones and 3 tenths.
The mass of each toy car is 2·3 kg.

Question ❷: 31 ÷ 10 = 3·1

Question ❸: The largest answer is 5·4 (54 ÷ 10 = 5·4).
The smallest answer is 1·4 (14 ÷ 10 = 1·4).

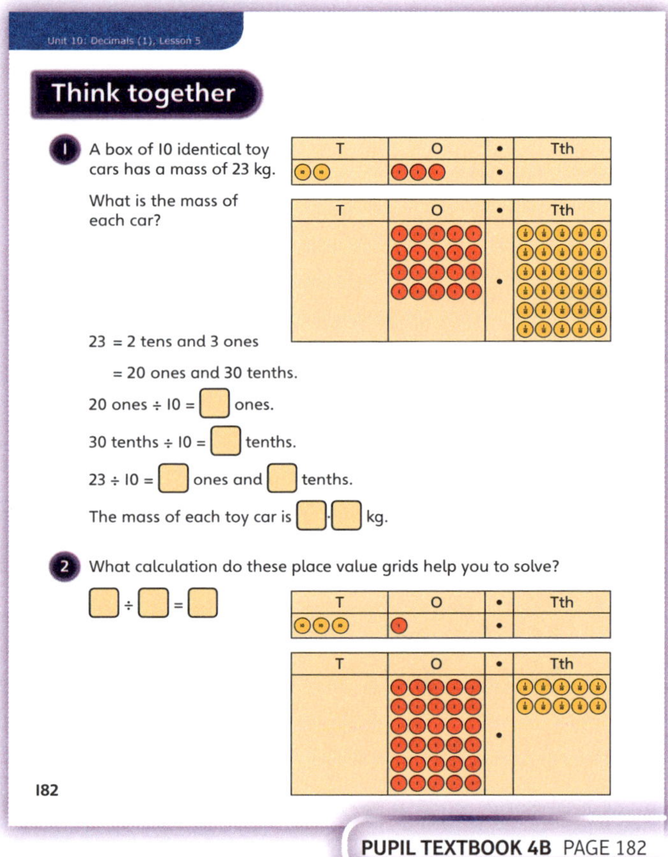

PUPIL TEXTBOOK 4B PAGE 182

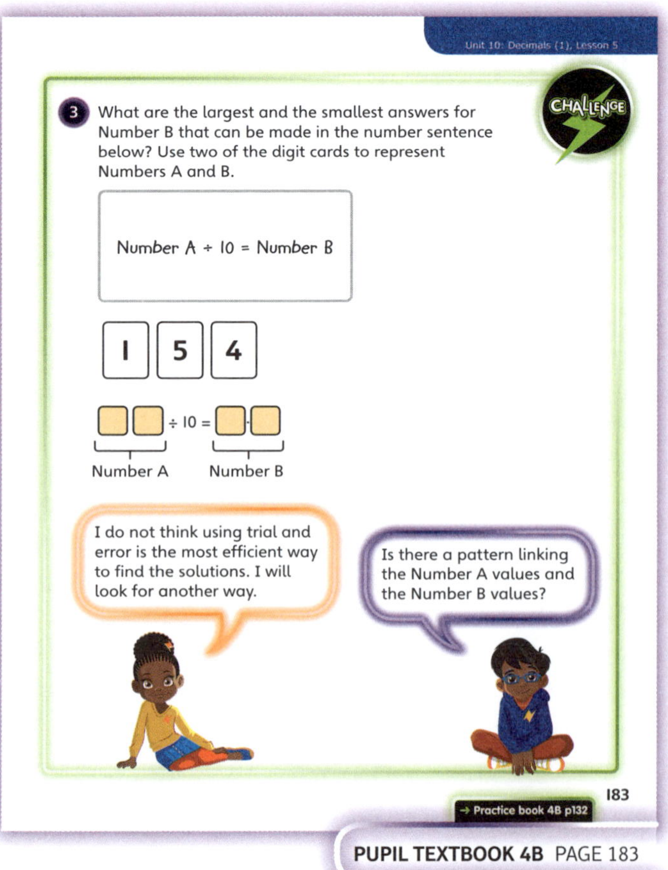

PUPIL TEXTBOOK 4B PAGE 183

Practice

WAYS OF WORKING Independent thinking

IN FOCUS In question ❶, children are provided with scaffolding, until part c), to help them to work through the procedure required to divide a 2-digit number by 10. Question ❷ requires children to create the calculation represented by a bar model. Question ❺ presents division calculations where the answer is shown before the equals sign, to challenge their understanding.

STRENGTHEN Children who are finding it difficult to follow all the steps of the calculation, either with or without counters, can use scaffolds, similar to those provided in question ❶ (for example, _ tens = _ ones, _ ones ÷ 10 = _ ones, and so on). This will allow them to work independently.

DEEPEN Ask children to write equations that meet other criteria, similar to those presented in question ❼; for example, 3·9 < _ ÷ 10 < 4·9.

ASSESSMENT CHECKPOINT Children should be able to explain their calculations and identify those that are not correct. They should be able to spot patterns within the calculations and to explain the conceptual steps when required to do so.

ANSWERS Answers for the **Practice** part of the lesson appear in the separate **Practice and Reflect answer guide**.

Reflect

WAYS OF WORKING Independent thinking

IN FOCUS Children are required to make links with the previous lesson and generalise about the similarities and differences between dividing a 1-digit number and dividing a 2-digit number by 10. Some children may have spotted patterns within the calculations, but they should be encouraged to explain the conceptual similarities as well.

ASSESSMENT CHECKPOINT The quality of children's written responses will allow you to assess the security of their understanding. Children who find it difficult to make links between the two lessons may have learnt a procedure to find the solution, without conceptually understanding the process.

ANSWERS Answers for the **Reflect** part of the lesson appear in the separate **Practice and Reflect answer guide**.

After the lesson ⏸

- Do children understand how to regroup the tenths in order to divide by 10?
- Do they recognise that the numbers remain the same, but change position, when dividing by 10?
- Are all children able to make links between this lesson and the previous lesson?

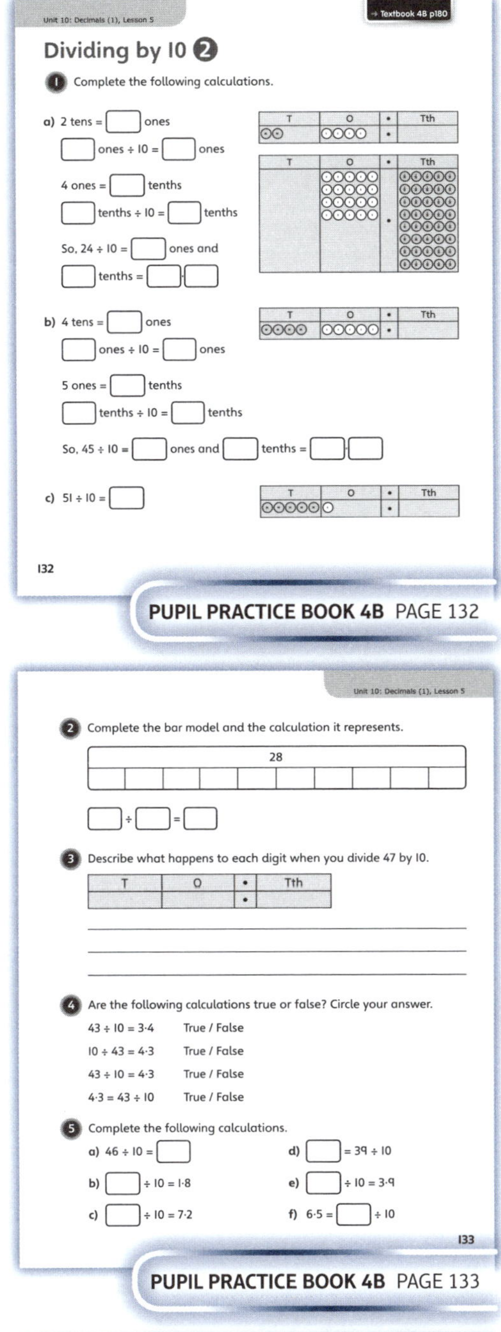

PUPIL PRACTICE BOOK 4B PAGE 132

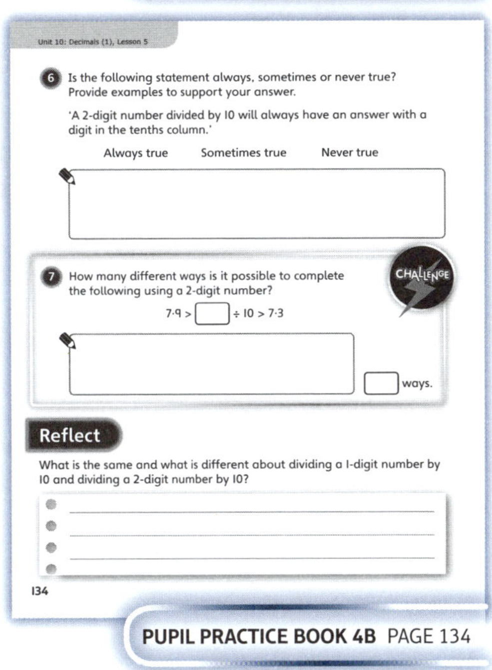

PUPIL PRACTICE BOOK 4B PAGE 133

PUPIL PRACTICE BOOK 4B PAGE 134

Hundredths 

Learning focus

In this lesson, children will understand that a hundredth as a decimal is 0·01 and will use a hundredths grid to make the connection between hundredths and tenths.

Small steps

→ Previous step: Dividing by 10 (2)
→ **This step: Hundredths (1)**
→ Next step: Hundredths (2)

NATIONAL CURRICULUM LINKS

Year 4 Number – Fractions (Including Decimals)
- Recognise and write decimal equivalents of any number of tenths or hundredths.
- Count up and down in hundredths; recognise that hundredths arise when dividing an object by one hundred and dividing tenths by ten.

ASSESSING MASTERY

Children can interpret a representation of a number on a hundredths grid and record this as a fraction and a decimal. They can explain the value of each digit in a number written to two decimal places and explain how the number of hundredths links to the number of tenths, where appropriate.

COMMON MISCONCEPTIONS

Children may find it difficult to count beyond nine hundredths and may not know how to record this as a decimal. Ask:
- *What are 10 hundredths the same as? What would it look like if 10 hundredths counters were placed on a hundredths grid? How can you write 1 tenth?*

STRENGTHENING UNDERSTANDING

Use a hundredths grid and counters to help children make links between the fractional and decimal notations of the same number. Create a game where the number rolled on a dice is the number of counters that must be added to the hundredths grid. Ask children to record their answers as a fraction and a decimal, to strengthen links.

GOING DEEPER

To deepen their understanding of how hundredths are made up of smaller parts, ask children to complete equations such as 0·15 = _ + _; or ask them to complete a part-whole model with 0·15 as the whole, experimenting with different values for the parts.

KEY LANGUAGE

In lesson: hundredths, hundredths column, tenths, equivalent, decimal, fraction

Other language to be used by the teacher: regroup

STRUCTURES AND REPRESENTATIONS

hundredths grid, place value grid

RESOURCES

Mandatory: laminated hundredths grid, place value counters

Optional: part-whole model

In the eTextbook of this lesson, you will find interactive links to a selection of teaching tools.

Before you teach 

- Are all children secure in their understanding of tenths?
- Do children remember the vocabulary of hundredths from their work on fractions in Unit 8?

Discover

 Pair work

ASK

- Question **1** a): *How many parts has the whole been split into?*
- Question **1** b): *Can the fraction that is covered by plain counters be represented using tenths?*

IN FOCUS Question **1** requires children to write a decimal number to two decimal places and use the hundredths column in a place value grid. Children should be encouraged to make links between the hundredths and tenths columns with the help of the hundredths grid to understand their relationship to each other.

PRACTICAL TIPS Create a game similar to that in the picture, where the number rolled on a dice is the number of counters that must be added to a hundredths grid. Ask children to record their answers as a fraction and a decimal, to strengthen links.

Alternatively, children can play a game in pairs where each player begins with 50 hundredths represented as counters on a hundredths grid. Each player rolls one or two dice and they must give their opponent the number of hundredths they roll. Each player must identify the number of hundredths they have at the beginning of each turn.

ANSWERS

Question **1** a): $\frac{1}{100}$ is covered by striped counters. This can be written as 0·01.

Question **1** b): $\frac{10}{100}$ is covered by plain counters. This can be written as 0·10 or 0·1.

Share

 Whole class teacher led

ASK

- Question **1** a): *Why do you think the hundredths column is named as it is?*
- Question **1** b): *How is the number of hundredths linked with the number of tenths in a number?*

IN FOCUS Question **1** a) introduces the hundredths column and how to record $\frac{1}{100}$ as a decimal. Although children have come across $\frac{1}{100}$ as a fraction before, this is the first time they see it as a decimal. Discuss the place value grid and the names of the columns in relation to the value of the digit in each column. In question **1** b), it is important that children understand that 10 hundredths is the same as 1 tenth. The use of counters on the hundredths grid should help children to understand this concept.

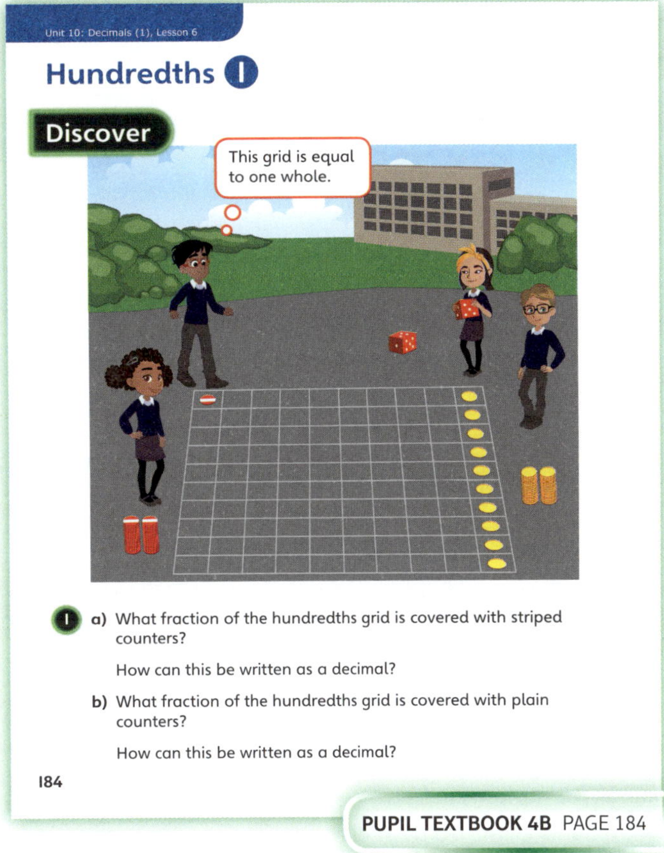

PUPIL TEXTBOOK 4B PAGE 184

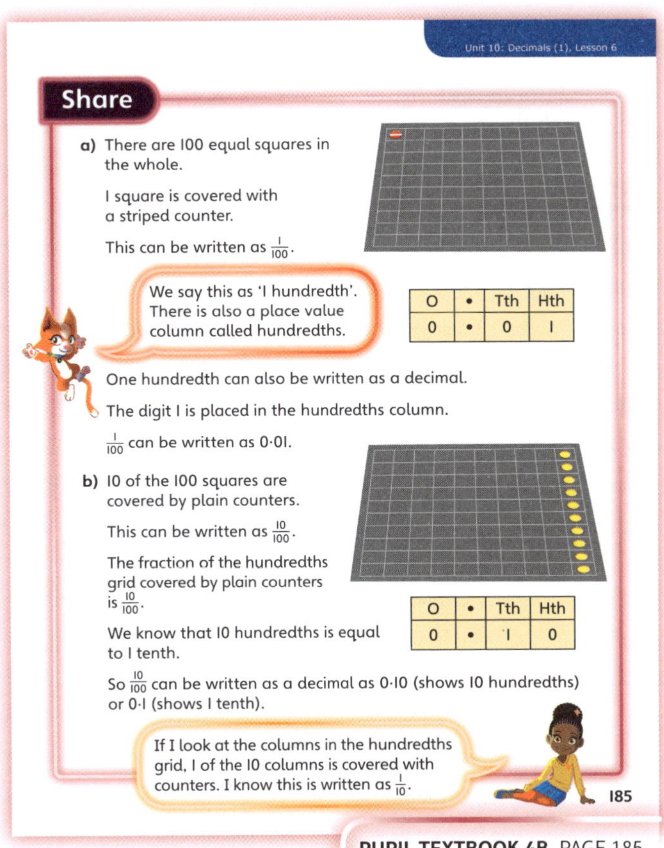

PUPIL TEXTBOOK 4B PAGE 185

Think together

 Whole class teacher led (I do, We do, You do)

ASK

- Question **1**: *How can counters on a hundredths grid help you to understand the concept of hundredths?*
- Question **2**: *What is 10 hundredths equivalent to?*

IN FOCUS Question **3** continues to use the context of counters on a hundredths grid to support understanding of hundredths and to interpret given values. Children are required to re-create Amelia's method of using counters to represent numbers on the grid and are asked to count upwards in hundredths, while considering how each step is written as a fraction and decimal. It is important for children to be comfortable representing hundredths in these different ways, to deepen their understanding of decimals.

STRENGTHEN Children should be able to use place value counters which show 0·01 and place these on a hundredths grid. They should understand the numbers they are presented with and how each number will change when more hundredths are added to it.

DEEPEN Children could explore counting upwards in 2 or more hundredths from different decimal numbers and write these steps as decimals, using a hundredths grid as required. This will deepen their understanding of decimals and the relatively small value of a hundredth in relation to 1.

ASSESSMENT CHECKPOINT Children should be able to count forwards in hundredths and write each step as a decimal. Using a hundredths grid to help them, children should be able to explain the link between the number of hundredths and the number of tenths in a number, where appropriate; they should be able to explain the value of each digit in a number written to two decimal places.

ANSWERS

Question **1** a): $\frac{7}{100}$

Question **1** b): 0·07

Question **2**: a) Children place 13 counters on a hundredths grid, one counter per square.

Question **2**: b) The digit 1 represents 1 row or column of 10 counters. The digit 3 represents 3 hundredths, shown by 3 counters in an incomplete row or column.

Question **3**: 4 hundredths, 5 hundredths, 6 hundredths, 7 hundredths, 8 hundredths, 9 hundredths, 10 hundredths, 11 hundredths.
0·04, 0·05, 0·06, 0·07, 0·08, 0·09, 0·1, 0·11

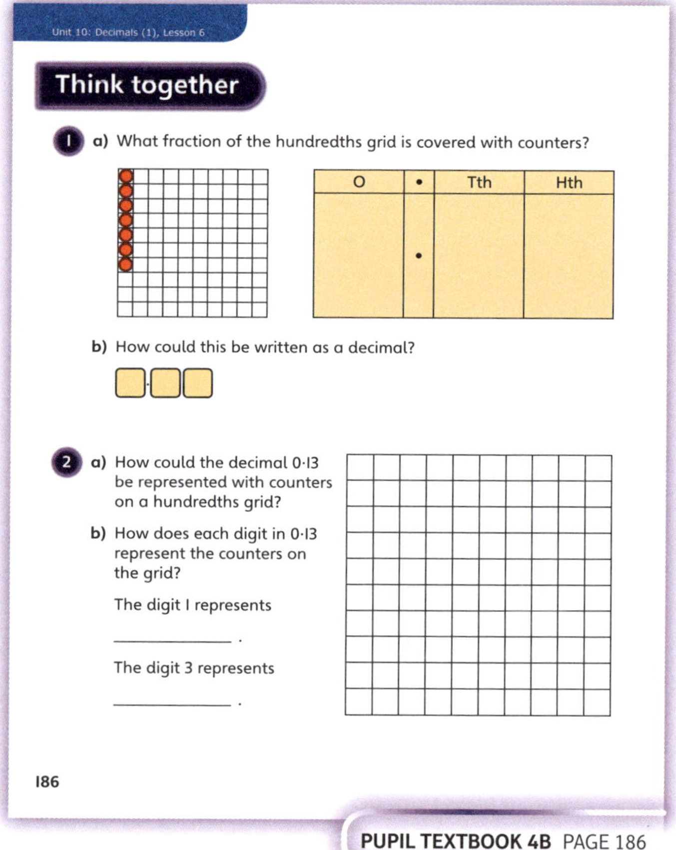

PUPIL TEXTBOOK 4B PAGE 186

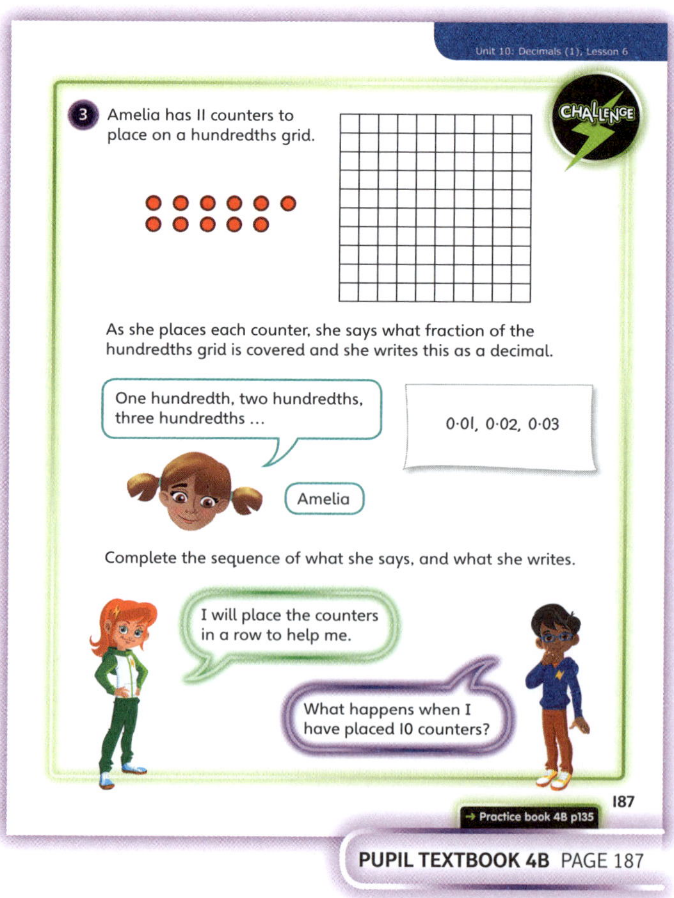

PUPIL TEXTBOOK 4B PAGE 187

Practice

 Independent thinking

 Question **1** requires children to work with fractional and decimal notations of the same number. They are provided with pictorial support to help with this. Question **2** starts to remove this scaffolding to check children's understanding.

 Children who are finding it difficult to move between fractional and decimal notation should be allowed to use place value counters that have the fraction $\frac{1}{100}$ on one side and the decimal 0·01 on the other side. Positioning these counters on a hundredths grid, and being able to turn them over, will help children to strengthen their understanding of equivalence.

 Children could be challenged to write how different numbers of hundredths are made up of smaller parts; for example, 0·15 = _ + _. Although this is not the main objective of the lesson, children could use the counters and move them into different parts to explore this concept. This will deepen their understanding of the relationship between tenths and hundredths.

 Question **4** requires children to interpret and write hundredths as words and numbers. Writing the numbers in this way will deepen children's understanding of the links between the fractional and decimal notations of the same numbers.

 Children should be able to explain the value of each digit in a number written to two decimal places. They should be able to make any number to two decimal places with resources and move between the fractional and decimal notations of the same number.

 Answers for the **Practice** part of the lesson appear in the separate **Practice and Reflect answer guide**.

Reflect

 Independent thinking

 In this section of the lesson, children are required to explain how they know that a fractional and decimal notation of the same number are equal to each other. Encourage them to use appropriate mathematical language within their answers.

 The quality of children's written responses will indicate the depth of their understanding of hundredths written as fractions and decimals.

 Answers for the **Reflect** part of the lesson appear in the separate **Practice and Reflect answer guide**.

After the lesson ⏸

- Are all children secure in their understanding that a hundredth as a decimal is 0·01 and how this is represented as a fraction?
- What concept within the lesson did children find the hardest to master?
- Do any children still need scaffolding for support?

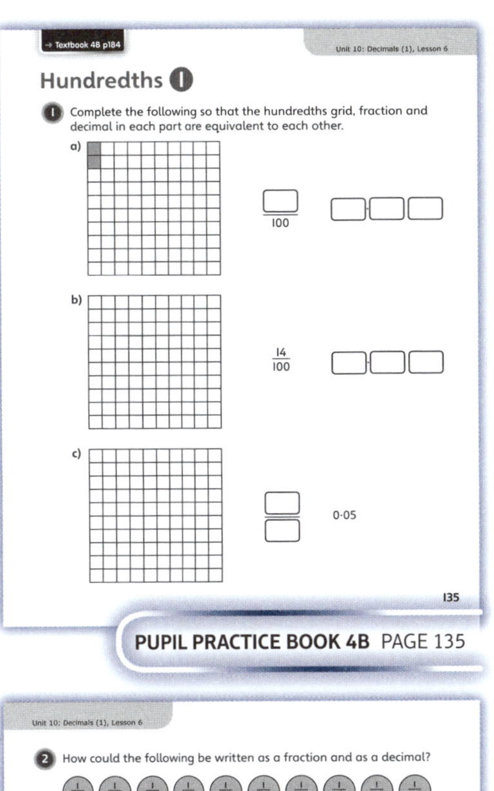

PUPIL PRACTICE BOOK 4B PAGE 135

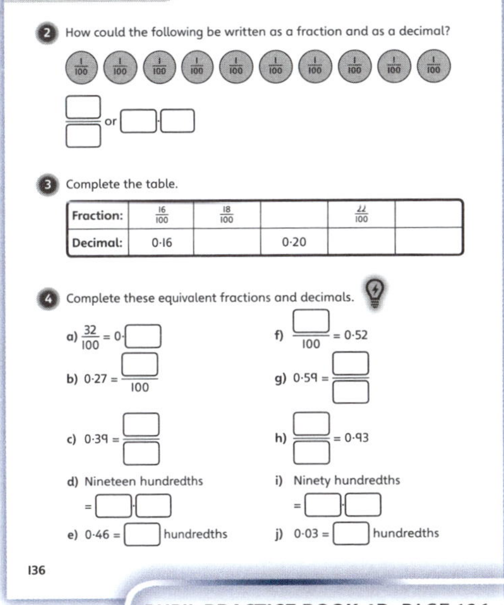

PUPIL PRACTICE BOOK 4B PAGE 136

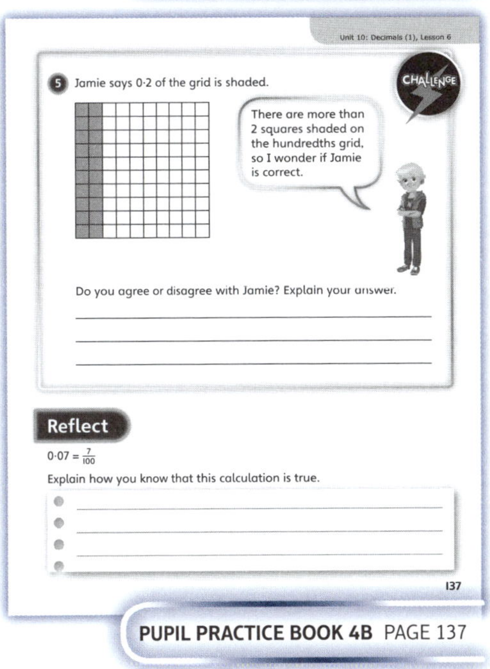

PUPIL PRACTICE BOOK 4B PAGE 137

Hundredths ②

Learning focus

In this lesson, children will practise writing hundredths as a decimal and counting forwards and backwards in hundredths from a given number.

Small steps

→ Previous step: Hundredths (1)
→ **This step: Hundredths (2)**
→ Next step: Hundredths (3)

NATIONAL CURRICULUM LINKS

Year 4 Number – Fractions (Including Decimals)
- Recognise and write decimal equivalents of any number of tenths or hundredths.
- Count up and down in hundredths; recognise that hundredths arise when dividing an object by one hundred and dividing tenths by ten.

ASSESSING MASTERY

Children can write any number of hundredths as a decimal and count forwards and backwards in hundredths from a given number. Using this understanding, they can count on to, or back from, 1 in an efficient way and use their knowledge of number bonds to 100 within this process.

COMMON MISCONCEPTIONS

When children are counting up from a number of hundredths to 1, they may try to make the number of tenths and hundredths amount to 10. Ask:
- *How many more hundredths do you need to make a multiple of 10? How many tenths are then needed to make 1?*

Children may confuse the value of the tenths and hundredths place value columns. Ask:
- *What is the value of the column the digit is in? How many parts out of 100 does this represent?*

STRENGTHENING UNDERSTANDING

Use counters and a hundredths grid with children who find it difficult to see the link between hundredths and tenths. They should be encouraged to complete the grid row by row to see when an additional tenth has been filled. Use prompts showing a pictorial representation of what each completed row represents in hundredths and tenths, alongside the amount shown as a decimal and a fraction.

GOING DEEPER

Provide children with a number of digit cards and ask them to see how many ways it is possible to make 1 out of two decimal parts. This could be recorded within a part-whole model, or within the equation $1 = \_ + \_$. This will help to deepen children's understanding of two decimal parts making a whole and how these parts must represent 100 hundredths.

KEY LANGUAGE

In lesson: hundredths, tenths, decimal

Other language to be used by the teacher: count on, count back, equivalent

STRUCTURES AND REPRESENTATIONS

hundredths grid, number line, place value grid

RESOURCES

Mandatory: place value counters

 In the eTextbook of this lesson, you will find interactive links to a selection of teaching tools.

Before you teach 

- Do all children work efficiently when recognising and writing decimal equivalents of hundredths?
- Do they understand the tenths and hundredths columns in a place value grid?

Discover

WAYS OF WORKING Pair work

ASK

- Question **1** a): *What resources could you use to replicate the context that you can see in the picture? How many parts of the whole are visible?*

IN FOCUS Question **1** a) builds on the work from the previous lesson, where children had to represent a number as a fraction and as a decimal. This lesson continues to build the foundations and provides practice for recognising and writing decimal equivalents of any number of tenths or hundredths.

PRACTICAL TIPS Use place value counters and a hundredths grid so children can see how a number of hundredths relates to a whole. When counting on or back from a decimal number in hundredths, use place value counters above the jumps on a number line to represent what is being added on each time.

ANSWERS

Question **1** a): $\frac{35}{100}$ or 0·35 of the floor has been carpeted.

Question **1** b): $\frac{43}{100}$ or 0·43 of the floor has been carpeted now.

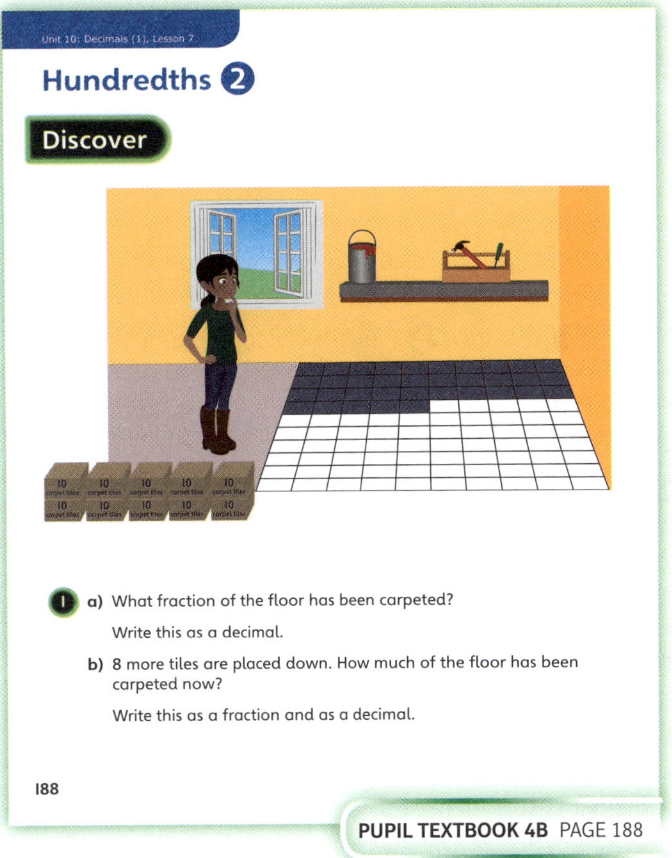

PUPIL TEXTBOOK 4B PAGE 188

Share

WAYS OF WORKING Whole class teacher led

ASK

- Question **1** a) and b): *Why have different models been used for each part of the question?*
- Question **1** b): *What number sentence could you write to show this calculation?*

IN FOCUS In question **1** a), children are prompted to identify the number of hundredths in an efficient way. Rather than counting in 1s to identify the total number of hundredths, they are shown how to identify the number of groups of 10 hundredths (the number of tenths), and then identify the remaining number of individual counters (the hundredths), to complete the number as a decimal.

In question **1** b), children count on from a decimal amount and use a number line to show this. When counting on, children can see the size of each jump and how this changes the number as a fraction and as a decimal.

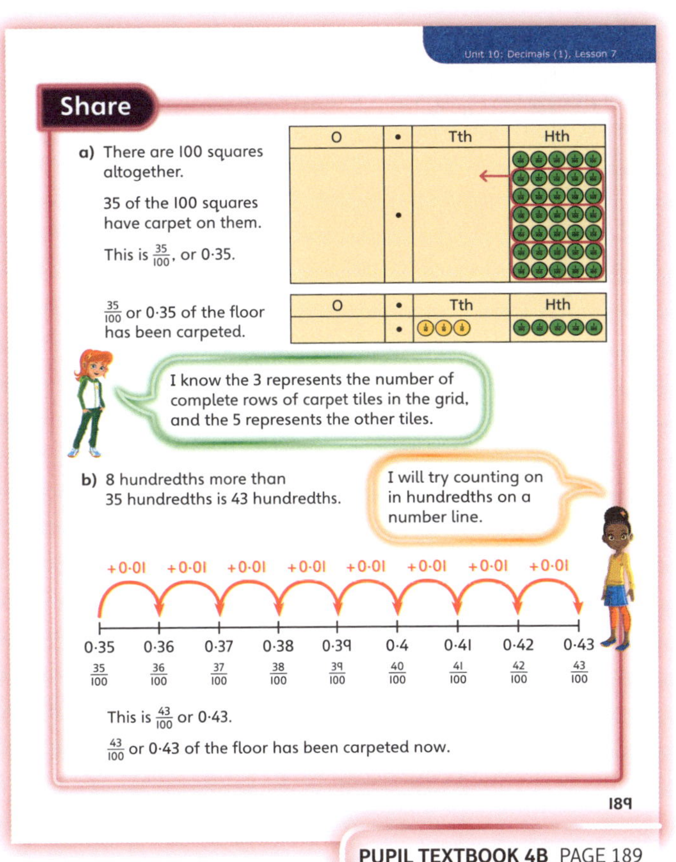

PUPIL TEXTBOOK 4B PAGE 189

Think together

WAYS OF WORKING Whole class teacher led (I do, We do, You do)

ASK

- Question **1**: *How many hundredths are in a whole?*
- Question **2**: *What method could you use to help you work out how many more squares Tom needs to shade?*

IN FOCUS Question **3** requires children to count on in hundredths from a known number and to use their understanding of working backwards to find solutions. This question allows children to practise calculating unknown quantities of hundredths as decimals.

STRENGTHEN Use place value counters to replicate the contexts in the questions. When children add or subtract hundredths from an amount, they should say the number they have made out loud, or write it down each time. This will help them to see when a complete tenth has been made and how, therefore, 0·3 is the same as 0·30.

DEEPEN To encourage children to see the relationship between the number of tenths and the number of hundredths within a number with two decimal places, children should identify these as different parts within the whole. These parts could be recorded within a part-whole model, or within an equation in the form 0·34 = _ + _. Children should be encouraged to write these in as many different ways as possible and to explain how they have chosen each part to show the depth of their understanding; answers could be 0·1 and 0·24, 0·2 and 0·14, 0·3 and 0·04, and so on.

ASSESSMENT CHECKPOINT Children should be able to explain the value of each digit in a number with two decimal places and should be encouraged to explain the position of these digits in a representation of the number. This will encourage them to think about the number of tenths and hundredths in a number, rather than only the number of hundredths.

ANSWERS

Question **1** a): 72 squares are shaded. This is $\frac{72}{100}$ or 0·72.

Question **1** b): 84 squares are shaded. This is $\frac{84}{100}$ or 0·84.

Question **2** a): 0·3, or $\frac{30}{100}$

Question **2** b): Tom needs to shade 4 more squares.

Question **3** a): Aki has lost 27 counters.

Question **3** b) This is $\frac{27}{100}$, or 0·27.

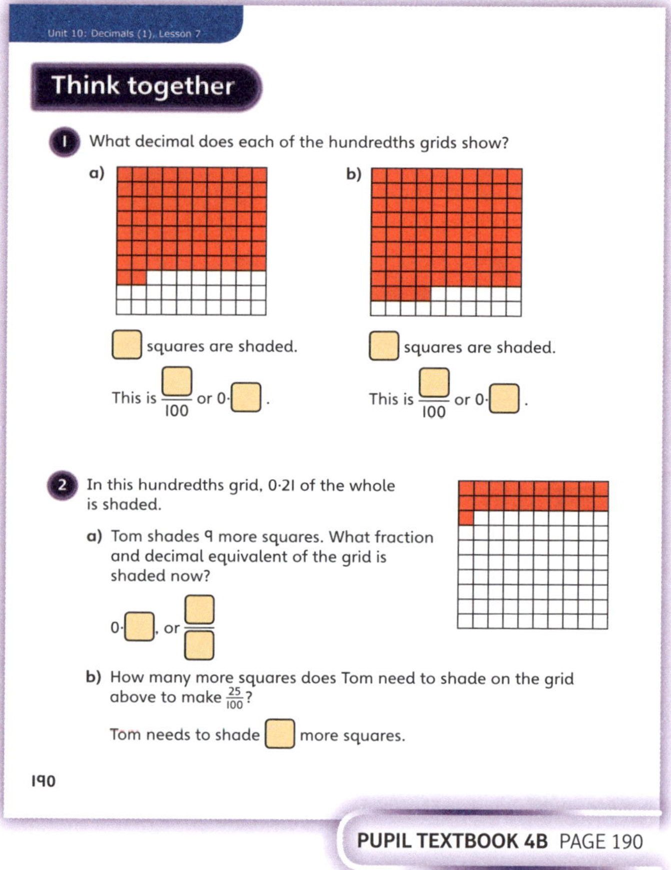

PUPIL TEXTBOOK 4B PAGE 190

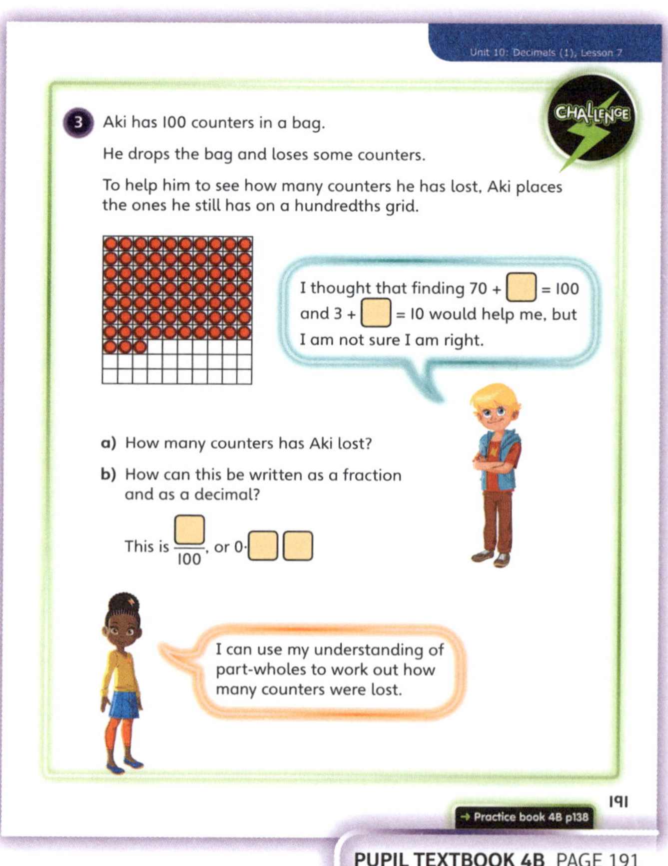

PUPIL TEXTBOOK 4B PAGE 191

Practice

WAYS OF WORKING Independent thinking

IN FOCUS For questions **2** and **3**, children must calculate the number of hundredths needed to make a whole. These questions encourage the use of known number facts within calculations, where appropriate, rather than always counting in steps of 1 hundredth.

STRENGTHEN Children may find it difficult to see the link between hundredths and tenths and so may not realise, for example, that 30 hundredths is the same as 3 tenths. Use counters and a hundredths grid, with the number of tenths and hundredths marked on it, and prompt children to fill the rows systematically.

DEEPEN Children could investigate what combinations of hundredths make 1, in the form $0·\_ + 0·\_ = 1$. Encourage them to find out how many different ways it is possible to complete this equation, using selected digit cards. This will deepen their understanding of how two decimal parts relate to each other to make a whole.

ASSESSMENT CHECKPOINT Children should be able to explain how they have worked, to ensure they are being efficient; they should also identify the number facts they have used to complete certain questions. They should be able to explain the value of each digit within a decimal number and identify where this value appears within different resources and representations.

ANSWERS Answers for the **Practice** part of the lesson appear in the separate **Practice and Reflect answer guide**.

Reflect

WAYS OF WORKING Independent thinking

IN FOCUS This section of the lesson provides children with a question that has multiple answers. Children should be encouraged to find all the possible ways to complete the equation and explain how they have worked to ensure they have found all the solutions.

ASSESSMENT CHECKPOINT Children should be able to explain the solutions they have found to satisfy the expression. If they can find all the solutions and work systematically, this shows that they have a strong understanding of hundredths and their relationship with tenths.

ANSWERS Answers for the **Reflect** part of the lesson appear in the separate **Practice and Reflect answer guide**.

After the lesson 

- How confident are children at describing the value of each digit within a number with two decimal places?
- Can children explain the difference between 0·4 and 0·04?
- Do any children still use inefficient methods to work out problems? Do they need ongoing support?

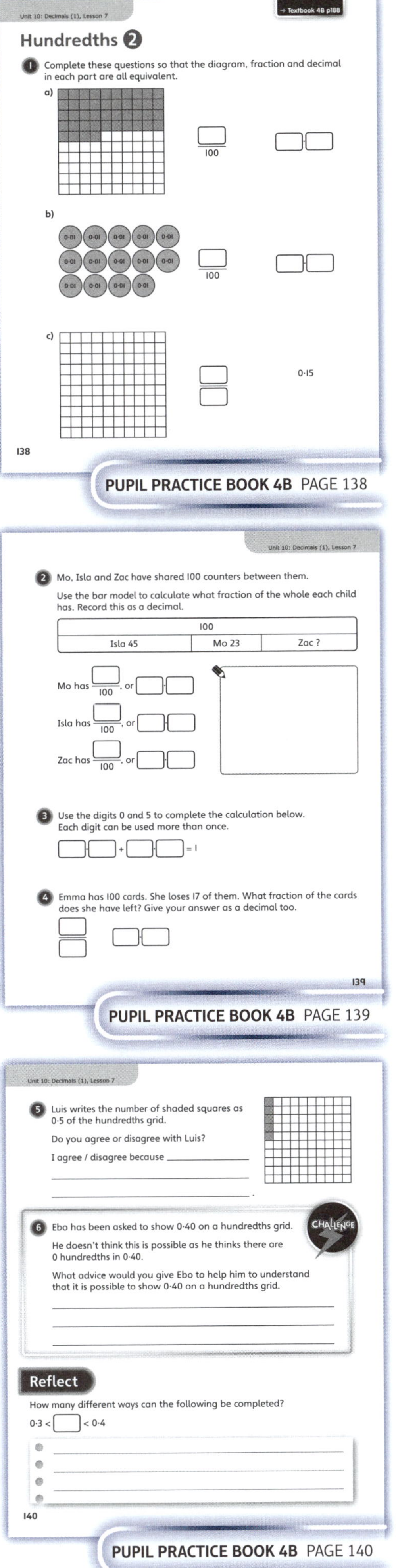

PUPIL PRACTICE BOOK 4B PAGE 138

PUPIL PRACTICE BOOK 4B PAGE 139

PUPIL PRACTICE BOOK 4B PAGE 140

Hundredths ③

Learning focus

In this lesson, children will build on their understanding and recognise that a number less than 1, with two decimal places, is a number of tenths plus some hundredths.

Small steps

→ Previous step: Hundredths (2)
→ **This step: Hundredths (3)**
→ Next step: Dividing by 100

NATIONAL CURRICULUM LINKS

Year 4 Number – Fractions (Including Decimals)
- Find the effect of dividing a one- or two-digit number by 10 and 100, identifying the value of the digits in the answer as ones, tenths and hundredths.
- Count up and down in hundredths; recognise that hundredths arise when dividing an object by one hundred and dividing tenths by ten.

ASSESSING MASTERY

Children can describe how a decimal number is made in different ways, identifying the number of tenths and hundredths in the number. They can use their understanding of counting on or back in tenths and then hundredths to identify the unknown part when presented with a part-whole model problem.

COMMON MISCONCEPTIONS

Children may confuse the tenths and hundredths place value columns and as a result calculate incorrectly. For example, with 0·43 they may believe the 4 represents 4 hundredths and the 3 represents 3 tenths. Ask:
- *What column is each digit in? Does this represent the number of hundredths or tenths?*

STRENGTHENING UNDERSTANDING

Use a labelled place value grid to help children who find it difficult to correctly distinguish between the tenths and hundredths columns. Give children opportunities to make different numbers with place value counters, and to write numbers based on the counter values they have been given.

GOING DEEPER

Children can investigate how numbers with two decimal places can be partitioned in different ways. The main focus of the lesson is to partition numbers into hundredths and tenths, but this can be taken further to deepen children's understanding; for example, in the calculation 0·24 = _ + _ all children should be able to write 0·2 and 0·04, but they could also write 0·19 and 0·05.

KEY LANGUAGE

In lesson: tenths, hundredths, part, whole

Other language to be used by the teacher: systematic, count on, count back, partition

STRUCTURES AND REPRESENTATIONS

part-whole model, number line, place value grid

RESOURCES

Mandatory: place value counters

Optional: dice, number line

 In the eTextbook of this lesson, you will find interactive links to a selection of teaching tools.

Before you teach 

- Are all children able to write hundredths as decimals?

Discover

WAYS OF WORKING Pair work

ASK

- Question **1** a): *Is there only one way to make 0·25?*
- Question **1** a): *What amounts have the children made? What does the digit to the right of the decimal point represent? What does the digit to the right of this digit represent?*
- Question **1** b): *What amount has Kate made?*

IN FOCUS Question **1** a) focuses on making a given number using place value counters. Question **1** b) introduces the fact that the same number can be made in different ways, as Kate can make 0·25 using a combination of tenths and hundredths counters.

PRACTICAL TIPS Use place value counters to practise making different numbers as well as recording the number shown by a given value of counters.

Use two dice to govern the number of tenths and hundredths a number must have and then write and make this number accordingly.

ANSWERS

Question **1** a): Reena has made 0·52 and Alex has made 0·25.

Question **1** b): Kate needs an additional 5 hundredths counters to make 0·25.

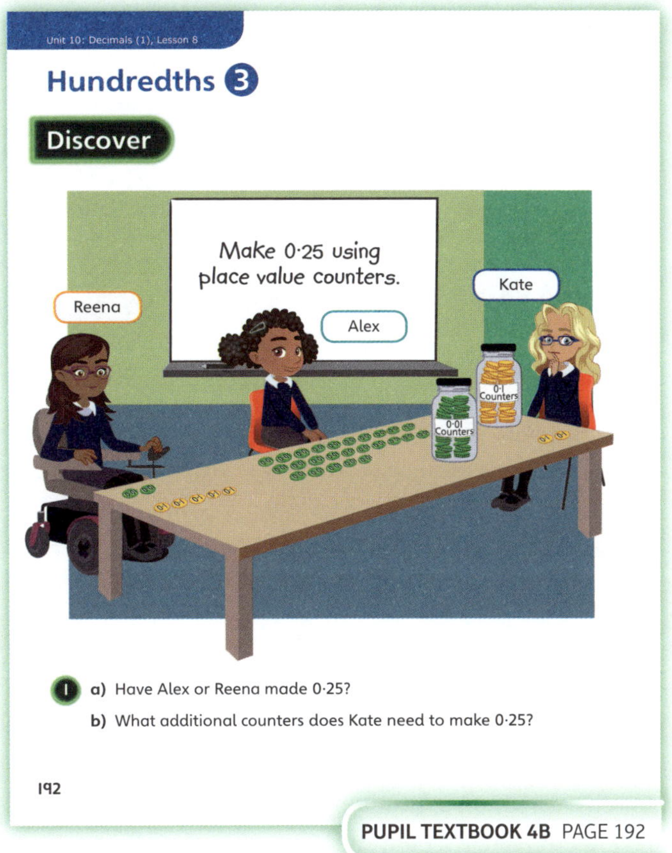

PUPIL TEXTBOOK 4B PAGE 192

Share

WAYS OF WORKING Whole class teacher led

ASK

- Question **1** b): *How is it possible to make the same decimal number in different ways?*

IN FOCUS Question **1** a) shows visually how counters can be used to make a given number. Children are reminded that 10 hundredths counters is the same as 0·1. This fact could be used to explore further ways that the same number could be made.

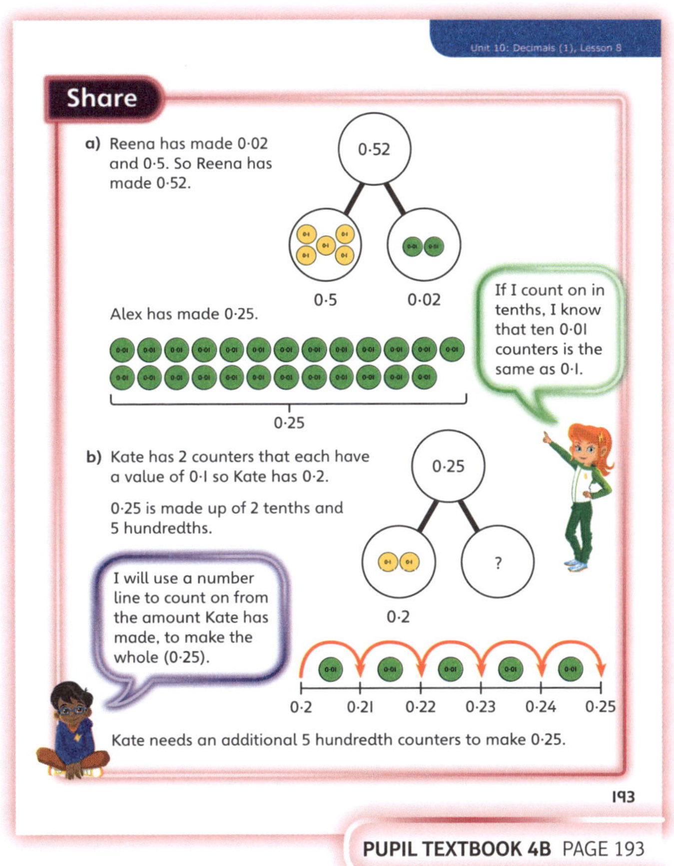

PUPIL TEXTBOOK 4B PAGE 193

Think together

 Whole class teacher led (I do, We do, You do)

ASK

- Question **1**: *How can a part-whole model be used to partition the same decimal number in different ways?*
- Question **3**: *How should you work to ensure that no possible solutions are missed out?*

IN FOCUS Question **2** explores how decimal numbers can be made in different ways. Children are given the whole and have to explore what the two parts could be. Children could be challenged to work systematically in order to maximise the number of solutions they can find.

STRENGTHEN Give children the opportunity to use counters to make or separate given numbers in different ways. Once children understand the concept, the counters should be removed so they can visualise the parts that make the whole number instead.

DEEPEN When children have found different combinations to make a decimal number that involve tenths and hundredths, they can be challenged to find alternative ways to do this; for example, initially finding 3 tenths and 5 hundredths = 0·35, then 2 tenths and 15 hundredths = 0·35 or 19 hundredths and 16 hundredths = 0·35, and so on.

ASSESSMENT CHECKPOINT Children should be able to describe how a decimal number can be made in many different ways. This should begin with the identification of the number of tenths and hundredths and then progress to partitioning the whole in different ways to form different parts. When using counters and the part-whole model, children should count in tenths and then hundredths within each part, in order to identify its value.

ANSWERS

Question **1** a): Richard has five 0·01 counters.
One 0·1 counter = ten 0·01 counters.
Richard needs 3 more 0·1 counters.

Question **1** b): Each jump = five 0·01 counters.

Question **2**: There are 46 different combinations, including: 0·4 + 0·03, 0·3 + 0·13, 0·2 + 0·23, 0·1 + 0·33, 0·01 + 0·42, 0·02 + 0·41, and so on.

Question **3**: 0·3 + 0·03 = 0·33; 0·3 + 0·4 = 0·7; 0·3 + 0·04 = 0·34; 0·03 + 0·4 = 0·43; 0·03 + 0·04 = 0·07; 0·4 + 0·04 = 0·44.

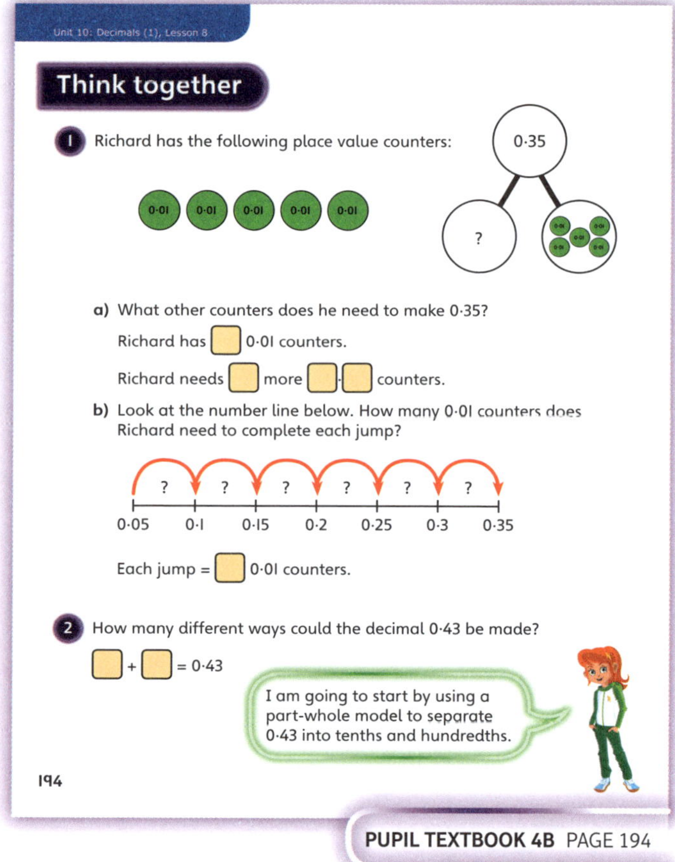

PUPIL TEXTBOOK 4B PAGE 194

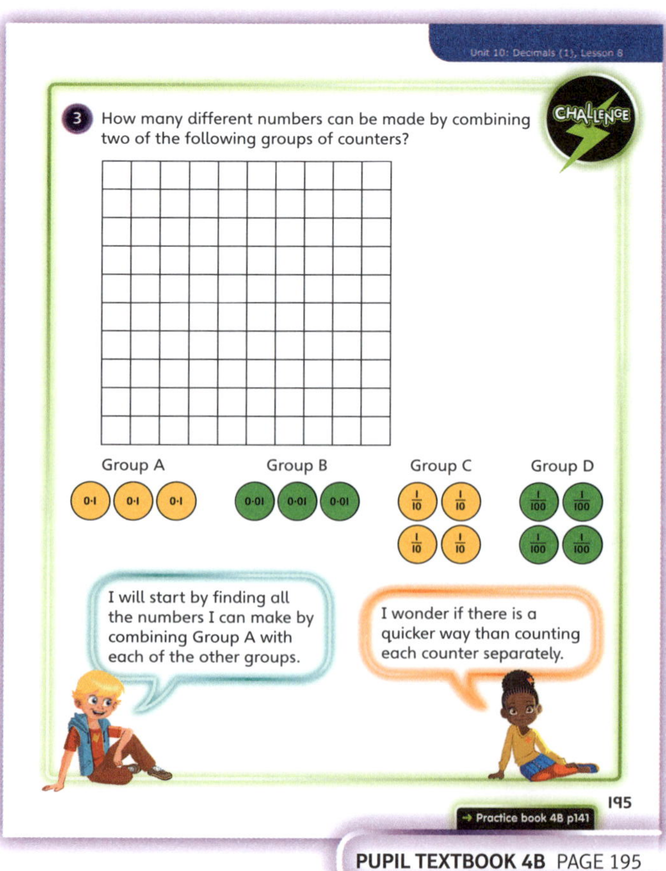

PUPIL TEXTBOOK 4B PAGE 195

Practice

WAYS OF WORKING Independent thinking

IN FOCUS Question ❷ requires children to complete part-whole models for numbers made from tenths and hundredths. The part-whole models provide scaffolds to help children with the process of combining parts to make the whole. In questions ❸ and ❹, these scaffolds are removed to add challenge, and children are required to identify different unknown values within the whole and the parts.

STRENGTHEN Children who do not know how to interpret a question, or get a question wrong, should use counters to understand it in more detail and identify their mistake. Eventually, children should be encouraged to try to solve questions without counters, in order to encourage visualisation, rather than counting in ones.

DEEPEN Children could investigate how a decimal number can be made with three parts and how changing one of these parts will affect the others. They could write their findings in the form _ + _ + _ = _. This will allow them to explore tenths and hundredths in a more complex context.

ASSESSMENT CHECKPOINT Children should be able to justify the choices they have made and to show, with resources, how they know that each of their answers is correct. They should be confident at using the language of tenths and hundredths and should be able to explain how changing one part of the whole affects the other part.

ANSWERS Answers for the **Practice** part of the lesson appear in the separate **Practice and Reflect answer guide**.

Reflect

WAYS OF WORKING Independent thinking

IN FOCUS This section of the lesson allows children to record how one number can be made in many different ways. As there are many different solutions, children should be encouraged to work in a systematic manner in order to show the logic of their thinking.

ASSESSMENT CHECKPOINT The more solutions children can find to the same question, the deeper their understanding is likely to be. Assessing the way that they work through their solutions will also give an indication of their understanding of tenths and hundredths as a whole, and how changing one part will affect the other part of a whole.

ANSWERS Answers for the **Reflect** part of the lesson appear in the separate **Practice and Reflect answer guide**.

After the lesson 

- How systematic are children when working through problems with multiple solutions?
- What were the barriers children encountered that stopped them finding all solutions to a problem?
- Can children explain how decimals can be made in different ways by making different partitions?

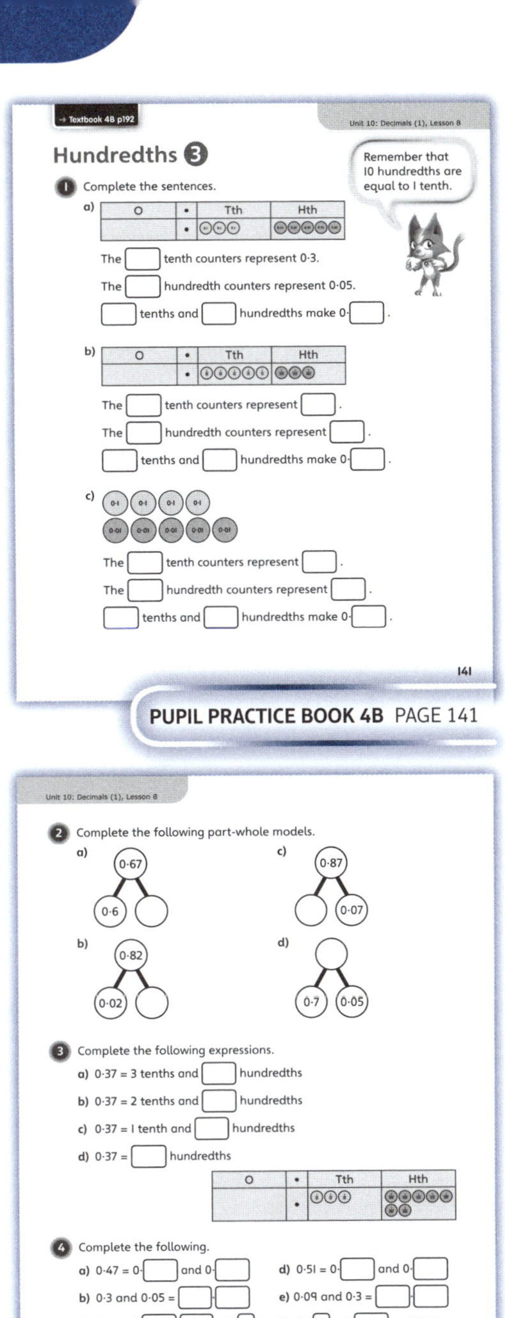

PUPIL PRACTICE BOOK 4B PAGE 141

PUPIL PRACTICE BOOK 4B PAGE 142

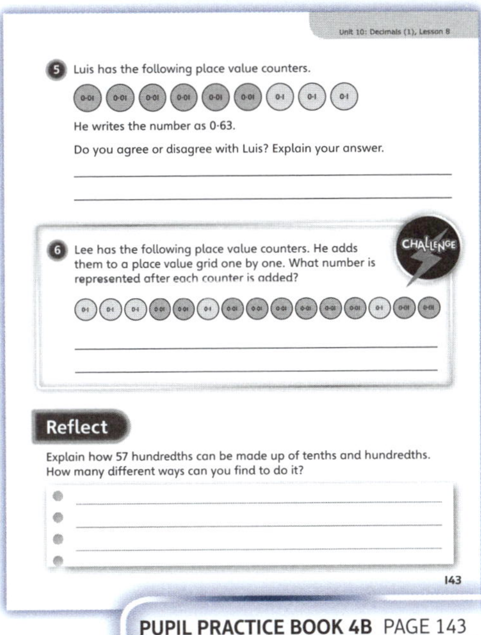

PUPIL PRACTICE BOOK 4B PAGE 143

Dividing by 100

Learning focus

In this lesson, children will divide 1- and 2-digit numbers by 100, building on their understanding of dividing by 10.

Small steps

→ Previous step: Hundredths (3)
→ **This step: Dividing by 100**
→ Next step: Dividing by 10 and 100

NATIONAL CURRICULUM LINKS

Year 4 Number – Fractions (Including Decimals)

Find the effect of dividing a one- or two-digit number by 10 and 100, identifying the value of the digits in the answer as ones, tenths and hundredths.

ASSESSING MASTERY

Children can explain the value of any digit in a number they are presented with throughout the lesson and can use this understanding to explain, and make, any 1- or 2-digit numbers 1 hundredth of their original size.

COMMON MISCONCEPTIONS

Children may identify patterns linking the initial and final numbers within calculations, before fully understanding the concept of dividing by 100. Children may also confuse the order of the digits in their answer. Ask:
- *Why does each digit move place value columns?*

STRENGTHENING UNDERSTANDING

Children may find it difficult to understand the concept of splitting an object into 100 parts and how this can be used to help to divide a number by 100. Use tens and ones counters to allow children to practise regrouping into tenths and hundredths. Children should not use this method throughout the lesson, as it is inefficient to have to count multiples of 100 counters repeatedly.

GOING DEEPER

Children should identify patterns linking the initial and final number within solutions and use this to help them to work efficiently. To deepen their understanding of the concept, ask children to find as many equations as they can that produce solutions meeting set criteria, for example: $0{\cdot}3 < \_ \div 100 < 0{\cdot}4$.

KEY LANGUAGE

In lesson: ones, tenths, hundredths, divide

Other language to be used by the teacher: equivalent, pattern, partition

STRUCTURES AND REPRESENTATIONS

hundredths grid, tenths grid

RESOURCES

Mandatory: place value grid, hundredths grids

Optional: place value counters

 In the eTextbook of this lesson, you will find interactive links to a selection of teaching tools.

Before you teach 

- What were the main misconceptions when children divided numbers by 10?
- Do children understand the relationship between tenths and hundredths?

Discover

 Pair work

ASK

- Question **1** a): *What calculation is the problem asking you to solve? What do you already know about decimals that could help you solve the problem?*
- Question **1** b): *What is different about this calculation? Will you use the same method?*

IN FOCUS Question **1** requires children to divide 1- and 2-digit numbers by 100 for the first time. They should build on their understanding and the method that was used to divide numbers by 10; encourage them to spot patterns within the calculations they complete. It is important that children understand the key concept of how the place value columns link and how this allows a number to be divided by 100.

PRACTICAL TIPS Use a number of tens and ones counters to allow practice at regrouping into tenths and hundredths. This will help children to understand the concept in more detail. However, they should not use this method throughout the lesson, as it would be inefficient to have to count multiples of 100 counters repeatedly.

ANSWERS

Question **1** a): Each piece of pizza is 0·03 m long.

Question **1** b): Each piece of cake is 0·12 m long.

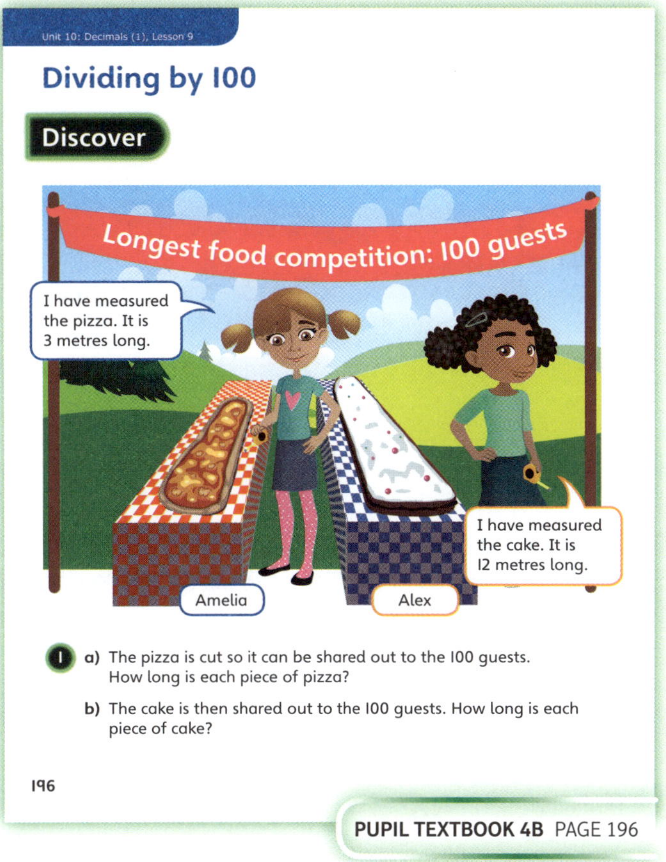

PUPIL TEXTBOOK 4B PAGE 196

Share

 Whole class teacher led

ASK

- Question **1** a): *How do you know 3 ones are the same as 300 hundredths? How does this help you to divide 3 by 100?*
- Question **1** b): *What knowledge of working with hundredths will you use to help you?*

IN FOCUS Question **1** a) provides children with a visual representation that shows the concept of dividing 1 object into 100 smaller parts; how each smaller part represents one hundredth of the original object; and how this can then be divided by 100. Place value counters are not used to represent the problem as the number of counters required would be too large.

There are two different methods presented for question **1** b). Discuss the similarities and differences between the two methods to help children see that when they have a calculation that involves tenths, they can split the full number up into hundredths and then work out how many tenths and hundredths they have; or they can separate the tenths first and then the hundredths, dividing each by 100 accordingly. Discuss why both methods are not possible for part a).

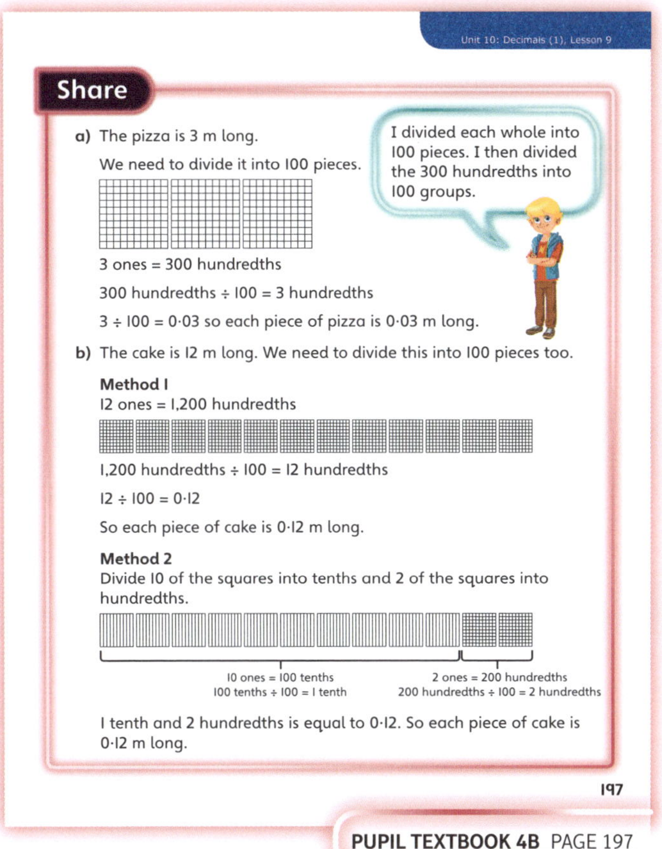

PUPIL TEXTBOOK 4B PAGE 197

Think together

WAYS OF WORKING Whole class teacher led (I do, We do, You do)

ASK

- Question **2** b): *What is the same and what is different about the two calculations?*
- Question **3**: *Can you spot patterns in the solutions that could help you to work more efficiently?*

IN FOCUS Question **3** encourages children to identify patterns within the solutions in order to find the answer to Mo's problem. Children should explain how they have reached their answer and how they have ensured that their method is efficient.

STRENGTHEN Children who are finding it difficult to spot patterns linking the starting number to the solution, should write a series of initial and final numbers in a place value grid. This will help them to identify that each digit remains the same, but moves two place value columns to the right.

DEEPEN Building on children's ability to spot patterns in solutions, ask them to find as many different solutions as they can to complete $0.3 < \_ \div 100 < 0.4$. Encourage children to find all possible ways to complete the equation and to explain how they know that they have found all the solutions.

ASSESSMENT CHECKPOINT Children should be able to explain the concept of changing a value of ones to be a value of hundredths. They should then use this knowledge to divide a number by 100. Children should also be able to identify patterns linking the starting number and the solution; they should understand that each digit becomes 1 hundredth of the size and therefore moves two place value columns to the right.

ANSWERS

Question **1**: 4 ones = 400 hundredths
400 hundredths ÷ 100 = 4 hundredths
$4 \div 100 = 0.04$
The mass of each plate is 0.04 kg.

Question **2** a): $16 \div 100 = 0.16$

Question **2** b): $60 \div 100 = 0.6$ ($60 \div 10 = 6$ ones, which divided by 10 again = 0.6)

Question **3**: $145 \div 100 = 1.45$

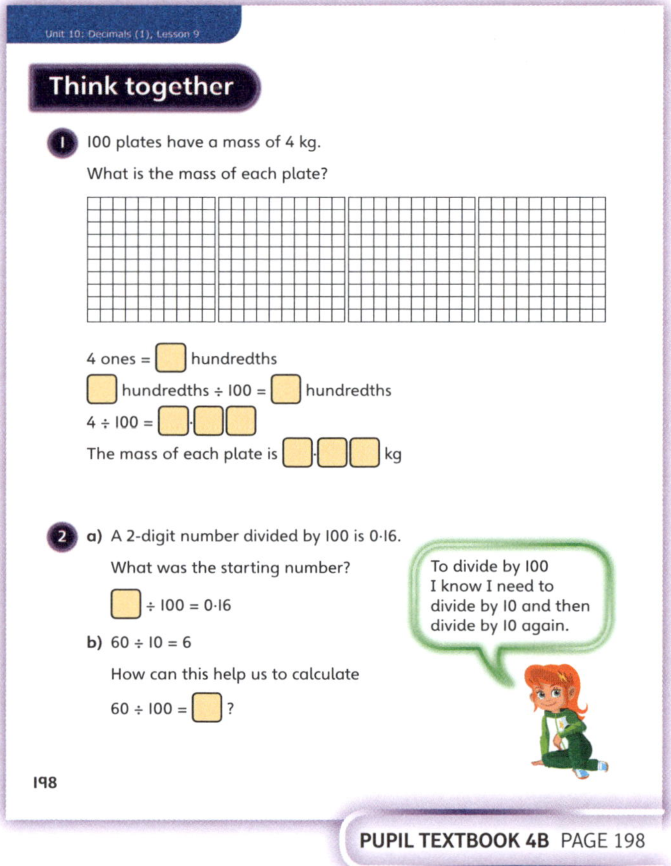

PUPIL TEXTBOOK 4B PAGE 198

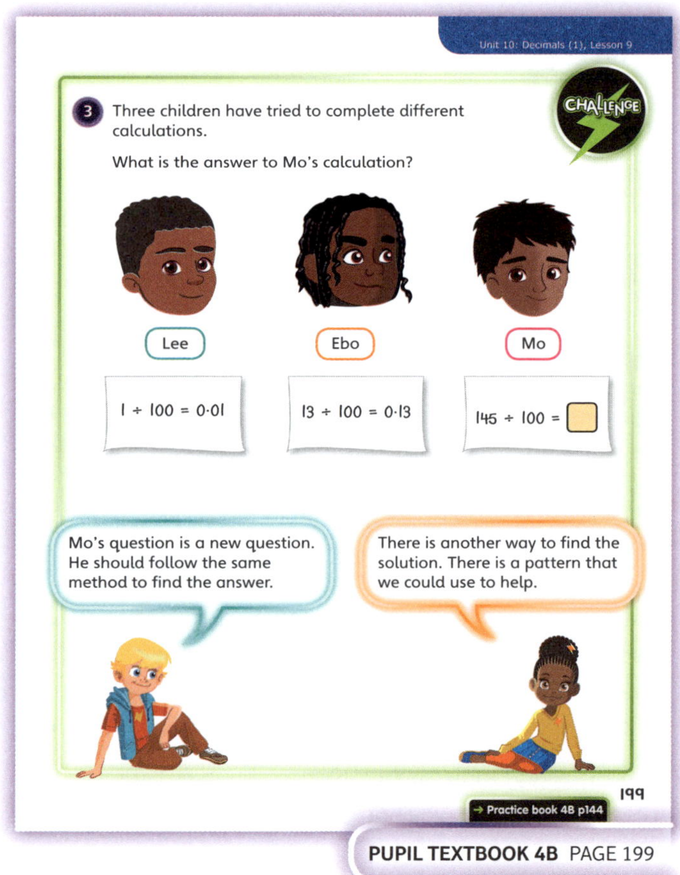

PUPIL TEXTBOOK 4B PAGE 199

Practice

WAYS OF WORKING Independent thinking

IN FOCUS Question ➊ provides children with step-by-step scaffolding to help them to understand the concept of dividing by 100 using both methods introduced in **Share**. This scaffold allows children to build the foundations of knowledge, so they are able to apply and explain their understanding later in the lesson.

Question ➌ challenges children to work efficiently when dividing a number by 100, by looking for patterns in the solutions, as given in the prompts.

STRENGTHEN Children who find the concept of dividing by 100 difficult to understand, and do not know how to work through the problems independently, could be given a visual scaffold, similar to the one provided in question ➊. They could also go back to placing hundredths onto a place value grid to revise what was learnt in the previous lesson.

DEEPEN Children could be asked to create as many of their own equations as possible from given digit cards, dividing each starting number by 100, for example: *How many different equations is it possible to make and complete using the digits 0, 3, 4, 5 and 6 and the format _ ÷ 100 = _ , where the initial number is a 1- or 2-digit number?*

ASSESSMENT CHECKPOINT Children should be able to explain the value of any digit in a number that they are presented with throughout the lesson. They should use this understanding to explain and make any 1- or 2-digit numbers 1 hundredth of their original size.

ANSWERS Answers for the **Practice** part of the lesson appear in the separate **Practice and Reflect answer guide**.

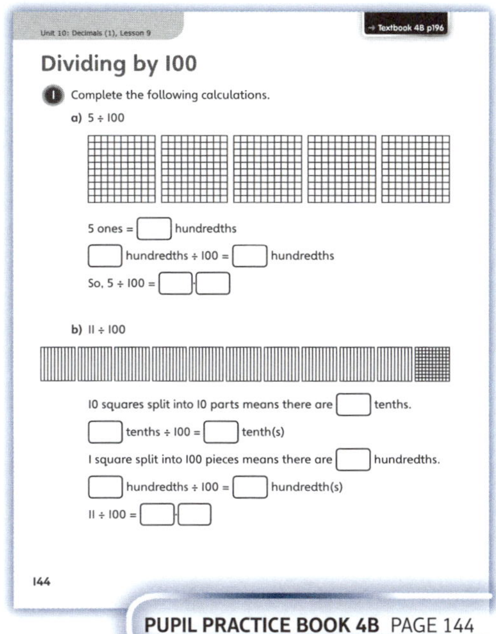

PUPIL PRACTICE BOOK 4B PAGE 144

PUPIL PRACTICE BOOK 4B PAGE 145

Reflect

WAYS OF WORKING Independent thinking

IN FOCUS Children are required to consider the patterns that they have used to find the solutions when dividing 1- and 2-digit numbers by 100: this concept was initially presented to them pictorially. Children should share their thoughts with the class, using the opportunity to write these independently.

ASSESSMENT CHECKPOINT The explanation given will allow you to assess whether children understand the concept of dividing a number by 100, or whether they have simply identified a pattern and used this to find the solutions.

ANSWERS Answers for the **Reflect** part of the lesson appear in the separate **Practice and Reflect answer guide**.

After the lesson ⏸

- Are all children secure in their understanding of dividing a number by 100?
- Can all children explain the concept of dividing by 100, and identify patterns to find solutions?

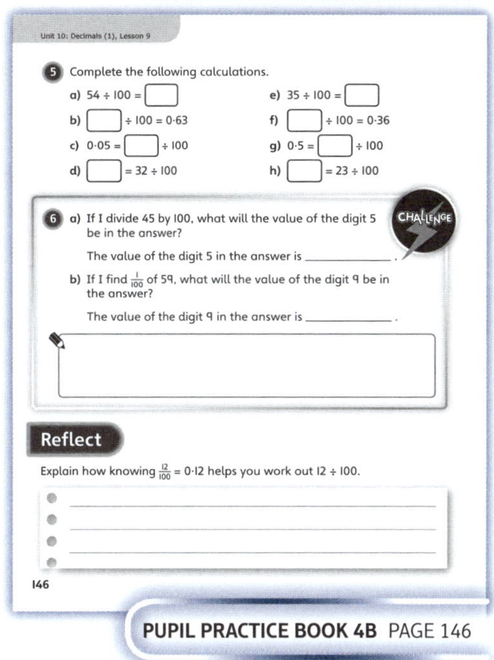

PUPIL PRACTICE BOOK 4B PAGE 146

Dividing by 10 and 100

Learning focus

In this lesson, children will divide numbers by 10 and 100 and see the connection between dividing by 10 and then 10 again, and dividing by 100.

Small steps

→ Previous step: Dividing by 100
→ **This step: Dividing by 10 and 100**
→ Next step: Making a whole

NATIONAL CURRICULUM LINKS

Year 4 Number – Fractions (Including Decimals)

Find the effect of dividing a one- or two-digit number by 10 and 100, identifying the value of the digits in the answer as ones, tenths and hundredths.

ASSESSING MASTERY

Children can divide 1- and 2-digit numbers by 10 and 100 and describe the effect of making each digit either 1 tenth or 1 hundredth of its original value. Children can make links between dividing the same number by 10 and 100 and can use different methods appropriately.

COMMON MISCONCEPTIONS

Children may confuse dividing by 10 and dividing by 100 and may just use a short-cut to find the solution. Ask:
• *Is the solution the required fraction of the original amount?*

Children may confuse the units that they are calculating during multi-step problems. Ask:
• *What is the value of each digit within each calculation?*

STRENGTHENING UNDERSTANDING

Use two cardboard versions of hundredths grids to strengthen children's understanding that dividing by 100 is the same as dividing by 10 and then by 10 again. Ask children to divide the first square into 100 pieces by cutting out each square individually; ask them to divide the other square by cutting it into 10 strips, and then cutting each strip into 10 pieces. Discuss what is the same and what is different about each method and the result.

GOING DEEPER

Children could be given number cards or a dice and be asked to use these to make a number to divide by 100. Alternatively, children could make a number with two decimal places and be asked to find the number that was divided by 100 or 10 to make this number. This will give an opportunity for extended learning.

KEY LANGUAGE

In lesson: ones, tenths, hundredths, divide, partition
Other language to be used by the teacher: equivalent, pattern

STRUCTURES AND REPRESENTATIONS

part-whole model, bar model, hundredths grid, place value grid

RESOURCES

Mandatory: place value counters

Optional: cardboard hundredths grid, scissors

 In the eTextbook of this lesson, you will find interactive links to a selection of teaching tools.

Before you teach 

• Are children able to distinguish between dividing by 10 and by 100?

Discover

 Pair work

- Question **1** b): *What is the same and what is different when dividing by 10 and 100? What have you used previously to help with dividing by 10 and 100?*

 Question **1** b) requires children to divide a quantity, previously divided by 10, by 100. Some children may identify that the answer to part a) could be used to find the answer to part b), but the focus is to return to the method used in the previous lesson.

 Give children two cardboard hundredths grids to help them understand that dividing by 100 is the same as dividing by 10 and then by 10 again. They could divide the first grid into 100 pieces and divide the other grid by cutting the square into 10 strips, and then cutting each strip into a further 10 pieces. They should then compare what is the same and what is different about the methods and the results.

Use place value grids to help children to visualise how the digits in the solutions change position when they divide by 10 and 100.

Question **1** a): Each table would receive 3·2 kg of cake.

Question **1** b): Each person would receive 0·32 kg of cake.

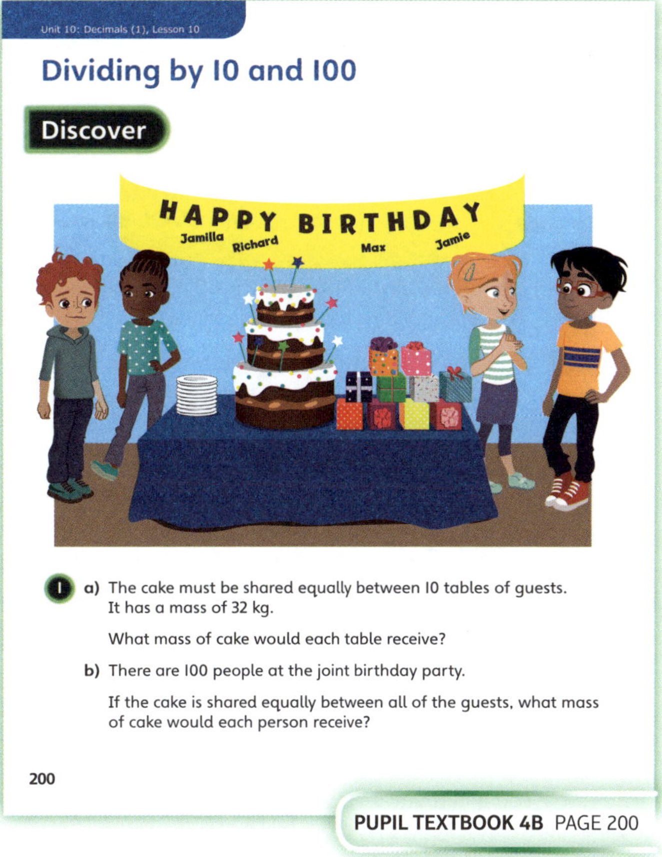

PUPIL TEXTBOOK 4B PAGE 200

Share

 Whole class teacher led

- Question **1**: *Is question **1** a) linked to question **1** b)?*
- Question **1** b): *What do you notice about the answer to this question in relation to part a)?*

 Question **1** b) uses the same method that children will have used in the previous lesson. Rather than multiplying each digit by different amounts to make it possible to divide by 100 based on their place value, the whole number (in this case 32 ones) is multiplied by 100 to make 3,200 hundredths, which can then be divided by 100. Explain this method and then compare it with the method used in part a). When both methods have been explained, facilitate discussions about the similarities and differences between the methods.

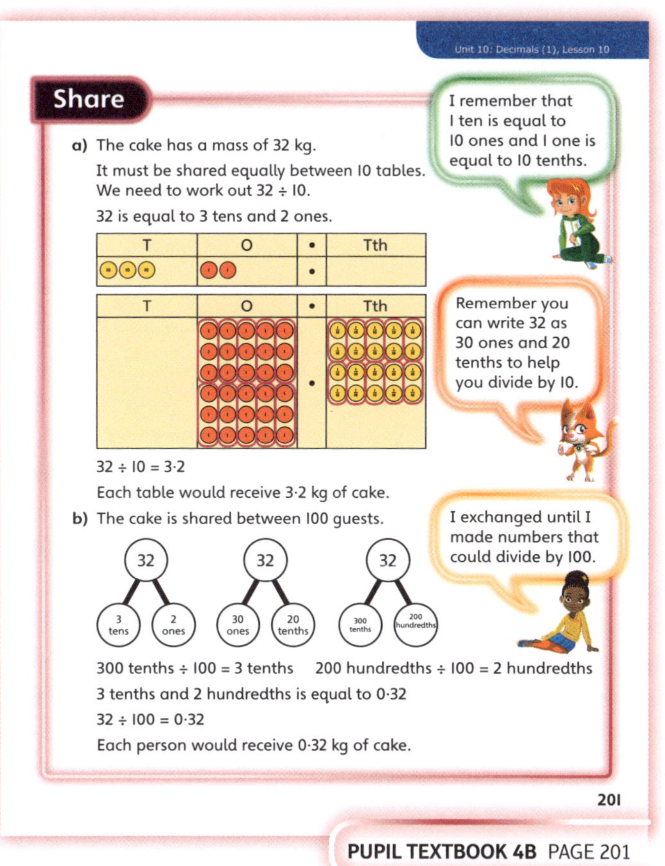

PUPIL TEXTBOOK 4B PAGE 201

Think together

WAYS OF WORKING Whole class teacher led (I do, We do, You do)

ASK

- Question **1**: *How is dividing by 100 linked with dividing by 10?*
- Question **2** b): *Why is it important to consider the value of each digit within each number?*

IN FOCUS Question **3** presents children with two different methods to complete the same calculation. The characters' responses should be used to consolidate the concept that dividing by 10 and then 10 again is the same as dividing by 100. Discuss that using a short-cut method is not necessarily the most efficient way to solve a problem and that children need to be careful when deciding where digits are placed.

STRENGTHEN Use place value counters, part-whole models and place value grids to act out the explanations given within question **3**. Children may also find these visual and concrete resources helpful to solve other questions.

DEEPEN To extend learning, children can explore whether it is always, sometimes or never true, that a 2-digit number divided by 100 will result in a solution with two decimal places. They can then try to explain why some solutions have one decimal place and some have two.

ASSESSMENT CHECKPOINT Children should be able to explain how dividing a number by 10 and then 10 again is the same as dividing a number by 100. They should be able to give examples to explain this concept and identify when it is easier to complete a calculation in one or two steps. They should also be able to identify and explain when an error has been made in their or others' working out.

ANSWERS

Question **1**: 600 tenths ÷ 100 = 6 tenths
400 hundredths ÷ 100 = 4 hundredths
64 ÷ 100 = 6 tenths and 4 hundredths
64 m ÷ 100 = 0·64 m
Each chair is decorated with 0·64 m of ribbon.

Question **2** a): Luis's hundredths grid is correct.

Question **2** b): Luis is correct.

Question **3** a): Zac has divided 20 by 10 and then 10 again. Reena has divided 20 by 100 using regrouping.

Question **3** b): Zac's method: 90 ÷ 10 = 9
9 ÷ 10 = 0·9
Reena's method: 90 ones = 900 tenths
= 9,000 hundredths
9,000 hundredths ÷ 100
= 90 hundredths = 0·9

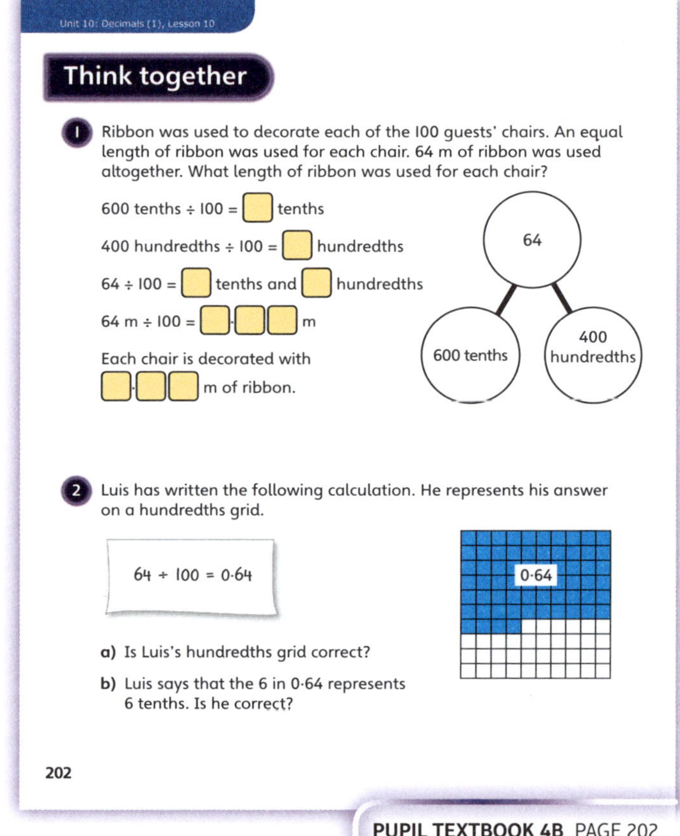

PUPIL TEXTBOOK 4B PAGE 202

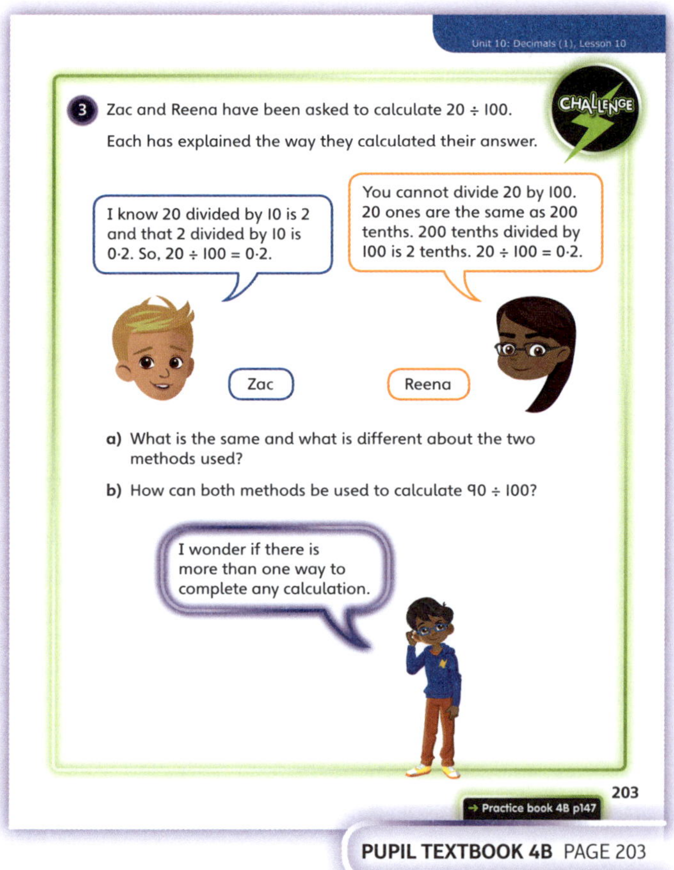

PUPIL TEXTBOOK 4B PAGE 203

Practice

WAYS OF WORKING Independent thinking

IN FOCUS Question ❶ requires children to use their understanding of dividing by 10 and by 100. The question includes sentence scaffolds to support children in the early stages of practice. Draw attention to how the digits change in place value.

Questions ❹ and ❺ allow less-structured practice for children to work out missing parts of equations, using pattern-spotting and other methods they have learnt. They should be encouraged to make links between calculations (especially in question ❹) to help them.

STRENGTHEN Use place value counters to divide an answer by 10, using regrouping, if children find it difficult to understand how the solution to a number divided by 10 can help them to calculate the same number divided by 100.

DEEPEN Children could roll a 9-sided dice twice, to make a number with two decimal places, and then find the number that was divided by 100 to make this number. This provides an opportunity for children to work backwards, which will strengthen their understanding of the relationship between the beginning and final number.

THINK DIFFERENTLY Question ❻ requires children to work backwards from a given solution to calculate what starting number has been divided by 100.

ASSESSMENT CHECKPOINT The strength of children's written response to question ❻, and the accuracy of the language used within their explanations, will allow you to assess the level of their understanding of working backwards from a given solution.

ANSWERS Answers for the **Practice** part of the lesson appear in the separate **Practice and Reflect answer guide**.

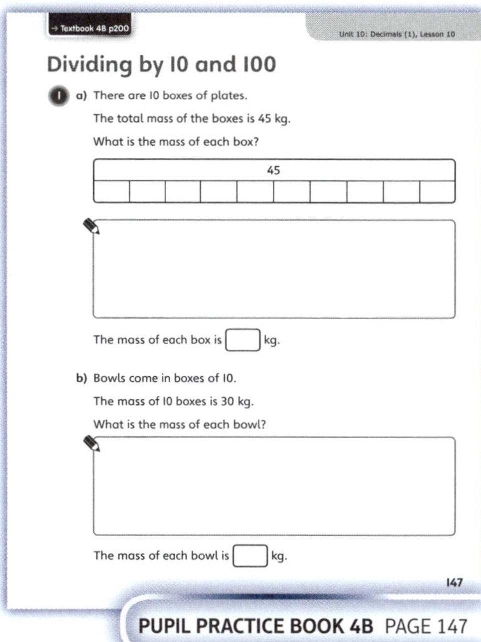

PUPIL PRACTICE BOOK 4B PAGE 147

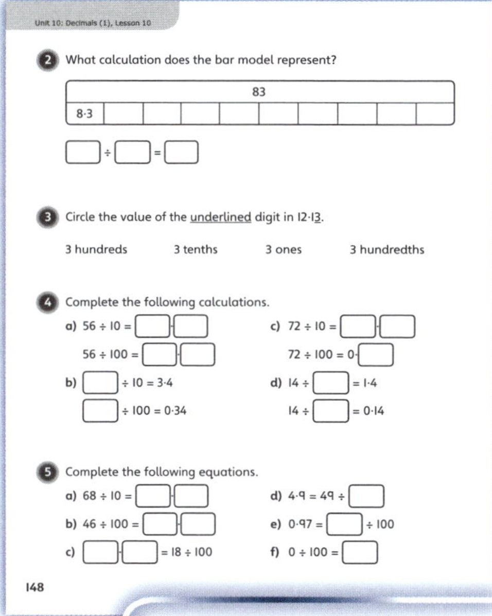

PUPIL PRACTICE BOOK 4B PAGE 148

Reflect

WAYS OF WORKING Independent thinking

IN FOCUS Children are required to make links between dividing a number by 10 and dividing the same number by 100. They should have the opportunity to write their explanations individually and then share these with the class.

ASSESSMENT CHECKPOINT The strength of children's written responses and the accuracy of the language used within their explanations will allow you to assess their understanding of the links between methods when dividing by 10 and 100.

ANSWERS Answers for the **Reflect** part of the lesson appear in the separate **Practice and Reflect answer guide**.

After the lesson ❚❚

- Are all children secure in their ability to divide whole numbers by 10 and 100 and record decimal answers accordingly?
- How confident are children at making links between calculations?

PUPIL PRACTICE BOOK 4B PAGE 149

End of unit check

Don't forget the *Power Maths* unit assessment grid on p26.

WAYS OF WORKING Group work adult led

IN FOCUS This **End of unit check** will allow you to focus on children's understanding of decimals to two decimal places (including how decimals relate to tenths and hundredths) and dividing 1- and 2-digit numbers by 10 and by 100. The questions within the **End of unit check** cover the full range of questions in the unit and are designed to draw out particular misconceptions or misunderstandings.

- Question **2** helps to assess whether children understand place value within decimal numbers.
- Questions **3** and **4** assess children's understanding of dividing both 1- and 2-digit numbers by 10. Question **4** will test whether children realise that both digits become one tenth of their original size.
- Question **5** assesses whether children have understood the link between fractional and decimal representations.
- Question **7** assesses whether children know the difference between dividing by 10 and dividing by 100.
- Question **9** is a SATs-style question.

ANSWERS AND COMMENTARY Children will be able to identify the value of any digit within a number up to two decimal places and will be able to count in tenths and hundredths, writing each step accurately. Children will also divide 1- and 2-digit numbers by 10 and 100, writing the solutions as decimal numbers.

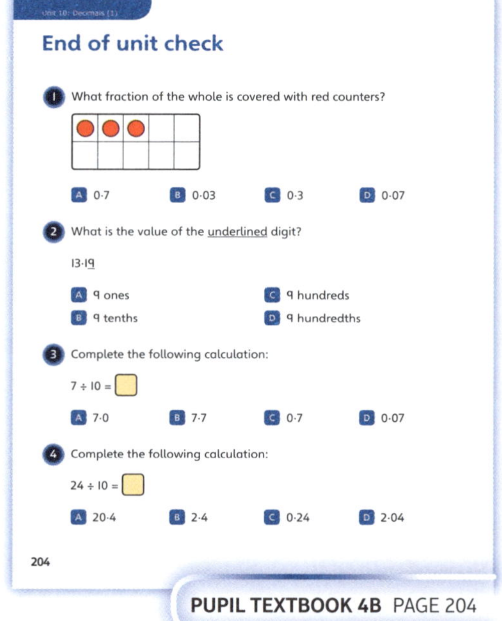

PUPIL TEXTBOOK 4B PAGE 204

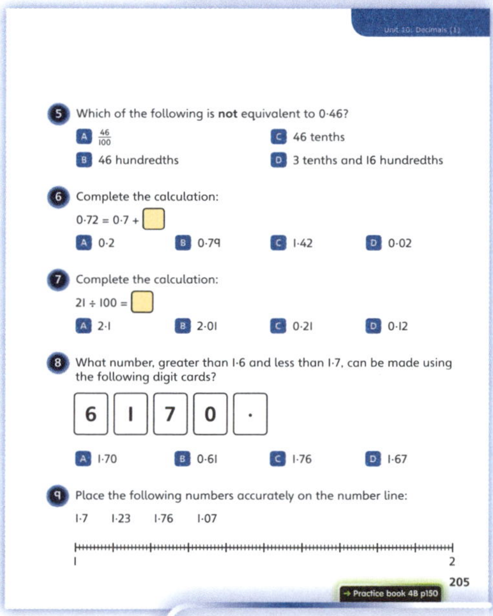

PUPIL TEXTBOOK 3B PAGE 205

Q	A	WRONG ANSWERS AND MISCONCEPTIONS	STRENGTHENING UNDERSTANDING
1	C	B and D suggest confusion of tenths and hundredths.	Children who find dividing by 10 difficult should be given place value counters and place value grids to help them understand the concept. Ask:
2	D	A, B or C suggest misunderstanding of place value.	• *How can counters be represented as a calculation?*
3	C	D shows a division of 100 rather than 10.	
4	B	D shows a misunderstanding between tenths and hundredths.	Children who find it difficult to identify the value of a digit within a decimal number should practise writing and making the number with place value counters on a place value grid. Ask:
5	C	B suggests a confusion between tenths and hundredths.	
6	D	A suggests interpreting the value of the 2 in 0·72 incorrectly.	• *How many tenths and hundredths does your grid have?*
7	C	A suggests confusing dividing by 10 and by 100.	
8	D	A suggests not understanding that 1·70 is not less than 1·7.	
9		Assess the accuracy of children's placement of the numbers.	

My journal

WAYS OF WORKING Independent thinking

ANSWERS AND COMMENTARY In question **1**, children use digit cards to make as many of the following numbers as possible: 3·41, 34·1, 3·14, 31·4, 4·31, 43·1, 4·13, 41·3, 1·34, 13·4, 1·43, 14·3

In question **2**, they can use any of the representations they have been exposed to during the unit to show one of the numbers they have created. For example, they could create a picture of place value counters, a number line, and a part-whole model with the amount partitioned into different parts.

Children who have difficulty thinking of more than one representation should be prompted to look back through their book. Ask:

- *How can you represent your number on a number line? Where will the line start and finish?*
- *How can you represent your number in a part-whole model? What will be the whole number?*
- *How can you represent your number in a place value grid? How many counters will you need for the tenths column and how many for the hundredths column?*

Power check

WAYS OF WORKING Independent thinking

ASK

- *How confident would you feel to teach someone else about decimals?*
- *Can you partition a number into tenths and hundredths?*
- *How confident are you dividing a number by 10 and 100?*

Power play

WAYS OF WORKING Pair work

IN FOCUS Children should work in pairs to see who can travel the furthest through the grid, as a result of the outcome of the spinner. Children should take it in turns to spin the spinner and should check each other's calculations for each step of the journey.

ANSWERS AND COMMENTARY The calculations children have to do will vary, based on the outcome of their spins. They should be challenged to explain their journey and state what decimal is added or subtracted at each stage. If they cannot explain each step, this suggests that they do not understand how to add or subtract tenths and hundredths to or from a number.

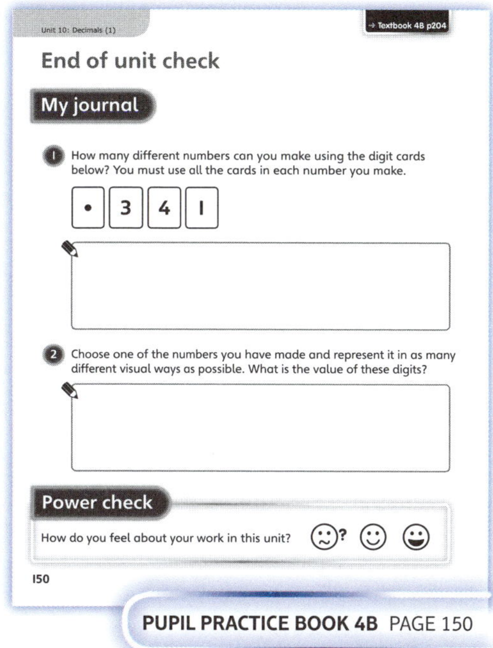

PUPIL PRACTICE BOOK 4B PAGE 150

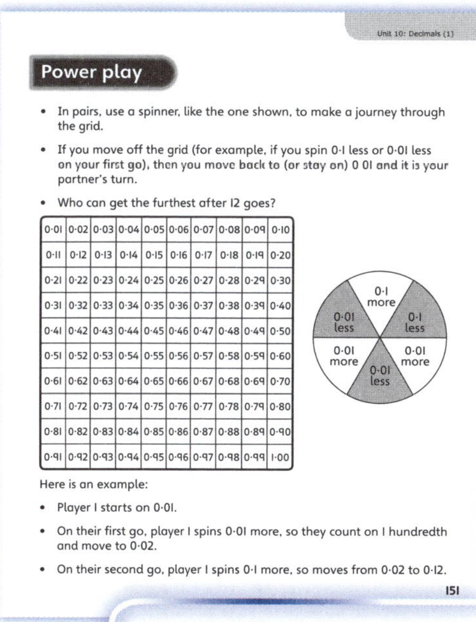

PUPIL PRACTICE BOOK 4B PAGE 151

After the unit ⏸

- Are all children secure in their mastery of tenths and hundredths and how they relate to one another?
- Can all children add and subtract tenths and hundredths?

Strengthen and **Deepen** activities for this unit can be found in the *Power Maths* online subscription.

Published by Pearson Education Limited, 80 Strand, London, WC2R 0RL.

www.pearsonschools.co.uk

Text © Pearson Education Limited 2018
Edited by Pearson, Little Grey Cells Publishing Services and Haremi Ltd
Designed and typeset by Kamae Design
Original illustrations © Pearson Education Limited 2018
Illustrated by Laura Arias, Nigel Dobbyn, Adam Linley and Nadene Naude at Beehive Illustration.
Cover design by Pearson Education Ltd
Back cover illustration © Diago Diaz and Nadene Naude at Beehive Illustration.

Series Editor: Tony Staneff
Consultants: Professor Liu Jian and Professor Zhang Dan

The rights of Tony Staneff, Belle Cottingham, Jonathan East, Caroline Hamilton, Rebecca Holland, Stephen Monaghan and Paul Wrangles to be identified as authors of this work have been asserted by them in accordance with the Copyright, Designs and Patents Act 1988.

First published 2018

21 20 19 18
10 9 8 7 6 5 4 3 2 1

British Library Cataloguing in Publication Data
A catalogue record for this book is available from the British Library

ISBN 978 0 435 19019 4

Printed in Slovakia by Neografia

www.activelearnprimary.co.uk

Note from the publisher
Pearson has robust editorial processes, including answer and fact checks, to ensure the accuracy of the content in this publication, and every effort is made to ensure this publication is free of errors. We are, however, only human, and occasionally errors do occur. Pearson is not liable for any misunderstandings that arise as a result of errors in this publication, but it is our priority to ensure that the content is accurate. If you spot an error, please do contact us at resourcescorrections@pearson.com so we can make sure it is corrected.